Sprole.
1st class.. 1868 to 1869

A

COURSE OF INSTRUCTION

IN

ORDNANCE AND GUNNERY;

PREPARED FOR THE USE OF THE

CADETS

OF THE

UNITED STATES MILITARY ACADEMY.

BY

BREVET-COL. J. G. BENTON,
MAJOR ORDNANCE DEPT.,
LATE INSTRUCTOR OF ORDNANCE AND SCIENCE OF GUNNERY, MILITARY ACADEMY, WEST POINT.

THIRD EDITION, REVISED AND ENLARGED.

NEW YORK:
D. VAN NOSTRAND, 192 BROADWAY.
1867.

PREFACE TO THIRD EDITION.

In Part I. of this work, the author has endeavored to describe briefly the principal articles of our Army Ordnance *Matériel*, and to state the general principles of their construction and operation. It is thought that Part II. contains the data and formula necessary to determine with practical accuracy the movement and effects of rifle as well as smooth-bore projectiles. The Appendix contains short descriptions of some of the most noted modern cannon and projectiles, and a tabular statement of some of the principal experiments made in England with armor plates.

The works which have been principally consulted, and in many instances quoted *verbatim*, are PIOBERT'S *Cours d'Artillerie*, *Ordnance Manual*, Reports on Powder, Cannon, Metals, and Small-arms, &c., by officers of the Ordnance Department, U. S. A., and HOLLEY'S Ordnance and Armor.

CONTENTS.

CHAPTER I. — PAGE.

Gunpowder ... 7

CHAPTER II.

Projectiles ... 71

CHAPTER III.

Cannon ... 104

CHAPTER IV.

Artillery Carriages ... 212

CHAPTER V.

Machines and Implements ... 248

CHAPTER VI.

Small-Arms ... 270

CHAPTER VII.

Pyrotechny ... 342

CHAPTER VIII.

Science of Gunnery ... 382

CHAPTER IX.

Loading, Pointing and Discharging Fire-Arms ... 435

CHAPTER X.

Different Kinds of Fires ... 450

CHAPTER XI.

Effects of Projectiles ... 471

CHAPTER XII.

Employment of Artillery ... 488

CHAPTER XIII.

Tables of Multipliers ... 505

Tables of Fire ... 516

APPENDIX ... 525

INDEX ... 537

ORDNANCE AND GUNNERY.

PART I.

CHAPTER I.

GUNPOWDER.

1. **General theory.** Gunpowder and the compositions of pyrotechny are the means used, in modern warfare, to propel projectiles, explode mines, destroy ships and buildings, and furnish light and signals for the operations of an army at night. They are simply mechanical mixtures of substances which give out light, heat, and gas in their combustion, or chemical union with each other.

The two classes of substances generally used for these purposes are the *nitrates* and *chlorates* on one hand, and *charcoal*, *sulphur*, *antimony*, &c., on the other. The former class contains a large amount of oxygen, which is a strong supporter of combustion; and the latter embraces those substances which have a powerful affinity for it.

Explosion is a phenomenon arising from the sudden enlargement of the volume of a body, as in the case of combustion, when a solid body is rapidly converted into one of vapor or gas. If this change of state be accompanied by the development of a large amount of heat, the explosive effect will be very much increased.

Gunpowder is an explosive substance, formed by the mechanical mixture of *nitrate of potassa*, *sulphur*, and *charcoal*.

The parts performed by these ingredients in the explosion will be best understood by an examination of the following table:

COMPOSITION OF GUNPOWDER.

BEFORE COMBUSTION.		AFTER COMBUSTION.
3 parts of carbon,	3 carbon,	3 carbonic acid (gas).
1 part of nitrate of potassa,	6 oxygen,	
	1 nitrogen,	1 nitrogen (gas).
	1 potassium,	1 sulphide of potassium (solid).
1 part of sulphur,	1 sulphur,	

A gunpowder can be made of nitrate of potassa and charcoal alone; but it is not so strong as when sulphur is present; besides, the substance of the grain is friable, has considerable affinity for moisture, and rapidly fouls the arms in which it is used.

Theoretically, sulphur does not contribute directly to the explosive force of gunpowder by furnishing materials for gas; but, by uniting with the potassium, it affords a large amount of heat, and prevents the carbonic acid from uniting with the potassa and forming a solid compound—the carbonate of potassa. It is to the heat and carbonic acid thus formed that gunpowder mainly owes its explosive force.

The strength of gunpowder, or amount of work which a certain quantity is capable of performing in a given time, depends on the mass of the powder and the velocity with which its gaseous particles are evolved.

This velocity of evolution of the gaseous particles, or "quickness," depends on the purity, proportion, and incorporation of the ingredients, and on the size, form, and density of the grains. These will be discussed in the following pages, under the heads of *Materials*, *Fabrication*, *Mechanical Effects*, and *Chemical Properties* of gunpowder.

MATERIALS.

SALTPETRE.

2. **Description.** *Saltpetre*, *nitre*, or *nitrate of potassa*, is composed of 53.45 of nitric acid, and 46.55 of potassa, or $KO+NO_5$. It crystallizes in colorless six-sided prisms; has a cooling, saline, and slightly bitter taste, and deflagrates with more violence than any other nitrate when thrown on burning charcoal. It is anhydrous; but its crystals often contain water mechanically confined. It is not deliquescent in common air (a very important quality in an ingredient of gunpowder), but is so in an atmosphere nearly saturated with moisture. It is insoluble in the oils and pure alcohol.

It is decomposed when strongly heated, and oxygen is evolved at first; finally nitrogen is given off, and peroxyde of potassium remains. When heated with combustible materials, nitre is completely deprived of its oxygen; it is consequently much used as an oxydizing agent. This is the part which it plays in gunpowder.

The *solubility* of nitre increases rapidly with the temperature.

100 parts of water at 32° dissolve 13.32 of nitre.
" " " 64.4° " 29.00 "
" " " 113° " 74.00 "
" " " 212° " 246.15 "

Hence, a hot saturated solution, on cooling, deposits the greater portion of the salt which had dissolved.

3. **Sources.** Nitrate of potassa, in connection with more or less of the nitrates of lime and magnesia, is obtained from several sources, among which may be enumerated calcareous caves, certain soils in warm climates, artificial nitre beds, and the mortar of stables or other buildings long occupied by animals, in which cases it generally occurs as an efflorescence. It is also found in the tobacco, sunflower, beet-root, cornstalk, and other plants.

The *caves* occurring in certain porous limestones are often found to contain large quantities of the nitrates of lime, potassa, and magnesia, deposited in the loose materials at the bottom, efflorescing from the sides, or even contained in the pores of the rock itself. Many of the limestone caves in Kentucky, Virginia, Tennessee, &c., abound in nitrates. In Madison county, Kentucky, there is a cave 1,936 feet long by 40 wide, which contains the nitrates of potassa and lime, mingled with the earthy matter at the bottom. One bushel of the earth yields, by double decomposition with carbonate of potassa, from three to ten pounds of nitre.

The greater part of the nitre used in England, and this country, is derived from the *soil* in various parts of the East Indies. It occurs in the same manner in various warm countries, as Egypt, Spain, &c. In the vicinity of Monclova, Mexico, it occurs in veins, or mines, in a state of great purity.

It appears to be generated spontaneously at the depth where the soil retains its moisture; and when dissolved by rains, the subsequent evaporation, by capillary attraction, causes it to rise to the surface, where it is deposited as a crust.

To obtain nitre from this source, the earth is removed to a certain depth, and treated with water, which dissolves the soluble salts. The solution is then transferred to large reservoirs, when it soon evaporates by solar heat, and deposits large crystals of nitre. This is known in commerce as *rough*, or *crude saltpetre.* The mother waters are rejected; but as they contain a large quantity of the nitrates of lime and magnesia, they might still afford some nitre if they were mixed with salts of potassa.

In the north of Europe, where nitre does not occur as a natural product, various artificial processes have long been employed to obtain it; and similar methods were much used in France during the Revolution, when that country could not be supplied from Spain and other countries. The *nitre-beds* were mostly used.

Nitre-beds. These were made by placing loosely on a floor of wood or clay, a layer, of three or four feet thick, of a mixture of earth, calcareous matter—such as marls, calcareous sands, mortar from stables, &c., and various animal products—such as blood, urine, stable manure, &c. Vegetable matter was found to be useful, probably furnishing potassa. The whole was placed under a shed, and occasionally moistened with additional quantities of blood or urine, and in about two years it was fit for lixiviation. In Prussia, the materials are placed in parallel *walls* about seven feet high, and three or four feet thick, which arrangement is found to be more convenient, and to occupy less space than the beds.

Mortar exposed to decomposing animal matter, in moist, warm places, becomes considerably charged with nitrates. In consequence, the mortar in old stables is often found to be rich in nitrates, and may be used to obtain nitre directly, or it may be advantageously mixed with the materials for nitre-beds.

The *lye of nitrified substances* contains nitrate of potassa, but especially nitrates of lime and magnesia, and also chlorides of sodium and calcium. The nitrates of lime and magnesia may be converted into nitrate of potassa by means of the carbonate of potassa; but on account of the increased value of this substance, the process now generally adopted is as follows, viz.: first, the nitrates of lime and magnesia are converted into the nitrate of soda, by means of the sulphate of soda, and then by the chloride of potassium, the nitrate of soda is converted into the nitrate of potassa.

The nitrate of potassa thus obtained, like that obtained from the soils of warmer climates, is called *rough saltpetre*, and contains from 15 to 25 per cent. of foreign matter, principally chlorides of sodium and potassium. These are separated by the process of *refining*.

4. **Refining.** The *refining of nitre* is founded on its rapidly-increasing solubility with elevation of temperature, while the solubility of the chlorides of sodium and potassium is nearly uniform. (For the details of this process see Ordnance Manual.)

If, after nitre has been refined, it be desired to preserve it for future use, it is fused in iron pots, and cast into cakes weighing about seventy pounds. This method has the advantage of reducing its volume, and expelling the water of crystallization; but it requires a little more work to pulverize it afterward in making gunpowder.

As the United States are in a great measure dependent on the East Indies for this important material of war, it has been the policy of the government to purchase yearly a certain quantity of rough saltpetre, refine and fuse it, and store it away in the arsenals for future use. The quantity now on hand in the arsenals amounts to several millions of pounds.

5. **Tests.** The *test of rough saltpetre* is founded on the fact that a solution of nitrate of potassa, saturated at a certain temperature, may be left in contact with an additional quantity of saltpetre at the same temperature, without sensibly dissolving any of it; while under the same circumstances it can dissolve sea salt, and many other soluble salts.

To a pound of rough saltpetre add a pint of water, saturated with pure saltpetre; stir the mixture for ten minutes with a glass rod, and decant the liquor on a filter; wash the saltpetre a second time in the same manner, with half a pint of the saturated solution, and pour the whole on the filter; let it drain, and then dry it perfectly by placing it first on a bed of some absorbent matter, such as ashes or lime, and then by evaporation in a glass vessel over a gentle fire. The saturated solution having taken up only the foreign salts, what remains on the filter (allowing two per cent. for earthy matter and the saltpetre left by the saturated water), is the quantity of pure saltpetre contained in the pound of rough. As the changes of temperature during the operation may affect the quantity of pure saltpetre remaining on the filter, it is proper to perform a corresponding operation at the same time and under the same circumstances, on a like quantity of pure saltpetre; the gain or loss thus ascertained will show the correction to be made in the former result.

Test of pure saltpetre. For powder, saltpetre should not contain more than 1-3000th of chlorides. To test this, dissolve 200 grains of saltpetre in the least possible quantity (say 1,000 grains) of distilled water; pour on it 20 grains of a solution of nitrate of silver, containing 10 grains of the nitrate to 1,032 grains of water, that being the quantity required to decompose 200-3000ths of a grain of muriate of soda; filter the liquid and divide it into two portions; to one portion add a few drops of the solution of the nitrate of silver; if it remains clear, the saltpetre does not contain more than 1-3000th of muriate of soda; to the other portion add a small portion of the solution of the muriate of soda; if it becomes clouded, the saltpetre contains less than 1-3000th. By using the test liquor in very small quantities, the exact proportion of muriate of soda may be ascertained. At the refinery of Paris it does not exceed 1-18,000th of the saltpetre; and this degree of purity is attained also at the refinery of Messrs. Dupont. Saltpetre for the best sporting powder is refined a second time, and contains not more than 1-60,000th part of chlorides.

6. **Chlorate of Potassa.** Other oxydizing substances, such as the chlorate of potassa and nitrate of soda, may be used in the manufacture of gunpowder; but for this purpose, they are inferior to the nitrate of potassa. The chlorate of potassa is a substance which parts with its oxygen easily, and makes a powder, which has been

found by experience, to give at least double the range with the mortar eprouvette, of that made with nitrate of potassa, but from its great quickness, resembles the fulminates in its destructive effects on the gun. Besides, it is more costly than nitrate of potassa, renders the powder liable to explode by slight causes, and gives a residue which rapidly corrodes iron.

Its use in the laboratory is chiefly confined to the preparation of colored fires and cannon primers.

The nitrate of soda is found as an extensive deposit in the soils of some portions of Peru and northern Mexico. It is cheaper than nitrate of potassa, and for the same weight affords a greater amount of nitric acid, or oxygen. Its affinity for moisture constitutes a serious objection to its use in the manufacture of a gunpowder for war purposes, or one that is to be preserved for any length of time.

The nitrate of soda may be used in obtaining the nitrate of potassa by decomposing it with carbonate of potassa—the potash of commerce.

CHARCOAL.

7. **Nature of charcoal.** Charcoal is the result of the incomplete combustion or distillation of wood. Its composition and properties vary with the nature of the wood, and mode of distillation employed.

Charcoal obtained from light wood is the best for gunpowder, as it is more combustible and easy to pulverize, and contains less earthy matters.

Willow and *poplar* are used for this purpose in the United States, and the black alder in Europe. The wood must be sound, and should not be more than three or four years old, and about one inch in diameter; branches larger than this should be split up. It is cut in the spring, when the sap runs freely, and is immediately stripped of its bark. The smaller branches are used for fine sporting powder.

The operation of charring may be performed in pits, but the method now almost universally used in making charcoal for gunpowder is that of *distillation*. For this purpose the wood is placed in an iron vessel, generally of a cylindrical form, to which a cover is luted; an opening with a pipe is made to conduct off the gaseous products, and the wood is thus exposed to the heat of a furnace. The progress of distillation is judged of by the color of the flame and smoke, and sometimes by *test-sticks*, which are introduced through tubes prepared for the purpose.

8. **Properties.** The charcoal thus obtained should retain a certain degree of elasticity, and should have a brown color, the wood not being entirely decomposed; it retains the fibrous appearance of the wood, and the fracture is iridescent. As it readily absorbs 1-20th of its weight of moisture, which diminishes its inflammability, it should be made only in proportion as it is required for use. Wood generally contains 52 per cent. of carbon, but distillation furnishes not more than 30 to 40 per cent. of charcoal.

The *specific gravity* of charcoal triturated under heavy rollers, is about 1.380; but in sticks, as it comes from the charring cylinders, it rarely exceeds .300.

The properties of charcoal vary much with the *temperature* employed in the preparation. If wood be merely heated until it ceases to give off vapor, a true charcoal is obtained; but by raising the temperature to redness or whiteness, its properties will be much changed, as is shown in the following table:

	When not heated to redness.	Heated to redness.	Heated to whiteness.
For electricity,	Non-conductor.	Good conductor.	Excellent conductor.
" heat.	Very bad conductor.	Good conductor.	Excellent conductor.
Combustibility.	Easy.	Less easy.	Difficult.

If sufficient heat be applied to drive off all the volatile matters in six hours, a black charcoal, containing from 28 to 33 per cent. of carbon, will be obtained. If the heat be reduced so as to prolong the distillation to twelve hours, the charcoal will have a yellowish brown color, and will contain from 38 to 40 per cent. of carbon.

Charcoal *inflames* at about 460° Fahrenheit. A black coal strongly calcined takes fire quickly, but is easily extinguished. A brown charcoal takes fire slowly, but burns strongly and rapidly. As it is desirable to have charcoal for gunpowder very combustible, it must therefore be prepared at a low temperature, and must be light. In distillation, the heat is kept below redness.

9. **Accidents.** When recently prepared charcoal is pulverized and laid in heaps, it is liable to absorb oxygen with such rapidity as to cause spontaneous combustion. This has been the cause of serious accidents at powder-mills; and hence it is important not to pulverize charcoal until it has been exposed to the air for several days. In Prussian powder-mills, pulverized

2

charcoal is kept in a fireproof room, in iron vessels, as a precaution against accidents.

When charcoal has not absorbed moisture, and is mixed with oxydizing substances, it may be inflamed by violent shocks, or by friction. This is the principal cause of the accidents which occur in the preparation of explosive mixtures which contain charcoal.

10. **Combustibility.** For the purpose of comparing the *combustibility* of charcoals made of different materials, a certain quantity of each is thoroughly mixed with nitre, in the proportion of 1 part of the former to 5 of the latter, and driven compactly into an iron tube about .25 inches in diameter; the weight and length of the filled tube are taken, and the duration of the combustion is ascertained by a pendulum or chronometer. The length of composition burned in a second of time is called the velocity of combustion, and is taken as the measure of the combustibility of that particular kind of charcoal. The amount of residue is ascertained by subtracting the weight of the tube and residue after burning from that of the filled tube before burning, and again subtracting this difference from the weight of the composition originally in the tube. The velocity of combustion is independent of the diameter of the tube and of the material of which it is made; but it varies slightly with the pressure used in driving the composition, and very much with the degree of trituration of the materials. The following tables contain some of the results thus obtained, viz.:

60 parts of nitre and 12 of charcoal. Black Charcoals.	Velocity of combustion.	Percentage of residue.
Charcoal of Hemp, . .	.31 inch.	16.6
" Grape Vine, . .	.26 "	27.7
" Pine, . . .	.18 "	41.6
" Black Alder, . .	.16 "	33.3
" Spindle Tree, .	.15 "	37.5
" Hazel, ...	.13 "	41.6
" Chestnut, . .	.14 "	50.0
" Walnut, . .	.11 "	45.8
" Coke . . .	.06 "	62.5
" Sugar, . . .	.04 "	66.6
Charcoal made by distilling Black Alder, and conducted so as to give,		
For 100 p'ts wood, 40 p'ts charcoal,	.14 inch.	39.0
" " 30 "	.16 "	37.0
" " 25 "	.15 "	33.3
" " 15 "	.12 "	35.0

The following table shows the influence of trituration and proportion of ingredients :

Mixture.	Parts of charcoal to 60 of nitre.	Charcoal made of Hemp.				Charcoal made of Pine.	
		6 hours' trituration.		4 hours' trituration.		6 hours' trituration.	4 hours' trituration.
		Velocity.	Per ct. of residue.	Velocity.	Pr. ct. of residue.	Velocity.	Velocity.
$\frac{1}{8}$	$8\frac{1}{2}$	.10 in.	58.0	.08 in.	55.0	.09 in.	.07 in.
$\frac{1}{7}$	10	.12	45.0	.10	43.0	.15	.09
$\frac{1}{6}$	12	.31	16.6	.17	26.0	.18	.12
$\frac{1}{5}$	15	.39	13.0	.27	17.0	.35	.20
$\frac{1}{4}$	20	.56	12.0			.39	.27
$\frac{1}{3}$	30	.65	11.0			.59	.53
$\frac{1}{2}$	60	.14					

SULPHUR.

11. **Properties.** Pure sulphur is of a citron-yellow color, and shining fracture; it crackles when pressed in the hand. The specific gravity of native sulphur is 2.033; that of sulphur refined by sublimation 1.900; its specific gravity is diminished by trituration. When heated, it melts at 226° into a thin, amber-colored liquid; if the temperature be then raised to about 400° it becomes dark and thick; but if heated still further, up to 800°, its boiling point, it becomes again thin and limpid. It begins to pass off in vapor at 115°, and if heated rapidly, inflames at 370°. It is insoluble in water, but soluble in oils and slightly so in alcohol.

Sulphur is generally found in great quantities in the neighborhood of volcanoes; it may also be obtained from metallic ores (pyrites) and other sources. Most of that used in the United States comes from Sicily through the French refineries.

Crude sulphur, as extracted by the first sublimation from the ore, contains about 8 per cent. of earthy matter. It is purified by a second sublimation, from which it is collected in the form of powder, called the *flowers of sulphur;* or, it is melted and run into moulds, making *roll brimstone.* It may be also refined, but not so thoroughly, by being simply melted and skimmed.

Pure sulphur is entirely consumed in combustion; and its purity is thus easily tested by burning about 100 grains in a glass vessel; the residue should not exceed a small fraction of a grain.

MANUFACTURE OF GUNPOWDER.

12. **Proportions of ingredients.**

	Nitre.	Charcoal.	Sulphur
By the atomic theory, . .	74.64	13.51	11.85
	(NO_5+KO)	$+ 3C$	$+ S.$
In the United States:			
For military service, . . .	76.00	14.00	10.00
For blasting and mining, .	62.00	18.00	20.00

The proportions of the ingredients of the earliest gunpowder known, differ but slightly from those now in use; and these, it will be seen, nearly agree with those called for by the theory of combining equivalents.

For the general purposes of artillery, slight variations in the proportions of the ingredients for powder are not found to affect its strength; but for blasting or mining purposes, a slower powder is found to answer nearly as well as a quick one, consequently the proportion of nitre is reduced much below that of gunpowder. Blasting powder is thus made cheap; but as it leaves a large amount of residuum, it cannot be advantageously used in fire-arms.

13. **Operations.** The several operations of fabricating gunpowder are:

1st. *Pulverizing;* which consists in reducing the ingredients to finely divided dust.

2d. *Incorporating;* which consists in bringing the particles of this dust into such intimate contact that each particle of powder shall be composed of one of each of the ingredients.

3d. *Compressing;* which gives strength and density to the substance of the powder, by converting the incorporated mixture into a cake which will not crumble in transportation.

4th. *Graining;* which breaks up the cake into small fragments or grains, and increases the surface of combustion.

5th. *Glazing;* which hardens the surface, to protect it from the action of moisture, and rounds the sharp angles of the grains to prevent the formation of dust in transportation.

6th. *Drying;* which frees the powder from the moisture introduced in certain operations of the fabrication.

7th. *Dusting;* which frees it from the dust, which would otherwise fill up the interstices and retard the inflammation of the charge.

The proportions of the ingredients, as well as the art of making gunpowder, vary in different countries, and even among the different manufactories of the same country.

The variations in the proportions are slight, however, and the differences in the modes of manufacture are principally confined to the more important operations of pulverizing, mixing, and compressing the composition. For French military powder, these operations are performed in the "pounding-mill," or a series of mortars and pestles. In Prussia the composition is pressed into cake by passing it between two heavy rollers, by means of an endless band of cloth, which receives the dust from a hopper. In England these operations are performed by the "rolling-barrel," "cylinder-mill," and "press." The superior strength and excellent preservative qualities of the English powder have led to the adoption of this mode of manufacture in the United States.

14. **Processes of manufacture.*** The buildings in which the different operations are carried on are separated from each other, and protected by trees or traverses as far as practicable.

Pulverizing. The saltpetre is usually pulverized sufficiently when it comes from the refinery. The charcoal is placed in large cast-iron barrels with twice its weight of zinc balls. The barrel has several ledges on the interior, and is made to revolve from 20 to 25 times in a minute. It is pulverized in 2 or 3 hours. The sulphur is placed in barrels made of thick leather stretched over a wooden frame, with twice its weight of zinc balls from .3 to .5 inches in diameter, and the barrel made to revolve about 20 times per minute. It takes one hour to pulverize the sulphur.

Incorporating. The ingredients having been weighed out in the proportions above given, the charcoal and sulphur are put together in a rolling-barrel similar to that in which the sulphur is pulverized, and rolled for one hour. The saltpetre is then added, and rolled for three hours longer. In some mills this operation is omitted. It is now taken to the cylinder, or *rolling-mill.* This consists of two cast-iron cylinders rolling round a horizontal axis in a circular trough of about 4 feet diameter, with a cast-iron bottom. The cylinders are 6 feet in diameter, 18 inches thick on the face, and weigh about 8 tons each. They are followed by a wooden scraper, which keeps the composition in the centre of the trough.

* *Vide* Ordnance Manual

A charge of 75 lbs. in some mills, and 150 lbs. in others, is then spread in the trough of the rolling-mill, and moistened with 2 to 3 per cent. of water, according to the hygrometric state of the atmosphere.

It is rolled slowly at first, and afterward from 8 to 10 revolutions of the roller per minute, for 1 hour for 50 lbs., and 3 hours for 150 lbs. of composition. A little water is added, as the process advances, if the composition gets very dry—which is judged of by its color.

When the materials are thoroughly incorporated, the cake is of a uniform, lively, grayish, dark color. In this state it is called *mill-cake.*

The quality of the powder depends much on the thorough incorporation of the materials, and burns more rapidly as this operation is more thoroughly performed.

The mill-cake is next taken to the press-house, to be pressed into a hard cake.

Pressing. The mill-cake is sprinkled with about 3 per cent. of water, and arranged in a series of layers about 4 inches thick, separated by brass plates. A powerful pressure is brought to bear on the layers, which are subjected to the maximum pressure for about 10 to 15 minutes, when it is removed. Each layer is thus formed into a hard cake about an inch thick.

Granulating. The cake is broken into pieces by means of iron-toothed rollers revolving in opposite directions, their axes being parallel and the distance between them regulated as required. *Fluted* rollers are sometimes used. The pieces are passed through a succession of rollers, each series being closer together, by which the pieces are broken into others still smaller, which pass over a sieve to another roller, the small grains passing through the sieve into a receiver below, until the whole is reduced to the required size. The various-sized grains are separated by the sieves between the different rollers.

Glazing. Several hundred pounds of the grained powder, containing from 3 to 4 per cent. of water, are placed in the glazing barrel, which is made to revolve from 9 to 10 times per minute, and in some mills from 25 to 30 times per minute. Usually from 10 to 12 hours are required to give the required glazing. In this operation the sharp angles are broken off, thereby diminishing the dust produced in transportation, and the surface of the grain receives a bright polish.

Drying. The powder is spread out on sheets stretched upon frames in a room raised to a temperature of 140° to 180° by steam-pipes or by a furnace. The temperature should be raised gradually, and should not exceed 180°, ventilation being kept up.

Dusting. The powder is finally sifted through fine sieves, to remove all dust and fine grains.

15. **Round powder.** In case of emergency, and when powder cannot be procured from the mills, it may be made, in a simple and expeditious manner, as follows: Fix a powder-barrel on a shaft passing through its two heads, the barrel having ledges on the inside; to prevent leakage, cover it with a close canvas glued on, and put the hoops over the canvas. Put into the barrel 10 lbs. of sulphur in lumps, and 10 lbs. of charcoal, with 60 lbs. of zinc balls or of small shot (down to No. 4, 0.014 in. in diameter nearly); turn it, by hand or otherwise, 30 revolutions in a minute.

To 10 lbs. of this mixture thus pulverized, add 30 lbs. of nitre, and work it two hours with the balls; water the 40 lbs. of composition with 2 quarts of water, mixing it equally with the hands, and granulate with the graining-sieve. The grains thus made, not being pressed, are too soft. To make them hard, put them into a barrel having 5 or 6 ledges projecting about 0.4 in. inside; give it at first 8 revolutions in a minute, increasing gradually to 20. The compression will be proportionate to the charge in the barrel, which should not, however, be more than half full; continue this operation until the density is such that a cubic foot of the powder shall weigh 855 oz., the mean density of round powder; strike on the staves of the barrel from time to time, to prevent the adhesion of the powder.

Sift the grains and dry the powder as usual. That which is too fine or too coarse is returned to the pulverizing-barrel.

This powder is round, and the grain is sufficiently hard on the surface, but the interior is soft, which makes it unfit for keeping, and may cause it to burn slowly. This defect may be remedied by making the grains at first very small, and by rolling them on a sheet or in a barrel, watering them from time to time, and adding pulverized composition in small proportions; in this way, the grains will be formed by successive layers; they are then separated according to size, glazed and dried.

It appears from experiments that the *simple incorporation* of the materials makes a powder which gives nearly as high ranges with cannon as grained powder.

The incorporated dust from the rolling-barrel may be used in case of necessity.

INSPECTION, PROOF, ETC.

16. **Proving powder.** Before powder for the military service is received from the manufacturer, it is inspected and proved. For this purpose, at least 50 barrels are thoroughly mixed together. One barrel of this is proved by firing three rounds from a musket, with service-charge, if it be musket powder; if cannon or mammoth powder, from an 8-inch columbiad, with 10 lbs. and a solid shot of 65 lbs. weight and 7.88 inches in diameter; if it be mortar powder, from a 3-inch rifle-gun, with a charge of 1 lb. of powder and an expanding projectile weighing 10 lbs. The general character of the grain, and its freedom from dust, are noted.

General qualities. Gunpowder should be of an even-sized grain, angular and irregular in form, without sharp corners, and very hard. When new, it should leave no trace of dust when poured on the back of the hand, and when flashed in quantities of 10 grains on a copper plate, it should leave no bead or foulness. It should give the required initial velocity to the ball, and not more than the maximum pressure on the gun, and should absorb but little moisture from the air.

Size of grain. There are five kinds of powder in the U. S. land service, depending on the size of the grain, viz.: *Mammoth* for the 15-inch gun, *Cannon* for smaller sea-coast guns and mortars, *Mortar* for field and siege cannon, *Musket* for rifle-muskets, and *Rifle* for pistols.

The size of the grain is tested by standard sieves made of sheet brass pierced with round holes.

The diameters of the large and small holes are as follows, viz.:

For Mammoth, 0.9 inch and 0.6 inch; for Cannon, 0.31 inch and 0.27 inch; for Mortar, 0.1 inch and 0.07 inch; for Musket, 0.06 inch and 0.035 inch. Not more than 5 per cent. should remain on the large, nor pass through the small standard sieves.

Gravimetric density. Is the weight of a given measured quantity. It is usually expressed by the weight of a cubic foot in ounces.

This cannot be relied upon for the true density when accuracy is desired, as the shape of the grain may make the denser powder seem the lighter.

Specific gravity. The specific gravity of gunpowder must be not less than 1.75; and it is important that it should be determined with accuracy. Alcohol and water saturated with saltpetre have been used for this purpose; but they do not furnish accurate results. Mercury, only, is to be relied upon.

Mercury densimeter. This apparatus was invented by Colonel Mallet, of the French army, and M. Bianchi, and consists of an open vessel containing mercury, a frame supporting a glass globe communicating by a tube with the mercury in the open vessel, and joined at top to a graduated glass tube, which communicates by a flexible tube with an ordinary air-pump. Stop-cocks are inserted in the tubes above and below the glass globe, and a diaphragm of chamois-skin is placed over the orifice at the bottom of the globe, and one of wire-cloth over the upper orifice.

The operation consists as follows: Fill the globe with mercury to any mark of the graduated tube, by means of the air-pump; close the stop-cocks; detach the globe, full of mercury, and weigh it; empty and clean the globe; introduce into it a given weight of gunpowder; attach the globe to the tubes; exhaust the air till the mercury fills the globe and rises to the same height as before; shut the stop-cocks; take off the globe and weigh it as before. If we represent by a the weight of the powder in the globe, by P the weight of the globe full of mercury, by P' the weight of the globe containing the powder and mercury, and by D the specific gravity of the mercury.

The specific gravities of the powder and the mercury being proportional to the weights of equal volumes of these two substances, we have

$$a : P-P'+a : : d : D$$

hence

$$d=\frac{a\times D}{P-P'+a}$$

A mean of three results will give the true specific gravity. If the powder is good in other respects, the density may vary from 1.67 to 1.79.

Initial velocity. The initial velocity is determined by the Electro-Ballistic Pendulum. It should not be less than 1,050 feet for Mammoth, 1,225 feet for Cannon, 1,000 feet for Mortar and 975 feet for Musket powder.

Strain on the gun. The strain on the gun is determined by Major Rodman's pressure-piston, an instrument which is attached to the breech of the proof gun, and the principles of which are explained on page 152.

For Mammoth powder the pressure should not be more than 10,000 lbs., for Cannon, not more than 40,000 lbs., and for Mortar not more than 50,000 lbs. to the square inch.

Inspection report. The report of inspection should show the *place* and *date* of fabrication and of proof, the *kind* of powder and its general qualities, as the number of grains in 100 grs., whether hard or soft, round or angular, of uniform or irregular size, and if free from dust or not; the initial velocities obtained in each fire; the amount of moisture absorbed; and, finally, the height of the barometer and hygrometer at the time of proof.

17. **Packing.** Government powder is packed in barrels of 100 lbs. each. The barrels are made of well-seasoned white oak; and hooped with hickory or cedar hoops, which should be deprived of their bark to render them less liable to be attacked by worms. Barrels made of corrugated tin are undergoing trial, to test their fitness to replace those made of wood.

Marks on the barrels. Each barrel is marked on both heads (in white oil-colors, the head painted black) with the number of the barrel, the name of the manufacturer, year of fabrication, and the kind of powder,—*cannon*, (used for heavy cannon,) *mortar*, (used for mortars and field cannon,) or *musket*—the mean initial velocity, and the pressure per square inch on the pressure-piston. Each time the powder is proved, the initial velocity is marked below the former proofs, and the date of the trial opposite it.

18. **Analysis.** Whatever may be the mode of proof adopted, it is essential, in judging of the qualities of gunpowder, to know the mode of fabrication and the

proportions and degree of purity of the materials. The latter point may be ascertained by analysis.

In the first place, determine the quantity of water that the powder contains, by subjecting it to a temperature of 212°, in a stove or in a tube with a current of warm air passing over it, until it no longer loses in weight. The difference in weight, before and after drying, gives the amount of moisture contained in the powder.

To determine the quantity of saltpetre. In a vessel of tinned copper, like a common coffee-pot, dissolve 1,000 grains of powder, well dried before weighing, in 2,000 grains of distilled water, and heat it until it boils; let it stand a moment, and then decant it on a piece of filtering-paper, doubled exactly in the middle; this operation is repeated three times; at the third time, instead of decanting, pour the whole contents of the vessel on the filter; drain the filter, and wash it several times with 2,000 grains of water heated in the vessel, using in all these operations 10,000 grains of water. After passing through the filters, this water contains in solution all the saltpetre, the quantity of which is ascertained by evaporating to dryness. Dry the double filter with the mixture of coal and sulphur, and take the weight of this composition by using the exterior filter to ascertain the weight of that on which the composition remains; this weight serves to verify that of the saltpetre and to estimate the loss in the process.

To determine the quantity of charcoal directly. To separate the sulphur from the charcoal, subject the powder, either directly or after the saltpetre has been dissolved out, to the action of a boiling solution of the sulphide of potassium or sodium, which dissolves the sulphur and leaves the charcoal, the weight of which may be easily determined.

It is important that the sulphides of potassium and sodium used in dissolving the sulphur, should contain no free potassa or soda; for each of these alkalies would dissolve a part of the carbon—particularly of the brown coal.

The sulphide of carbon also dissolves the sulphur contained in powder, and may be used to determine the weight of charcoal which it contains.

The charcoal, separated from the saltpetre and sulphur, is dried with care and weighed, and should then be submitted to analysis in an apparatus used for burning organic matters. The composition of the charcoal may be judged of by comparing it with the results obtained in the analysis of charcoal of known quality used in the manufacture of powder.

To determine the quantity of sulphur directly. Mix and beat in a mortar 10 grains of dry powder, 10 of carbonate of potassa, 10 of saltpetre, and 40 of chloride of sodium; put this mixture in a vessel (capsule) of platinum or glass, on live coals, and, when the combination of the materials is completed and the mass is white, dissolve it in distilled water, and saturate the solution with nitric acid; decompose the sulphate which has been formed, by adding a solution of chloride of barium, in which the exact proportions of the water and the chloride are known. According to the atomic proportions, the quantity of sulphur will be to that of the chloride of barium used as 16. to 104.

19. **Hygrometric qualities.** The susceptibility of pow-

der to absorb moisture is due to the charcoal and the presence of deliquescent salts, principally chloride of sodium or common salt. The absorbent power may be judged of by exposing 1 lb. to the air in a moist place (such as a cellar which is not too damp) on a glazed earthen dish, for 15 or 20 days, stirring it sometimes so as to expose the surface better; the powder should be previously well dried, at the heat of about 140°. Well-glazed powder, made of pure material, treated in this way, will not increase in weight more than 5 parts in 1,000, or a half of one per cent.

20. **Quickness of burning.** The relative quickness of two different powders may be determined by burning a train laid in a circular or other groove which returns into itself, one half of the groove being filled with each kind of powder, and fire communicated at one of the points of meeting of the two trains; the relative quickness is readily deduced from observation of the point at which the flames meet.

21. **Restoring unserviceable powder.** When the quantity of water does not exceed 7 per cent., the powder may be restored by drying; this may be effected in the magazine, if it be dry, by means of ventilation, or by the use of the chloride of calcium for twenty or thirty days.

Quick-lime may be used; but the use of it is attended with danger; on account of the heat evolved in slaking.

When powder has absorbed from 7 to 12 per cent. of water, it may still be restored by drying in the sun or drying-house; but it remains porous and friable, and unfit for transportation: in this case it is better to work

3

it over. In service, it may be worked by means of the rolling-barrels, as described for making round powder.

When powder has been damaged with salt water, or become mixed with dirt or gravel, or other foreign substances which cannot be separated by sifting, or when it has been under water, or otherwise too much injured to be reworked, it must be melted down to obtain the saltpetre by solution, filtration, and evaporation.

22. **Storage, &c.** In the powder-magazines, the barrels are generally placed on the sides, three tiers high, or four tiers if necessary; small skids should be placed on the floor, and between the several tiers of barrels, in order to steady them; and chocks should be placed at intervals on the lower skids, to prevent the rolling of the barrels. The powder should be separated according to its kind, the place and date of fabrication, and the proof range. Fixed ammunition, especially for cannon, should not be put in the same magazine with powder in barrels, if it can be avoided.

Besides being recorded in the magazine book, each parcel of powder should be inscribed on a ticket attached to the pile, showing the entries and the issue.

23. **Preservation.** For the preservation of the powder, and of the floors and lining of the magazine, it is of the greatest importance to preserve unobstructed the circulation of the air, under the flooring as well as above. The magazine should be opened and aired in clear, dry weather, when the air outside is colder than that inside the magazine; the ventilators must be kept free; no shrubbery or trees should be allowed to grow so near as to protect the building from the sun. The moisture of a magazine may be absorbed by chloride of calcium,

suspended in an open box under the arch, and renewed from time to time; quick-lime, as before observed, is dangerous.

The sentinel or guard at a magazine, when it is open, should have no fire-arms; and every one who enters the magazine should take off his shoes, or put socks over them; no sword or cane, or any thing which might occasion sparks, should be carried in.

24. **Transportation.** Barrels of powder should not be rolled for transportation; they should be carried in hand-barrows, or slings made of rope or leather. In moving powder in the magazine, a cloth or carpet should be spread; all implements used there should be of wood or copper; and the barrels should never be repaired in the magazine.

When it is necessary to roll the powder, for its better preservation and to prevent its caking, this should be done with a small quantity at a time, on boards in the magazine yard.

In wagons, barrels of powder must be packed in straw, secured in such a manner as not to rub against each other, and the load covered with thick canvas.

EFFECTS OF GUNPOWDER.*

25. **History, etc.** The projectile arms of the ancients, such as bows, ballistas, and catapults, were operated by the same motive power—that of the spring.

Although large masses were thrown from these machines, the velocity imparted was feeble, as the springs rapidly lost their power, from being bent; and

* Vide PIOBERT'S *Cours d'Artillerie.*

the introduction of gunpowder, a more certain as well as powerful agent, gradually caused them to be superseded.

As before stated, the power of this agent is essentially due to the almost instantaneous development of expansive gases and heat by combustion; and although its properties were known for a long time, its use was at first confined to fireworks and incendiary compositions alone.

The advantage of using an agent capable of communicating great velocity to a projectile, arises not only from the intensity of the shock, the possibility of disabling a large number of men, and penetrating very resisting objects, but from the fact that it allows of the use of lighter machines, whereby the projectile can be directed with greater ease and certainty against its object.

Although the combustible nature of powder was known in Asia from the earliest times, and its properties were described by Marcus Græcus and Roger Bacon, its application to projectiles seems to have been a subsequent result of accident.

It is stated that about the year 1330, Berthold Schwartz, a monk of Fribourg, was engaged in making experiments with a mixture of saltpetre, sulphur, and charcoal, such as described by Marcus Græcus, and had left the mixture in a mortar, covered with a large stone, when it unexpectedly caught fire and exploded, throwing the stone to a distance with great force. The experiment was repeated, and with such success that military men saw at once that it could be applied to move large projectiles. Its progress as a projectile power, however, was comparatively slow, and it was only at the beginning of the 16th century that it was generally used for military purposes.

For a long time after its introduction, gunpowder was used in the form of dust, or "mealed powder," from which it derived its name; but it was found difficult to load small arms with gunpowder in this condition, on account of the moisture which sometimes collects in the bore after a few discharges. To overcome this difficulty, it was given a granular form, and received the name of "musket powder." It was soon discovered, however, that two parts of *grained* powder produced as much effect as three parts of mealed powder; but the larger fire-arms of the day had not sufficient strength to resist this increased force, and mealed powder continued to be used until the close of the 16th century.

At first, the ingredients of powder were converted into cake with a hand-pestle; a process which gave grains of very irregular size and shape. It was afterward discovered that the quality could be much improved by careful manipulation, without sensibly altering the proportions of the ingredients first established.

Any improvement in gunpowder which increases its strength, also increases its injurious effects on the arms in which it is used. It becomes necessary, therefore, to study the form and thickness of fire-arms, and the nature of the agent whose operations they are intended to restrain and direct.

It is impossible to embrace in a single glance the details of a phenomenon as complicated as the explosion of a charge of powder. The senses cannot detect the relations which exist between the partial operations of a phenomenon, where they are produced with such rapidity that they seem blended into one. In this case the only sure method of investigation is to separately study the different facts, and then unite them as a whole, borrowing from the physical sciences a thorough knowledge of the substances operated upon.

If the numerous circumstances which influence the results of the explosion of gunpowder, and the enormous expansive force which is developed in its limited duration, prevent us from accurately determining the measure of its effects, we can at least determine the limits between which this measure is included; which is sufficient for artillery purposes. From the results thus obtained were calculated the iron and bronze howitzers introduced to supersede those of Gribeauval's system. With less thickness of metal, these pieces were found to answer every requirement of service; a fact which tends to confirm the accuracy of the data from which they were constructed.

26. **Explosion.** The phenomenon of the explosion of powder may be divided into three distinct parts, viz.: *ignition*, *inflammation*, and *combustion*.

By ignition is understood the setting on fire of a particular point of the charge; by inflammation, the spread of the ignition from one grain to another; and by combustion, the burning of each grain from its surface to centre.

27. **Ignition.** Gunpowder may be ignited by the electric spark, by contact with an ignited body, or by a sudden heat of 572° Fahrenheit. A gradual heat decomposes powder without explosion by subliming the sulphur. Flame will not ignite gunpowder unless it

remains long enough in contact with the grains to heat them to redness. Thus, the blaze from burning paper may be touched to grains of powder without igniting them, owing to the slight density of the flame, and the cooling effect of the grains. It may be ignited by friction, or a shock between two solid bodies, even when these are not very hard. Experiments in France, in 1825, show that powder may be ignited by the shock of copper against copper, copper against iron, lead against lead, and even lead against wood; in handling gunpowder, therefore, violent shocks between all solid bodies should be avoided.

The time necessary for the ignition of powder varies according to circumstances. For instance, damp powder requires a longer time for ignition than powder perfectly dry, owing to the loss of heat consequent on the evaporation of the water; a powder, the grain of which has an angular shape and rough surface, will be more easily ignited than one of rounded shape and smooth surface; a light powder, more easily than a dense one; and a powder made of a black charcoal, more easily than one made of red, inasmuch as the latter is compelled to give up its volatile ingredients before it is acted on by the nitre.

28. **Combustion.** The *velocity of combustion* is the space passed over by the surface of combustion in a second of time, measured in a direction perpendicular to this surface.

The diameter of the largest-size grain of mortar-powder does not exceed 0.1 inch; the time of its combustion, therefore, is altogether too transient to be ascertained by direct observation. It may be deter-

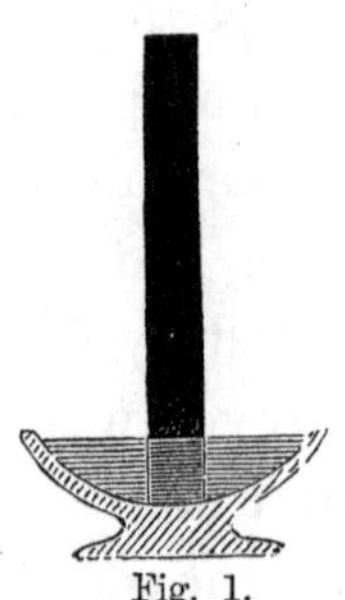
Fig. 1.

mined by compressing the composition into a tube and burning it, or by burning the "press-cake." In the latter case, take a prism of the cake about fourteen inches long and one inch square at the base. Smear the sides with hogs' lard, and place it on end in a shallow dish of water. The object of the lard is to prevent the spread of the flame to the sides; and the water is to prevent the lower end from being ignited by burning drops of powder. Set the upper end on fire, and note the time of burning of the column with a stop-watch beating tenths of seconds.

In either way it will be shown that the composition, if homogeneous, burns in parallel layers, and that the velocity of combustion is uninfluenced by the size of the column, or by the temperature and pressure of the surrounding gas.

The velocity of combustion of dry French war-powder is thus found to be 0.48 in., and of English powder, which American powder closely resembles, it is about 0.4 in.

It may be shown by direct experiment that the burning of a grain of powder in a fire-arm, is progressive, and that the size of the grain exerts a great influence on the velocity of the projectile, especially in short arms.

For this purpose take a mortar eprouvette and load it with a single fragment of powder weighing forty-six grains; fire it, and the ball will not be thrown out of the bore; divide the same weight into seven or eight fragments, and it will barely be thrown out of the bore; divide it into fifteen fragments, and it will be thrown about ten feet; fifty fragments will throw it about thirty feet; and the same weight of cannon-powder, about one hundred and seventy feet.

The progressive burning of powder is further confirmed by the fact, that burning grains are sometimes projected from a gun with sufficient force to perforate screens of paper, wood, and lead, at considerable distances. It is even found that they are set on fire in the gun, and afterward extinguished in the air before they are completely consumed. The large grains of powder used in the fifteen-inch columbiad are thrown out burning, to a distance of one hundred yards.

The velocity of combustion of powder varies with the *nature*, *proportions*, *trituration*, *density*, and *condition* of the ingredients.

Purity of ingredients. To secure the greatest velocity of combustion, it is necessary that the nitre and sulphur should be pure, or nearly so. This can always be effected by a proper attention to the prescribed modes of refining; but with the charcoal it is different, for the part which it plays in combustion depends upon certain characters which are indicated by its color and texture. The velocity of combustion will be greater for red charcoals than those that are black and strongly calcined; and for light and friable charcoals, than those that are hard and compact. It appears, in fact, to be nearly proportioned to the combustibility of the charcoals given in the tables on page 19.

Proportions. The proportions of the ingredients have a very great effect on the combustion; by varying them, all velocities between 0 and .55 inch can be obtained; the latter number can scarcely be exceeded. The proportions which give a maximum, appear to be comprised between the two following:

Nitre, 76.	Charcoal, 15.	Sulphur, 9.
" 76.	" 14.	" 10.

As it is often useful in preparing fireworks to know the proportions which will give a certain velocity of combustion, a table is given of a series of proportions of nitre, sulphur, and charcoal, and the corresponding velocities of combustion:

Sixty parts of nitre, compounded with certain proportions of sulphur and charcoal, gave the following velocities:

Parts of Sulphur.	Parts of Black Charcoal.							
	0	5	10	11	15	20	30	60
	Inch.	Inch.	Inch.	Inch.	Inch.	Inch.	Inch.	Inch.
0	.0	.02	.11	.14	.24	.34	.43	.07
5	.0	.05	.24	.30	.43	.47	.35	.00
8	.0	.06	.50	.51	.49	.41	.20	.00
10	.0	.08	.47	.49	.47	.39	.16	.00
15	.0	.11	.43	.44	.36	.35	.14	.00
20	.0	.16	.39	.40	.38	.30	.10	.00
30	.0	.27	.34	.33	.29	.21	.01	.00
60	.0	.00	.00	.00	.00	.00	.00	.00

It will be seen that the proportions 6—1—1 are among those that give the greatest amount of gas in a given time, other circumstances being equal; for the reason, that the weight burned during this time is greater, and because each unit of weight gives a greater volume of gas.

Trituration. Trituration of the ingredients increases the velocity of combustion; and this increase is much greater as the proportions approach those which give the greatest velocity. For the results of experiments on this point, see accompanying table:

Duration of trituration.	Velocity of combustion.			Remarks.
	Composition.			
	A.	B.	C.	
Hours.	Inches.	Inches.	Inches.	
0	.12	.13	.0189	Compositions dry.
1	.31	.25	.0192	Nitre. Ch'coal. Sulphur. Composition.
2	.38	.29	.0200	A, 75.00 12.5 12.50 Gunpowder.
3	.40	.32	.0204	B, 68.00 12.0 28.00 Fuze composition. C, 66.66 2.0 31.34 Port-fire "
4	.44	.34	.0212	
5	.46	.35	.0216	The nitre was taken as it came from the refinery. The sulphur and charcoal
10	.48	.37	.0236	had already been triturated in the rolling-barrels.

Density. For each set of proportions, the maximum velocity corresponds to a very small density. By increasing the density, the velocity is diminished; and more rapidly for quick compositions than slow ones. When in the form of a dust, gunpowder composition burns more slowly without compression than with it. For the results of experiments on the preceding compositions, see the following table; the trituration was extended to ten hours:

Density.	Velocity of combustion.			Remarks.
	Composition.			
	A.	B.	C.	
0.80	.360	.310		The pulverized composition is simply poured into a tube, and settled by striking lightly on a table.
1.00	.440	.410	.0319	The composition poured in as above, and compressed under a weight of 22 lbs. without shock.
1.20	.470	.390	.0295	Composition driven with a mallet weighing 2.2 lbs., falling through a height of 3.9 inches.
1.40	.480	.380	.0252	Same, save the height, which was 27 inches.
1.60	.890	.366	.0224	These densities were obtained by increasing the number of blows with the mallet for each ladleful of composition.
1.80	.443	.360	.0220	
2.00		.340		The density of a composition under the same pressure, increases with the trituration of the ingredients.
2.16		.330		

Moisture. By moistening the composition with pure water, alcohol, or vinegar, and then drying it completely, the velocity of combustion is increased. With pure water alone, this increase of velocity may amount to 0.1 of an inch. On the contrary, the velocity is diminished where oils, fatty or resinous substances, are added to the composition, or when it incloses water or other liquids.

Dry Powder, or one containing ½ per cent. of moisture, has a velocity of 0.48 in.
" " 1½ " " " 0.39 in.
" " 2½ " " " 0.33 in.

29. **Law of formation of gaseous products.** When the form and size of the grains and the velocity of combustion are known, we can ascertain, at any given mo-

ment, the amount of powder consumed, as the velocity is uniform and independent of the surface.

Spherical grain. Take a spherical grain of powder of homogeneous structure, one that will completely burn up in $\frac{1}{10}$ of a second. Apply fire at any point of its surface, the flame will immediately envelop it, and burn away the first spherical layer; if, for example, we suppose the time of this partial combustion be $\frac{1}{10}$ of the time required to burn up the entire grain, then the radius of the remaining sphere will be only $\frac{9}{10}$ of the first; but the volumes of spheres being to each other as the cubes of their radii, the primitive sphere will be to the one which remains after the burning of the first layer, as 1.0 is to 0.729, the cube of .9. Subtracting the second of these numbers from the first, we shall have 0.271, which expresses the difference of volumes of the two spheres, or the amount consumed in the first $\frac{1}{10}$ of time, compared to that of the entire grain. By making similar calculations on the other layers, we shall obtain the results contained in the following table:

Fig. 2.

Time of burning	0.000	.100	.200	.300	.400	.500	.600	.700	.800	.900	1.000
Decreasing radii	1.000	.900	.800	.700	.600	.500	.400	.300	.200	.100	0.000
Volumes of grain	1.000	.729	.512	.343	.216	.125	.064	.027	.008	.001	0.000
Volumes burnt	0.000	.271	.488	.657	.784	.875	.936	.973	.992	.999	1.000
Volumes burnt in each 0″.01	0.000	.271	.217	.171	.127	.091	.061	.037	.019	.007	0.001

It will be seen from this, that for equal intervals of time, those taken in the first period of combustion give forth very much larger amounts of gas than those taken in the last. If, instead of a sphere, we suppose the grain to be a *polyhedron* circumscribing a sphere, the burning layers being parallel, the decreasing grain will continue to be a similar polyhedron, circumscribing a sphere. The results given in the table will be strictly true for this case, as well as for grains of conical or cylindrical form, provided their bases be equal to their heights.

Fig. 3.

General formula. A general formula may be deduced to show the amount of gas developed at any instant of the combustion of a grain or charge of powder. For this purpose take a spherical grain of powder, and consider it inflamed over its entire surface.

Let t represent the time of burning, from the instant of ignition to the moment under consideration: t', the time necessary to burn from the surface to the centre, or total combustion: R, the radius of the grain.

Since the combustion passes over the radius R in the time t', the velocity of combustion is equal $\frac{R}{t'}$, and for the time t, it will pass over the space $t\frac{R}{t'}$ or $R\frac{t}{t'}$; the radius of the decreasing sphere will therefore be $R\left(1-\frac{t}{t'}\right)$. The volume of the grain of powder and that of the decreasing sphere are $\frac{4}{3}\pi R^3$ and $\frac{4}{3}\pi R^3\left(1-\frac{t}{t'}\right)^3$, respectively; and their difference, or the quantity of powder burned, will be equal to $\frac{4}{3}\pi R^3\left(1-\left(1-\frac{t}{t'}\right)^3\right)$.

The first factor of this expression represents the primitive volume of a grain of powder, and the other expresses the relation of the volume burned to the primitive volume.

The same expression will answer for all the grains of a charge of powder, if they are of the same size and composition; consequently, if we let A represent the volume or weight of the grains composing a charge of powder, the quantity remaining unburned after the time t will be represented by $A\left(1-\frac{t}{t'}\right)^3$; and the quantity burned, by $A\left(1-\left(1-\frac{t}{t'}\right)^3\right)$.

Although the grains of powder are not spherical, their sharp angles are partially worn away by rubbing against each other in glazing and in transportation; and the mode of fabrication and inspection reduces the variation in size within narrow limits; therefore, if we examine the influence which the actual form and size of the grains exercises over the phenomenon of combustion of powder, we shall find that the effect varies but slightly from that due to the spherical form.

Application to ordinary powder. Take a grain of oblong form, like that of a spheroid, or cylinder terminated by two hemispheres: it will present a greater surface than a spherical grain of the same weight, and consequently the amount of gas formed from it in the first instants of time, will be greater, and the duration of the combustion will be less. It can be shown, however, that so long as the size of the grains is kept within the regulation limits, this influence will be slight. To do this, take an oblong grain the cylindrical part of which has a diameter of .054 in., let it be terminated by two hemispheres, and have a total length of .097 in. (these being the minimum and maximum size of a grain of French cannon-powder, respectively); its weight will be about .07 grain, or $\frac{1}{210}$ of a *gramme*, and with a velocity of combustion of 0.48 it will take 0.056″ to burn up completely.

French war-powder is composed of grains of different weights, numbering about 310 to every *gramme*, or 15.4 grs. Troy. If, therefore, powder contain oblong grains of the size stated above, there must be others still smaller: if we suppose them to be in equal quantities, and the larger to be $\frac{1}{210}$ of the unit of weight, then the smaller must be equal to $\frac{1}{410}$ of the unit of weight; which would be equal to spheres with a radius of 0.028 inch. Comparing the quantities of gas developed in intervals of .008″, or about $\frac{1}{8}$ of the time necessary for the combustion of the smallest grains, we obtain the result in the following table:—

Kinds of grains of Powder.	Relation of the volume of powder burned, to the volume of the grains after a time of						
	0″.008	0″.016	0″.024	0″.032	0″.040	0″.048	0″.056
Elongated grains, diamr. .054 in.; length, 0.098 in.,—210 to the gramme, or 15.4 grs., . .	0.310	0.555	0.737	0.864	0.946	0.987	1.000
Spherical grains of 410 to the gramme, or .056 in. diameter,	0.357	0.616	0.794	0.907	0.968	0.994	0.999
Elongated and spherical grains as above, in equal quantities, forming a mixture of 310 to the gramme, ...	0.333	0.585	0.766	0.885	0.958	0.990	0.999
Spherical grains of 310 to the gramme, or 0.063 in. diameter,	0.330	0.580	0.758	0.875	0.948	0.985	0.998
Difference between mixed grains and spherical grains of the same mean weight, ...	0.003	0.005	0.008	0.010	0.010	0.005	0.001

The differences in the results do not much exceed $\frac{1}{100}$, and may be neglected in practice; we may accordingly consider all the grains of a charge of powder as spheres with radii corresponding to their mean weight. *This mean weight is an important element, and may be determined by counting the number of grains in a given charge, and dividing the weight of the charge by this number.*

In war-powder the largest portion of each grain is burned in the first two-tenths of the time required to consume the entire grain: as it has been shown that a grain of ordinary cannon-powder requires 0.1 second for its combustion, the largest portion of the grain will be burned in the first $\frac{2}{100}$ of a second. If we consider the velocity of the projectile on leaving a gun, and the time necessary to overcome its inertia in the first period of its movement, we shall see that a very large portion of each grain is burned up before the projectile leaves the gun. *If the size of the grain be increased, the effect will be to diminish the amount of gas evolved in the first instants of time, and to diminish the pressure on the breech.** This principle has been made use of lately to increase the endurance of large cannon.

29. **Inflammation.** When grains of powder are united to form a charge, and fire is communicated to one of them, the heated and expansive gases evolved, insinuate themselves into the interstices of the charge, envelop the grains and ignite them, one after the other.

* This idea has been carried out more fully in the experiments of Captain Rodman, by converting the powder into one or more cakes, which are perforated with numerous small holes for the passage of the flame. In this way a large portion of the powder is consumed on an increasing instead of a decreasing surface, and the amount of gas given out in the last moments will be greater than in the first; and thus the strain on the breech of a gun may be very much diminished without proportionately diminishing the velocity communicated to the projectile. For actual results obtained with this kind of powder, see Note appended to section 109.

This propagation of ignition is called *inflammation*, and its velocity the *velocity of inflammation*. It is much greater than that of combustion, and it should not be confounded with it.

The velocity with which inflamed gases move in a resisting tube, like a cannon, is very great. Hutton calculated it to be from 3,000 to 5,000 feet per second; and Robins determined it by experiment to be about 7,000 feet per second.

But when these gases are forced to pass through the interstices of powder, the resistance offered will considerably diminish the velocity of their expansion: it is found to vary with the form and size of the grains; and may be supposed to be reduced to 33 feet per second. The velocity of combustion, as before stated, is only .48 inch per second.

Although the velocity of inflammation of a train of powder can afford but an imperfect idea of this velocity in a gun, it may be interesting to study it.

The velocity of inflammation of a train of powder generally varies with the size of the grains, with the quantity of powder employed, and the disposition of the surrounding bodies, as will be shown by the following results of actual experiment.

The amount of powder in each train was about .11 lb. to the linear foot, and the time corresponding to the distances was one second.

On a plane surface in the open air,	7.87	feet.
In an uncovered trough,	8.13	"
In a linen tube,	11.38	"
In the same tube placed in the trough,	17.48	"
In the trough covered up,	27.88	"

These velocities are less than those obtained in fire-arms, for the reason that the powder is not only confined at the sides, but at one end, which was not the case in the experiment with the covered trough, where it could expand in both directions.

A velocity of more than three hundred feet can be obtained by burning quick-match inclosed in a cloth tube.

The size of the cross-section influences the velocity, as was shown by burning a train containing .062 lb. per foot in an open trough: the velocity was 5.77 feet, instead of 7.87 feet; and in a covered trough it was twenty feet, instead of 27.88. The velocity, therefore, increases with the cross-section of the train.

To determine the influence of the size of the grains on the velocity of inflammation, two trains were fired, one composed of fine grains, and the other of large ones; the velocity of the first was 8.2 feet, and the second was 7.54. This difference was due to the greater amount of gas developed by the small grains in the first instants of combustion.

The nature of the charcoal exerts an influence, the black being more favorable to inflammation than the red.

For a specific gravity of 1.3, the velocity is 7.5 feet.
" " " 1.6, " " 7.2 "
" " " 1.8, " " 6.2 "

Light powder is therefore found to be more inflammable than heavy.

If the grains be round the interstices are larger, and more favorable to the passage of the flame, and the inflammation of the mass. If they be sharp and angular, they will close upon each other in such a way as to reduce the interstices; and although the ignition of such grains may be more rapid, its propagation will be diminished.

It has been shown that, when powder is burned in an open train, fine powder inflames more rapidly than coarse; such, however, is not the case in fire-arms, owing to the diminution of the interstices. If a charge were composed of mealed powder, the flame could no longer find its way through interstices, and the velocity of inflammation and combustion would become the same. The velocity of inflammation of powder compressed by pounding is about .64 inch, while that of mealed powder in the same condition is only .45 in.

PRODUCTS, ETC., OF COMBUSTION.

30. **Nature of products.** Temperature and atmospheric pressure considerably influence the products obtained from burning gunpowder. When exposed in the open air to a temperature gradually increasing to 572° Fahrenheit, the sulphur sublimes, taking with it a por-

NOTE.—By compressing grain-powder under a hydrostatic press it may be converted into a solid cake, and be used in loading fire-arms, in place of the ordinary cartridge. No cement is required to unite the grains, as the pressure brings the particles of the surface of the grains within the limits of cohesive attraction, in the same way that artificial limestone is formed by compressing sand. As the pressure diminishes the interstices of the grains, it also diminishes the velocity of inflammation, and the rapidity with which the charge is converted into flame.

Experiments made at West Point, on some specimens of powder thus prepared by Dr. Doremus, of New York, showed that the pressure on the surface of the bore may be increased or diminished by diminishing or increasing the pressure on the cakes.

The cakes were covered by a water-proof, but highly inflammable varnish, which protected the powder from moisture, without apparently diminishing its inflammability.

tion of the carbon. This was shown by Saluces, who passed the volatilized products through a screen of very fine cloth, and found carbon deposited on it. Powder may be, therefore, completely decomposed by a gradual heat, without explosion; but when suddenly brought in contact with an ignited body, the temperature of which is at least 572° Fahrenheit, the sulphur has not time to sublime before explosion takes place

The proportions for war-powder for the United States service are seventy-six parts of nitre, ten of sulphur, and fourteen of carbon. By the atomic theory the proportions should be 74.64 nitre, 11.85 sulphur, 13.51 carbon. If we adopt these last proportions, the formula for gunpowder becomes

$$(NO^5+KO)+S+3C.$$

If we suppose the ingredients to be pure, and to arrange themselves under the influence of heat according to their strongest affinities, there will result one equivalent of nitrogen, three of carbonic acid, and one of sulphide of potassium, for

$$(NO^5+KO)+S+3C=N+3CO^2+SK.$$

The products are, therefore, solid and gaseous. Usually, powder contains a slight quantity of moisture; the ingredients are not absolutely pure, nor are they proportioned strictly according to their combining equivalents; it might be expected, therefore, that the actual would differ from the theoretical results.

The actual gaseous products obtained by combustion are, principally *nitrogen* and *carbonic acid*, sometimes *carbonic oxide*, a little *sulphuretted hydrogen*, *carburetted hydrogen*, and *nitrous oxide*. The solid products are, *sulphide of potassium*, *sulphate of potassa*, *carbonate of potassa* (mingled with a little *carbon*), and traces of *sulphur*.

When the sulphide of potassium comes in contact with the air, it is converted into sulphate of potassa, and gives rise to the white smoke which follows the explosion of gunpowder. A portion of the sulphide is sometimes condensed on the surface of the projectile, which accounts for the red appearance of shells, sometimes observed in mortar-firing.

The solid products are probably volatilized at the moment of explosion by the high temperature which accompanies the combustion; but, coming in contact with bodies of much lower temperature, they are immediately condensed. In chambered arms, small drops of sulphur may be observed condensed on the sides of the bore, which show that the sulphur has been volatilized; and we know that good powder burns on paper and leaves no trace. This fact, however, was most completely shown by the experiments of Count Rumford. This celebrated observer used a small eprouvette of great strength, which he partially filled with powder, and hermetically closed with a heavy weight. The powder was fired by heating a portion of the eprouvette to redness. Whenever the force was sufficient to raise the weight, the entire products escaped; when it was not, a solid substance was found condensed on the surface of the bore furthest from the source of heat.

31. **Temperature.** The temperature of the gaseous products of fired gunpowder has been variously estimated. Saluces determined by experiment that pure copper, which melts at a temperature of 4,622 Fahr., was not always melted by them; while brass, the melting-

point of which is about 3,900 Fahr., was invariably melted ; he was, therefore, induced to place their temperature at about 4,300 Fahr. As metals absorb a large amount of heat before melting, it is probable that the temperature of fired gunpowder is actually more than is here stated.

DETERMINATION OF THE FORCE OF GUNPOWDER.

32. **Absolute force.** The absolute force of gunpowder is measured by the pressure which it exerts when it exactly fills the space in which it is fired. Various experiments have been made to determine mechanically the absolute expansive force of fired gunpowder, but with widely different results. Robins estimated it at 1,000 atmospheres, Hutton at 1,800, D'Antoni from 1,400 to 1,900, and Rumford carried it as high as 100,000 atmospheres.* These discrepancies arise, in a great measure, from the very great difference which exists between the expansive force of the gases in the different moments of combustion, and from a want of coincidence in the observations.

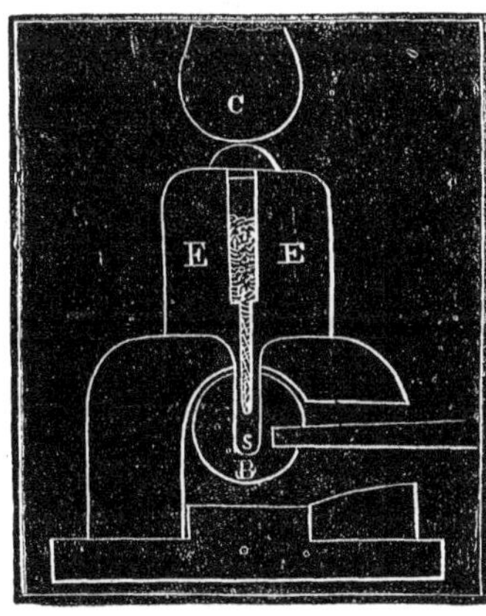

Fig. 5.

The apparatus used by Rumford to determine this point consisted, essentially, of a small eprouvette, E, capable of holding exactly 25 grains of powder. The orifice

* Rodman's experiments show the absolute pressure to be at least 13,333 atmospheres, or 200,000 lbs. to the square inch.

was closed with a heavy weight, and the powder was fired by heating the stem of the eprouvette, S, with a redhot cannon-ball, B. For the first trial, he filled the eprouvette with 25 grains of the best quality of dry powder, and rested upon the cover the knob, C, of a 24-pd. gun, whose weight was 8,081 lbs. Notwithstanding its great strength, the eprouvette was burst at the first fire into two pieces; and the 24-pdr. was raised. Rumford endeavored to show from the weight thus raised, that the pressure of the gases on the sides of the eprouvette was greater than 10,000 atmospheres. He further attempted to show, that as the tenacity of good iron is equal to 4,231 times the pressure of the atmosphere on the same surface, and as the surface of rupture was 13 times that of the bore, the force necessary to produce the rupture of the éprouvette must have been $13 \times 4{,}231$, or 55,003 atmospheres.

There are circumstances attending this experiment which should be taken into account, and which will very materially diminish this result. They are, the diminution of the tenacity of the iron, due to heating the eprouvette to produce explosion, and the incorrect method by which Rumford estimated the strength of a hollow cylinder subjected to a strain of expansion.

33. **Relation between density and force.** Experiments were continued, with a similar apparatus, to determine the relation between the density and the expansive force of fired gunpowder. The capacity of the eprouvette was nearly 25 grains. It was fired with various charges from 1 up to 18 grains; and the expansive force of each discharge was determined by the smallest weight necessary to close the orifice against the escape

of the gas. With the results of 85 trials a table was formed, from which a curve was constructed which expresses the relation between the density and expansive force of fired gunpowder, from 1 to 15 grains. By analogy and calculation, this curve was continued up to a charge of 24 grains; and for the density corresponding to this charge, the pressure was found to be 29,178 atmospheres.

This pressure is much greater than that developed in the explosion of projectiles and mines, owing to the low temperature of the surrounding surfaces, and the large amount of heat which they absorb. It is the same with cannon, for the most rapid firing does not raise the temperature of the bore above 210 Fahr., which is much below that of the eprouvette. Besides, the powder does not completely fill the space in rear of the ball; and, as powder burns progressively, this space is enlarged before the gases are completely developed, and consequently their density is diminished. There is also a loss of force by the escape of the gases through the windage and vent.

The following equation expresses the relation found to exist between the density and expansive force of charges of gunpowder, from 1 to 15 grains, fired in an eprouvette the capacity of which was 25 grains, or in other words, for charges in which the densities vary from .04 to .6:

$$p = 1.841(905d)^{1+0.362d};$$

in which p represents the pressure in atmospheres, and d the density of the inflamed products.

It will be seen from this equation, that the pressure

increases more rapidly than the density, since the exponent of the density is greater than unity.

The density of the gases is equal to the weight of the powder burned divided by the space occupied by the gases. By substituting this in the equation, we can determine the pressure exerted at any given instant of the combustion.

Although this relation is deduced for a particular kind of powder, it may be used for all service-powders and service-charges without serious error, since the actual amount of gaseous products is nearly the same for all, and the densities of the highest service-charges never exceed 0.6.*

34. **Force of powder when inflammation is instantaneous.** If the size, form, and density of the grains of a charge of powder, the velocity of combustion, and the

* The accuracy of Rumford's formula has been lately verified by a series of experiments made by Captain Rodman. The apparatus used by this officer consisted of a very thick cast-iron shell, to which was attached an indenting piston for determining the pressure on the inner surface, or powder cavity of the shell.

The following table shows the pressures calculated by the formula and the pressures obtained by the experiments, for three different densities:

Density.	Pressure by Rumford's Formula.	Pressure by Rodman's Experiments.
$d=\frac{1}{20}$	1,290 lbs.	1,066 lbs.
$d=\frac{1}{10}$	2,900 "	2,525 "
$d=\frac{1}{2}$	3,700 "	3,220 "

The lesser pressure obtained by Rodman's experiments may be in a great measure explained by the facts, that the shell was not heated, but fired with a friction tube, and that the gas was allowed to escape through the vent. Further experiments were made which show that so long as the volume of the charge bears the same proportion to the space in which it is fired, the pressure on the unit of surface remains the same, no matter what may be the amount of the charge. This follows also from Rumford's formula, since the value of p is not affected so long as d remains the same.

space in which it is contained, are known, we can determine the density of the gaseous products at any particular moment of combustion. For this purpose, take the case in which the inflammation of the whole charge is considered instantaneous, and let

P be the weight of the charge,
d' the density of the composition of which the powder is made,
V the space in which the gases expand,
t' the time of combustion of a single grain,
t the time since the combustion began,
d the density of the gases at a given instant.

According to section 29, the weight of powder remaining after a time, t, will be equal to $P\left(1-\frac{t}{t'}\right)^3$, and the volume will be equal to $\frac{P}{d'}\left(1-\frac{t}{t'}\right)^3$; the weight of gaseous products evolved will be equal to $P\left(1-\left(1-\frac{t}{t'}\right)^3\right)$; and their density will be equal to this quantity divided by the space V, diminished by the space occupied by the powder unburnt at the end of the time t.

Or,

$$d=\frac{P\left(1-\left(1-\frac{t}{t'}\right)^3\right)}{V-\frac{P}{d'}\left(1-\frac{t}{t'}\right)^3}.$$

Let K represent the ratio of the weight of powder which would fill the space V, to the weight of the charge P, and D the gravimetric density, or weight of a unit of volume of powder, we shall have the equation,

$$\frac{DV}{P}=K, \text{ or } \frac{V}{P}=\frac{K}{D};$$

and the formula for the density of the gaseous products becomes,

$$d=\frac{1-\left(1-\frac{t}{t'}\right)^3}{\frac{K}{D}-\frac{1}{d'}\left(1-\frac{t}{t'}\right)^3}=D\frac{1-\left(1-\frac{t}{t'}\right)^3}{K-\left(1-\frac{t}{t'}\right)^3\frac{D}{d'}}. \qquad (1)$$

If the charge fill the entire space V, $K=1$, and

$$d=D\frac{1-\left(1-\frac{t}{t'}\right)^3}{1-\left(1-\frac{t}{t'}\right)^3\frac{D}{d'}}.$$

When the grains are consumed, $t=t'$, and $d=\frac{D}{K}$; and if $K=1$, $d=D$.

Having determined the mean density of the gaseous products at any instant of the combustion, we can determine the pressure exerted on the enclosing surfaces by means of Rumford's formula

$$P=1.841(905d)^{1\times 0.362d}.$$

This value of P supposes that the entire charge is inflamed at the same time—a supposition that is not strictly correct, except for small and lightly-rammed charges. When the charge is large, and well-rammed, as in cannon, it is necessary to take into consideration the time of inflammation.

35. **Density when the inflammation is not instantaneous.** In a majority of cases the preceding formulas will give the relation between the density and expan-

sive force of gunpowder, without sensible error; but when the grains are small, and the charge is compressed by ramming, the interstices are diminished in size, and the inflammation is comparatively less rapid; besides, the size and form of the charge exert an influence which increases with its length. It is proposed, therefore, to modify the formulas, and adapt them to the most general case, by considering the inflammation progressive.

Take a charge of powder, of any form whatever, and consider it ignited at the point *A*, the inflammation will reach the surface of the concentric zone *B*, the radius of which is tv, in the time t, v being the velocity of inflammation. There will be portions of the charge situated within this zone which the flame will not have reached; others in which the combustion is completed; and others, between these two, in which the inflammation is completed, but the combustion is only partially completed. See figure 7.

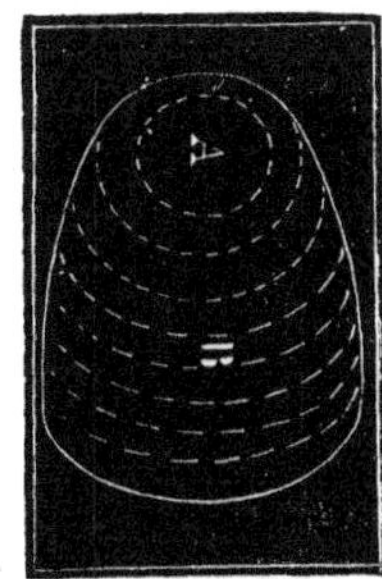

Fig. 6.

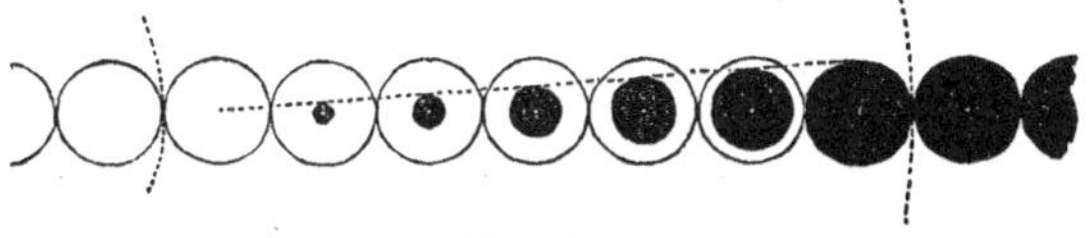

Fig. 7.

The extent of the inflamed zones being determined by the form and dimensions of the charge, exerts a great influence on the development of the gases, and consequently on their density.

If the velocities of inflammation and combustion be known, the quantity of gas formed from each zone can be calculated, and the question becomes one of analysis.

In this calculation, the integral limits which refer to the extent of the zones are determined by the surface of the charge; and those which refer to the progress of the combustion of the grains will be the point of ignition and the surface of inflammation; or, if θ be the time necessary for the flame to reach the surface of the zone, the radius of which is x, the time of partial combustion of a grain of this zone will be $t-\theta$, and its complete combustion is expressed by the relation $t=t'+\theta$.

For this zone the density of the gaseous products at the instant of inflammation will be $d=0$, and when completely consumed $d=D$.

The intermediate values may be determined by formula (1)

$$d=\frac{D\left(1-\left(1-\frac{t}{t'}\right)^3\right)}{K-\left(1-\frac{t}{t'}\right)^3\frac{D}{d'}};$$

by substituting $t-\theta$ for t, and supposing $K=1$, should the charge completely fill the space in which it is burned. Integrating between the determined limits, we obtain the mean density of the gases developed.

The solution of this question, in a general sense, is very difficult, and requires the aid of the differential calculus. There are particular cases, however, where the solution is not difficult; for instance, where the charge is of cylindrical form and is placed at the bottom of the bore of a gun.

36. **Calculation of the density of a charge of cylindrical form.** Although the charge of a gun is ignited at the rear and upper portion, we may consider that all portions of the circular layer at the bottom are inflamed

at once, and that the inflammation spreads by parallel layers throughout its extent. The space at the bottom of the bore, and the escape of gas through the vent, favor this supposition.

Let L represent the total length of the charge, and θ' the time necessary for the inflammation to pass over this length. Let us assume that $\theta'=nt'$, t' being the time necessary for the combustion of a single grain of the charge; n, therefore, is the ratio of these times.

The velocity of inflammation will be $\frac{L}{\theta'}$, or $\frac{L}{nt'}$; and $\frac{Lt}{nt'}$ will represent the portion of the charge inflamed in the time t. The length of the charge which will be consumed (and no portion can be entirely consumed unless $t>t'$) will be $(t-t')\frac{L}{nt'}$; and the thickness of the burning layer will be the difference between these two quantities, or $\frac{L}{n}$; which is constant.

If the area of a section of the charge, perpendicular to its axis, be taken as the unit of surface, the volumes may be represented by their lengths. Divide the length of the burning portion into a number, h, of smaller sections, the length of one of the smaller sections will be equal to $\frac{L}{nh}$; if h be very large, the grains of each very small section may be considered in the same stage of combustion, and the radii of the consumed layers in each grain of the small sections will be represented in parts of the primitive radius, as follows:—

For the 1st, 2d, 3d. . . $h-2$, $h-1$, h, sections.

$$\frac{h}{h}, \frac{h-1}{h}, \frac{h-2}{h}, \ldots \frac{3}{h}, \frac{2}{h}, \frac{1}{h}.$$

The radii of the burning grains will be,

$$o, \frac{1}{h}, \frac{2}{h}, \ldots \frac{h-3}{h}, \frac{h-2}{h}, \frac{h-1}{h};$$

and the corresponding volumes of the unburnt portions will be represented by.

$$o, \left(\frac{1}{h}\right)^3, \left(\frac{2}{h}\right)^3, \ldots \left(\frac{h-3}{h}\right)^3, \left(\frac{h-2}{h}\right)^3, \left(\frac{h-1}{h}\right)^3.$$

The volumes burned will be represented by,

$$1, \quad 1-\left(\frac{1}{h}\right)^3, \; 1-\left(\frac{2}{h}\right)^3, \ldots 1-\left(\frac{h-1}{h}\right)^3.$$

If D represent the gravimetric density of the powder, the weight of each small section will be $\frac{L}{nh}D$, and the weight of the gaseous products in all the sections will be

$$\frac{L}{nh}D\left\{h-\frac{1^3+2^3+3^3+4^3\ldots+(h-1)^3}{h^3}\right\};$$

but we know in general terms that

$$1^3+2^3+3^3\ldots z^3, \text{ or } \Sigma z^3=\left(z\left(\frac{z+1}{2}\right)\right)^2;$$

therefore the sum of the weights of the gases formed will be,

$$\frac{L}{nh}D\left\{h-\frac{1}{h^3}\left(\frac{h(h-1)}{2}\right)^2\right\}=\frac{LD}{n}\left\{1-\left(\frac{h-1}{2h}\right)^2\right\}.$$

If we suppose h, the number of sections, to be infinite, the above expression will reduce to

$$\frac{LD}{n}(1-\tfrac{1}{4})=\tfrac{3}{4}\frac{LD}{n}.$$

The portion of the charge entirely consumed being equal to $\frac{t-t'}{nt'}L$, its weight will be $\frac{t-t'}{nt'}LD$, and the total weight of gaseous matter developed will be,

$$\frac{t-t'}{nt'}LD+\tfrac{3}{4}\frac{LD}{n}=\frac{LD}{nt'}\left(t-\frac{t'}{4}\right).$$

The space which they occupy is equal to the volume of the inflamed portion of the charge, diminished by the volume of the unburned grains at the end of the time t; the volume of the burning powder is $\frac{L}{n}$, and its weight is $\frac{L}{n}D$. The weight of the portion burned being equal to $\tfrac{3}{4}\frac{LD}{n}$; that which remains unburned will be equal to $\tfrac{1}{4}\frac{LD}{n}$, and the density of the grains being d', their volume will be equal to $\tfrac{1}{4}\frac{LD}{nd'}$. The volume into which the gases expand will consequently be equal to

$$\frac{tL}{nt'}-\tfrac{1}{4}\frac{LD}{nd'}.$$

Finally, the mean density of the gases at the instant t, will be,

$$d=\frac{\frac{L}{nt'}D\left(t-\frac{t'}{4}\right)}{\frac{tL}{nt'}-\tfrac{1}{4}\frac{LD}{nd'}}=D\frac{t-\frac{t'}{4}}{t-\frac{t'D}{4d'}}.$$

From this it will be seen that the density is independent of the velocity of inflammation and length of the charge. The formula, however, can only be applied from the instant $t=t'$ to that in which $t=\theta'-t'$, that is to say, so long as there exists a portion of the charge in which the combustion is ceasing on its posterior surface, and commencing on its anterior surface.

Without committing a serious error, we can, however, apply the formula when $t=\frac{1}{2}t'$, because, in taking the sum of the cubes $0+1^3+2^3+3^3+\ldots+(h-1)^3$ from 1 to $(h-1)^3$ it will only be necessary to take it from $\left(\frac{h}{2}\right)^3$ to $(h-1)^3$, which makes an error equal to $1^3+2^3+3^3\ldots+\left(\frac{h}{2}-1\right)^3$, or $\frac{1}{16}$ of the total sum, as may be seen by replacing h by $\frac{h}{2}$ in the expression $\left\{\frac{h(h-1)}{2}\right\}^2$.

If the section of the charge, instead of being equal to the section of the bore of the gun, is only $\frac{1}{K}$, the gases being developed freely in a space K times greater, the density D will be diminished in an inverse ratio, and we shall have

$$d=\frac{D}{K}\frac{t-\frac{t'}{4}}{t-\frac{t'D}{4Kd'}}$$

37. **Application to practice.** Thus it will be seen that the density, and consequently the expansive force of fired gunpowder can be determined at each instant of combustion, either in the case in which the inflammation is considered instantaneous, or when considered progressive.

The accuracy of the formulas was verified in France some years since, in the course of a series of experiments to determine the influence which the size and density of grains of powder exert upon the initial velocity of a projectile.

There were six different sizes of grains tried, viz.:—.26 in., .21 in., .18 in., .15 in., .10 in. (cannon), .05 in. (musket); of each size there were six different densities, viz.:—1.3, 1.4, 1.5, 1.6, 1.7, 1.8, and four different modes of manufacture, making 144 varieties of powder in all. The instruments used were the ballistic pendulum, the 4-pdr. gun pendulum, the mortar eprouvette, and the infantry musket.

The results of calculation and direct experiment show a remarkable agreement, and may be summed up as follows, viz.:—

1. With the 4-pdr. gun the high densities gave greater velocities when combined with the smallest grains, and *vice versa*, the low densities gave greater velocities when combined with the largest grains. The grains which gave the highest velocities possessed medium size and density, or a density of 1.5 combined with a diameter of 0.18 in.

2. With the mortar eprouvette, which fired a smaller charge than the 4-pdr. gun, the fine-grained powder gave almost invariably greater velocities than the coarse. For a grain of .1 in. (or mortar size), the lowest densities gave the best results, and for a grain of .05 in. (or musket size), the highest densities gave the best results.

3. With the infantry musket, and a still smaller charge, the superiority of fine grains was more marked for all densities, and particularly so for the least.

It would appear from the foregoing, that the size and

density of grains of powder should be increased in a certain ratio with the weight of the projectile to be moved, the size of the charge, and the length and diameter of the bore in which the powder is burned; and it is for this reason that five different powders have been adopted for the military service of this country (see page 27).

GUN-COTTON.

38. **Gun-cotton, or pyroxile.** The action of nitric acid on such vegetable substances as saw-dust, linen, paper, and cotton, is to render them very combustible. In their natural state these substances are almost entirely composed of *lignine*, the constituents of which are *oxygen*, *hydrogen*, and *carbon*; nitric acid furnishes *nitrogen*, a substance which enters into the composition of nearly all explosive bodies.

Gun-cotton was discovered by Prof. Schönbein, and published to the world in 1846. His method of preparing it consists in mixing three parts of sulphuric acid, sp. grav. 1.85, with one part of nitric acid, sp. gr. 1.45 to 1.50; and when the mixture cools down to between 50° and 60° Fahr., clean rough cotton, in an open state, is immersed in it; when soaked, the excess of acid is poured off, and the cotton pressed tightly to remove as much as possible of what remains. The cotton is then covered over and left for half an hour, when it is again pressed, and thoroughly washed in running water to remove all free acid. After being partially dried by pressure, it is washed in an alkaline solution made by dissolving one ounce of the carbonate of potash in a gallon of water. The free acid being thus expelled, it is placed in a press, the excess of alkaline solution expelled, and the cotton left nearly dry. It is then washed in a solution of pure nitrate of potash, one ounce to the gallon, and being again pressed, is dried at a temperature of from 150° to 170°.

The sulphuric acid has no direct action on lignine, its use in the preparation of pyroxile being to retain the water abstracted from the cotton, and prevent the solution of the compound, which would take place, to a greater or less extent, in nitric acid alone.

Cotton, in its conversion into an explosive substance, increases very considerably in weight, owing to the formation of a new and distinct chemical compound.

Gun-cotton, when properly prepared, explodes at a temperature of about 380° Fahr. It will not, therefore, ignite gunpowder, when loosely poured over it.

Under ordinary circumstances, the electric spark will not explode it; but if the fluid be retarded in its progress by being passed over the surface of a string moistened with common water, and in contact with the cotton, explosion will follow.

From the experiments of Major Mordecai, made at Washington arsenal, in 1846, the following facts regarding the use of this substance in the military service, were ascertained:—

1. The *projectile force*, when used with moderate charges in musket or cannon, is equal to that of about twice its weight of the best gunpowder.

2. When compressed by hard ramming (as in filling a fuze), it burns slowly.

3. By the absorption of moisture, its force is rapidly diminished, but it is restored by drying.

4. Its *explosive force*, or bursting effect, is in a high degree greater than that of gunpowder. In this respect the nature of gun-cotton assimilates much more to that of the fulminates than to gunpowder. It is, consequently, well adapted for many purposes in mining.

5. Gun-cotton, well prepared, leaves no perceptible stain when a small quantity is burnt on white paper.

6. It evolves little or no smoke, as the principal residue of its combustion is water and nitrous acid; the latter is made sensible by its odor, and by its effects on the barrel of a gun, which will soon be corroded by it, if not wiped after firing.

7. In consequence of the quickness and intensity of action of gun-cotton, when ignited, it cannot be used with safety in our present fire-arms. An accident of service, such as that of inserting two charges into a musket before firing (which frequently occurs), would cause the bursting of the barrel; and it is probable that the same result would be produced by regular service charges, repeated a moderate number of times.

NOTE.—The following is said to be the improved method of preparing gun-cotton for the Austrian service: The cotton is immersed in the strongest preparation of acid—one part of nitric to three parts of sulphuric—and is permitted to remain in the acid bath for forty-eight hours. It is then washed in a stream of running water for four, six, or eight weeks, so as to remove every trace of free acid, which is not the case with gun-cotton made elsewhere.

It is prepared for service by spinning it into a thread, which is wound around a stick to form it into a cartridge of the required size. Thus prepared, the inflammation, which depends on the compactness of the mass of fibres, is reduced to a safe limit for cannon, and at the same time the means afforded of rendering the explosive effect of the different cartridges uniform.

It is stated that the Austrian Government is so far pleased with this substitute for gunpowder, as to have forty batteries of rifled field artillery prepared expressly for its use.

CHAPTER II.

PROJECTILES.

39. **Definition.** A projectile is intended to reach and strike, pass through, or destroy, a distant object; the effect of a projectile varies with its form and the material of which it is composed.

To destroy an object against which it is thrown, a projectile should have certain hardness and tenacity; if it be softer and less tenacious than the object, it will spread out laterally, or break into pieces, and presenting a greater surface, will meet with greater resistance, and consequently penetrate less than if it had preserved its primitive form. Great density is also favorable to penetration, inasmuch as it gives a projectile a greater mass for an equal surface.

40. **Materials.** *Stone, lead, wrought and cast iron* are materials, each possessing peculiar advantages for projectiles, according to the circumstances under which they are fired, and the objects against which they are used.

Stone. Stone projectiles were used before the invention of gunpowder, and very generally after it, until the year 1400, when the French made them of cast iron.

The defects of stone as a material for projectiles, are a want of density and tenacity, which requires it to be used in large masses, and fired with comparatively small charges of powder. The effect of stone balls against the walls of ancient cities was very great, but against modern fortifications, where the walls are sustained by large masses of earth, their effect is very slight. Until quite lately, bronze guns, throwing stone balls of enormous calibre, were used by the Turks in defending the passage of the Dardanelles. It is stated that when the English fleet, under Admiral Duckworth, forced the passage of these straits, a stone ball weighing 800 lbs. struck and nearly destroyed the English admiral's ship, and that one hundred men were killed and wounded by it.

Lead. Lead as a material for projectiles, possesses the essential quality of density; but it is too soft to be used against very resisting objects, since it is flattened even against water.

From its softness and fusibility, large projectiles of this material are liable to be disfigured, and partially melted, by the violent shock and great heat of large charges of powder. Its use is chiefly confined to small-arms and case-shot, which are generally directed against animate objects. These defects of lead may be corrected, in a measure, by alloying it with tin, antimony, &c.

Wrought iron. When great strength and density, combined, are required in a projectile, wrought iron may be used, but it is generally attended with considerable expense.

Cast iron. The introduction of cast iron, for large projectiles, was an important step in the improvement of artillery, as it unites in a greater degree than any other material, the essential qualities of hardness, strength, density, and cheapness; it is exclusively used for this purpose in the United States' service.

Compound. Compound projectiles are sometimes made so as to combine the good and correct the bad qualities of different metals. Thus, at the siege of Cadiz, cast-iron shells filled with lead, forming projectiles of great strength and density, were thrown from mortars to a distance of three miles and three quarters.

For rifle-cannon, projectiles are made occasionally of cast iron, and covered with a soft coating of lead, or other soft metal, to obviate the serious results that might arise from the wedging of the flanges in the grooves of the gun. Such is the construction of Armstrong's projectile in England, and Sawyer's and others, in this country.

In the rifle-cannon lately used by the French army in Italy, it is stated that the flanges which projected into the grooves of the bore were made of tin.

Considerable success has also been attained in uniting cast iron and wrought iron, and cast iron and soft metal, in such manner as to attain the strength of one metal, and the softness and expansibility of the other.

41. **Classification.** Projectiles may be classified according to their form, as *spherical* and *oblong.*

Spherical projectiles. Spherical projectiles are commonly used in smooth-bored guns, and for this purpose possess certain advantages over those of an oblong form: 1. They present a uniform surface to the resistance of the air as they turn over in their flight. 2. For a given weight they offer the least extent of surface to the resistance of the air. 3. The centres of figure and inertia coincide. 4. They touch the surface of the bore at only one point; they are therefore less liable to wedge in the bore, and endanger the safety of the piece.

5. Their rebounds on land and water being certain and regular, they are well suited to rolling and ricochet firing.

Oblong projectiles. The great improvements which have been made within the last few years in the accuracy and range of cannon and small-arms, *consist simply* in the use of the oblong instead of the spherical form of projectile.

The superiority of the oblong form has been long known, and for many years used in the sporting rifles of this country; but serious obstacles have always stood in the way of its general adoption into the military service.

To attain accuracy of flight and increase of range with an oblong projectile, it is necessary that it should move through the air in the direction of its length. Though experience would seem to show that the only sure method of effecting this, is to give it a rapid rotary motion around its long axis by the grooves of the *rifle*, numerous trials have been, and are now being made, to produce the same effect with smooth-bored arms.

Centre of gravity, &c. One of the simplest plans used for this purpose, is to place the centre of gravity, or inertia, in advance of the centre of figure, or resistance. If these points be situated in the long axis of the projectile, as they should be, the forces of propulsion and resistance, which act in opposite directions, will cause it to coincide with the line of flight.

This plan was tried on a hollow projectile in the time of Louis XIV., by dividing the cavity into two compartments by a partition, and filling the front one with bullets, and the rear one with powder; but the flight of these projectiles was uncertain and irregular, and it was observed that some of them burst in the air, and that others struck the object sidewise.

Another plan of this kind, proposed by Thiroux, is to make the projectile very long, with its rear portion of wood, and its point of lead or iron, somewhat after the manner of an arrow; but it does not appear that that method has ever been submitted to the test of practice.

Chain-ball. To arrest the motion of rotation of an oblong projectile, thrown under high angles, and with a moderate velocity, it has been proposed to attach a light body to its posterior portion, by means of a cord or chain, which will offer a resistance to the flight of the projectile, and cause it to move with its point foremost.

Nail-ball. This is a round projectile, and has an iron pin projecting from it to prevent it from turning in the bore of the piece.

Grooved balls. Attempts have also been made to give an oblong projectile a motion of rotation around its long axis, by cutting spiral grooves on its base for the action of the charge, or by cutting them on the forward part for the action of the air. These plans have not succeeded in practice, for the reason, perhaps, that the projectile naturally turns over end for end, and the air and charge do not act on the grooves with sufficient promptness, energy, and certainty to prevent it.

42. **Solid shot.** Projectiles may be further classified according to their structure and mode of operation, as *solid*, *hollow* and *case shot.*

Solid projectiles. Solid projectiles are used in *guns* and *small-arms*, and produce their effect by impact alone. When used in heavy guns they are known as *solid shot*, *round shot*, or *shot*. They are made of cast iron, and on account of their great strength and density, and the comparatively large charges of powder with which they are fired, are used when great range, accuracy, and penetration are required. They are the only projectiles that can be used with effect against very strong stone walls, or floating batteries covered with wrought-iron plates. Solid shot for guns are classified according to their weight, which, in the United States' land service, is as follows, viz.:

Field service, 6 and 12 pounders.

Siege service, 12, 18, and 24 pounders.

Sea-coast service, 32 and 42 pounders.

Solid shot for columbiads are classified according to the diameter of the bore, as 8 and 10 inch solid shot.

43. **Bullets.** The object of small-arms is to attain animate objects; their projectiles are, therefore, made of lead, and are generally known as *bullets*. They are both round and oblong; but in consequence of the great improvements that have been made of late, in adapting the principle of the rifle to small-arms, the oblong ball is now very generally used in all military services, the round bullet being chiefly retained for use in case-shot.

Round bullets. Round bullets are denominated by the number contained in a pound; this method is often used to express the calibre of small-arms; as, for instance, the calibre of the old musket was 17 to the pound, and the rifle was 32. In 1856, these two calibres were replaced by one of 24 to the pound, that of the new rifle musket. The number is sometimes prefixed to the word *gauge*, in which case the rifle-musket would be called a 24-*gauge gun*. This mode, however, is principally used to designate sporting arms.

The oblong bullet is denominated by its diameter and weight; for instance, the new rifle-musket ball has a diameter of 0.58 in., and weighs 540 grains.

Oblong bullet. The oblong bullet at present used in the United States' service, is composed of a cylinder surmounted by a conoid—the conoid being formed of the arcs of three circles. The cylinder has three grooves cut in it, in a direction perpendicular to its axis, to hold the grease necessary for lubricating the bore of the piece in loading, and possibly to guide the bullet in its flight, after the manner of the feathers of an arrow.

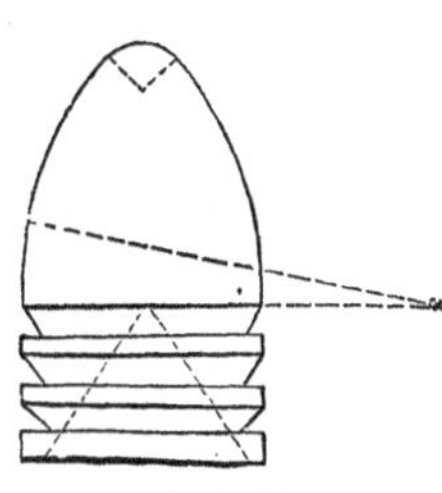

Fig. 9.

A conical cavity is formed in the bottom, in which the gas of the charge expands, and forces the sides of the bullet into the grooves or rifles of the gun. From these grooves it receives a rotary motion around its long axis, which prevents it from turning over in its flight.

44. **Shells.** Under the head of *hollow shot* are included shells for guns, howitzers, and mortars, and hand and rampart grenades. These projectiles are all made of cast iron; and for guns and field howitzers their calibres are expressed by the weight of the equivalent solid shot, as 12, 24, and 32 pound shells; and for all other howitzers and mortars, by the diameter of the bore of the piece, as 8 and 10 inch shells.

Shells have less strength to resist a shock, they are therefore fired with a smaller charge of powder, than solid shot. Their weight, and consequent mean density, is generally about *two-thirds* that of a solid shot of the same size.

Shells act both by impact and explosion, and are used against animals and such inanimate objects as will not cause them to break on striking.

The principal parts of a spherical shell are: 1. The *cavity*—the shape of which is similar to and concentric with the exterior. The use of the cavity is to contain a bursting charge of powder, if the object be merely to destroy by explosion; or a bursting charge and incendiary composition, if the object be to destroy by explosion and combustion together. The size of the cavity should be as large as possible, to produce the greatest explosive effect; but as the shell should have sufficient strength to resist the shock of the discharge, and sufficient weight to overcome the resistance of the air, the size of the cavity will necessarily be subordinate to these conditions, which fix the thickness of the metal. 2. The *fuze-hole*, which is used in inserting the bursting charge, and to hold the fuze which communicates fire to it. As the presence of the fuze-hole diminishes the effect of the bursting charge, the diameter of its orifice should be as small as possible. 3. The *ears* are two small recesses made near the fuze-holes of all shells larger than a 42-pounder, for the purpose of inserting the "hooks," and lifting the shells up to the bore of the piece in loading. A small hole is sometimes made in the upper hemisphere of shells, for the purpose of charging them after the fuze is driven; but late improvements in the construction of the fuze allow it to be dispensed with, so that the powder can now be poured directly through the fuze-plug, and the charging can be deferred until the moment of loading.

Fig. 10 represents a mortar-shell, and fig. 11 a shell used in a gun or sea-coast howitzer. The mortar-shell is fired with a lighter charge of powder than the gun-shell, and has therefore less thickness of metal.

The fuze-hole of the gun-shell, is reinforced with metal, so that the fuze will not be driven in by the force of the discharge. This reinforce serves, in a measure, to compensate for the metal taken out of the fuze-hole, and renders the shell more concentric.

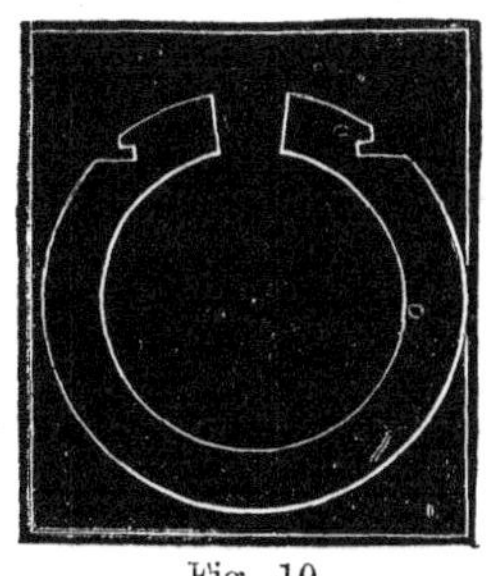

Fig. 10.

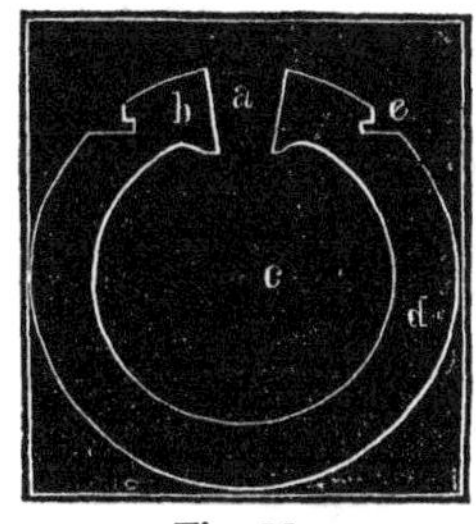

Fig. 11.

a. . Fuze-hole.
b. . Reinforce.
c. . Cavity.
d. . Sides, or thickness of metal.
e. . Ears.

45. **Grenades.** The hand grenade, as its name indicates, is a projectile thrown from the hand, against troops in mass.

The particular projectile used for this purpose, in our service, is the 6-pounder spherical case-shot.

Rampart Grenade. Rampart grenades are intended to be rolled down the rampart of a work, to protect a breach against the attack of a storming column. Shells of any size will answer for this purpose, and particularly those which are unserviceable for ordinary purposes.

Grenades are filled with a bursting charge, and are

armed with a short fuze,* which is lighted by a match in the hands of the grenadier immediately before it is thrown. They act by the force of their explosion alone.

46. **Case-shot.** Case-shot are a collection of small projectiles enclosed in a case or envelope. The envelope is intended to be broken in the piece by the shock of the discharge, or at any point of its flight, by a charge of powder, enclosed within it; in either case, the contained projectiles continue to move on after the rupture, but scatter out into the form of a sheaf or cone, so as to cover a large surface and attain a great number of objects. These projectiles can only be used with effect against animate objects situated at a short distance from the point of rupture.

The three principal kinds of case-shot in use are *grape*, *canister*, and *spherical case-shot*, or *shrapnel.* They are adapted to all guns and howitzers below those of 10-inch calibre, and receive their name from the pieces in which they are used.

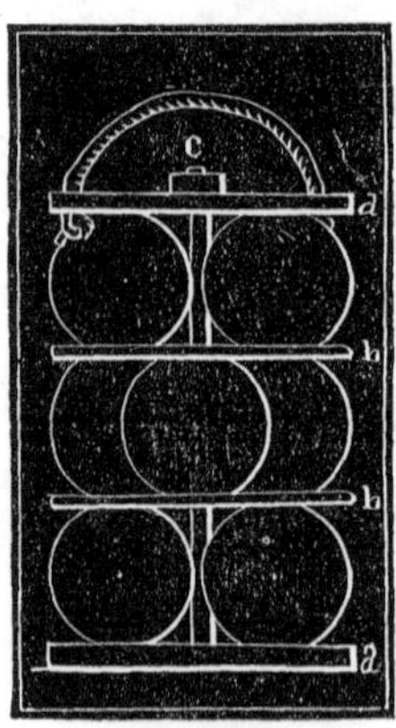

Fig. 12.

Grape-shot. A grape-shot is composed of nine small cast-iron balls, disposed in three layers of three balls each. Formerly the balls were held together by a covering of canvas and network of twine; but the present method is more simple and durable.

* Ketchum's hand grenade, which has lately been introduced into the American service, is a small, oblong percussion shell, which explodes on striking a slightly resisting object. To prevent accidents, the "plunger," or piece of metal which communicates the shock to the percussion cap is not inserted in its place until the moment before the grenade is thrown.

The parts of a stand of grape are, two plates, *a*, *a*, see Fig. 12, for the top and bottom layers; two rings, *b*, *b*, for the intermediate layer, and a screw-bolt, *c*, which passes through the plates and unites the whole. A handle is formed by passing a piece of rope-yarn through two holes in the upper plate, and tying the ends into knots to prevent them from pulling out.

Grape-shot are used in all except the field and mountain services.

*Canister-shot.** A canister-shot for a gun contains 27 small cast-iron balls, arranged in four layers, the top of 6, and the remainder of 7 each. A canister-shot for a howitzer contains 48 small iron balls, in 4 layers of 12 each. For the same calibre, it will be seen that the balls used in canister-shot are smaller than those used in grape-shot. The envelope is a tin cylinder, closed at the bottom by a thick cast-iron plate, and at the top by one of sheet-iron. The plates are kept in place by cutting the edges of the cylinder into strips about 0.5 inch long, and lapping them over the plates. To give more solidity to the mass, and prevent the balls from crowding upon each other when the piece is fired, the interstices are closely packed with sawdust. The handle is made of wire, and attached to the thin plate at the top.

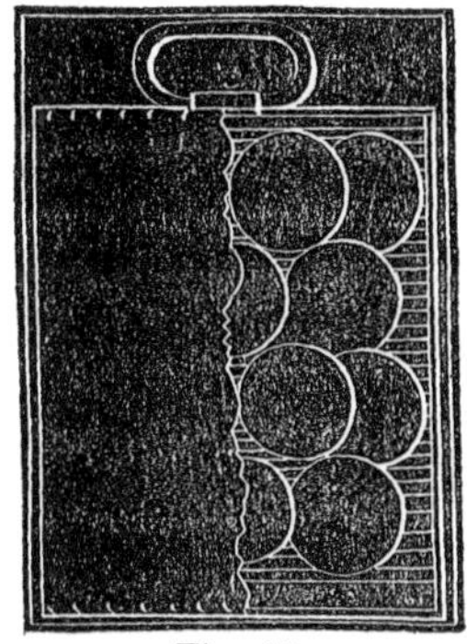

Fig. 13.

Canister-shot are used in the field, mountain, siege, and sea-coast services.

* The balls for canister for bronze rifle-guns are made of lead, or enclosed in a case of some soft material, to avoid injury to the surface of the bore.

It is stated that canister-shot were first used in the defence of Constantinople, about the middle of the 15th century.

Spherical case-shot. Though projectiles similar to spherical case-shot were used in France as early as the time of Louis XIV., the credit of perfecting them is due to Colonel Shrapnel of the British army. They were first successfully used by the English against the French, in the Peninsular war.

The envelope in the spherical case-shot, is a thin cast-iron shell, the weight of which, when empty, is about one half that of the equivalent solid shot. To prepare this shot, it is first filled with round musket-balls, 17 to the lb., and the interstices are then filled up by pouring in melted sulphur or resin; the object of which is to solidify the mass of bullets, and prevent them from striking, by their inertia, against the sides of the case and cracking it, when the piece is fired. A hole is bored through the mass of sulphur and bullets, to receive the bursting charge; and, in order not to displace too many bullets, and not to scatter them too far when the shot bursts, the bursting charge should only be sufficient to produce rupture.

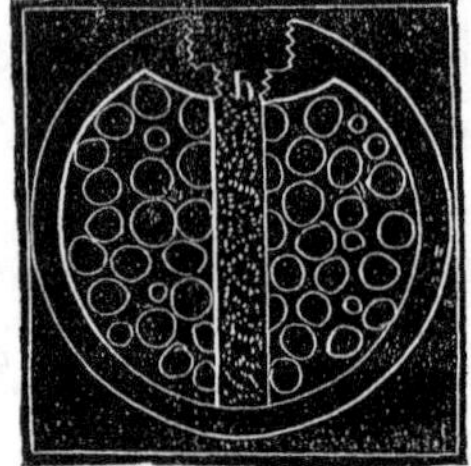

Fig. 14.

If the iron, of which the case is made, were always of suitable quality, and the cavity filled with bullets snugly packed in, there would be no necessity for sulphur to prevent accidents. In this case, it would not be necessary to remove any of the bullets, as the bursting charge would be disseminated through the interstices; and the difficulty, which now sometimes arises from their adhering to fragments of the case, would be entirely obviated.

To increase the effect of a small bursting charge, the lower portion of the fuze-hole, *b*, fig. 14, is partially closed, by screwing into it a disk perforated with a small hole for the passage of the flame from the fuze. The spherical case-shot mostly used for field service is the 12-pounder; it contains about 80 bullets; its bursting charge is 1 oz. of powder; and it weighs when finished 11.75 lbs.,—nearly as much as a solid shot of the same calibre.

The rupture of a spherical case-shot may be made to take place at any point of its flight; and in this respect it is superior to canister and grape-shot, which begin to separate the moment they leave the piece.

47. **Bar-shot.** Bar-shot consist of two hemispheres, or two spheres, connected together by a bar of iron; the motion of rotation which these projectiles assume in flight, renders them useful in cutting the masts and rigging of vessels; but, as they are very inaccurate, they are only employed at short distances. They are very little used, however, at the present day.

Chain-shot only differ from bar-shot in the mode of connection, which is a chain, instead of a bar.

48. **Percussion bullets.** Percussion bullets may be made by placing a small quantity of percussion powder, enclosed in a copper envelope, in the point of an ordinary rifle-musket bullet, or by casting the bullet around a small iron tube, which is afterward filled with powder and surmounted with a common percussion-cap. The impact of the bullet against a sub-

Fig. 15.

stance no harder than wood is found to ignite the percussion charge or cap, and produce an effective explosion.

These projectiles can be used to blow up caissons, and boxes containing ammunition, at very long distances.

49. **Carcasses.** Carcasses are shells which have three additional holes, of the same dimensions as the fuze-hole, pierced at equal distances apart in their upper hemispheres, with their exterior openings tangent to the great circle perpendicular to the axis of the fuze-hole. The object of a carcass is to set fire to wooden structures, by the flame of the burning composition which issues through the holes.

CHARGE OF RUPTURE OF SHELLS.

50. **Plane of rupture.** Suppose the cavity of the shell to be spherical, and concentric with the exterior. As soon as the enclosed charge of powder is inflamed, the gases developed expand into the cavity, and the expansive force increases until it is sufficient to overcome the tenacity of the metal, and produce rupture; which will take place in the direction of least resistance, or following a surface composed of lines normals to the two surfaces.

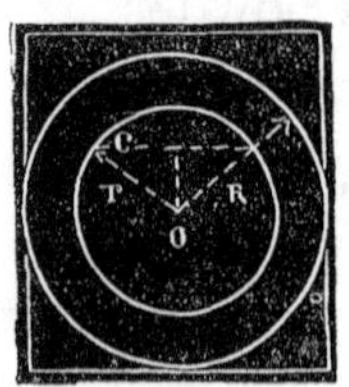

Fig. 16.

Let R be the radius of the exterior, and r the radius of the interior surface; o, the common centre of the two spheres; T, the tenacity of the material of which the sphere is composed; and p, the pressure on a unit of surface to overcome the tenacity of the metal.

Let C be the radius of the circle of rupture on the interior surface. From the known properties of gases, the pressure exercised on the area of this circle to produce rupture is equal to the components of all the normal pressures acting on the spherical segment of which it is the base, taken perpendicularly to the plane of this circle; therefore $\pi p C^2$ is the pressure of the gases which tends to break the sphere.

Under this supposition, rupture should follow the surface of the frustum of a cone of which this circle is the smaller base.

The surface of this frustum is equal to the difference of the surfaces of two cones whose common apex is at the centre of the sphere. The base of the smaller is $2\pi C$, and its slant height r; its surface therefore is equal to πCr. The surface of the larger cone, whose generatrix is the radius of the exterior sphere, will be to the smaller as R^2 is to r^2, and therefore $\pi Cr\frac{R^2}{r^2}$; their difference, or the area of the surface of rupture will be equal to $\pi Cr\left(\frac{R^2}{r^2}-1\right)$.

If the pressure of the gases acted normally to the surface of the fracture, or in the direction of the tenacity, this surface multiplied by T would give the total resistance, which should be equal to the pressure of the gases; but it acts obliquely, and to produce rupture should be increased by a quantity which depends on the law of increase of the resistance due to the angle which the pressure makes with the normal. Although we cannot measure this resistance, it must be admitted that the effort to overcome is greatest when the power is in the direction of the normal to the surface of rupture.

We shall, therefore, have the relation,

$$p\pi C^2 = T\pi Cr\left(\frac{R^2}{r^2}-1\right)+\delta.$$

Or,

$$p = T\frac{r}{C}\left(\frac{R^2}{r^2}-1\right)+\frac{\delta}{\pi C^2}.$$

In this expression the value of δ is unknown; but it is easy to be seen that it diminishes as the direction of the pressure approaches the normal, and when they coincide δ becomes 0. At the same time C increases, and the value of p diminishes, until C becomes equal to r, its maximum value. Therefore, the section of easiest rupture of a hollow sphere passes through a great circle, and the pressure which is *in equilibrio* with the tenacity of the metal, will be given by the foregoing formula, by making $C=r$, and $\delta=0$; it will then become,

$$p = T\left(\frac{R^2}{r^2}-1\right) = T\left(\left(\frac{R}{r}\right)^2-1\right).$$

When the pressure is less than this value of p, the sphere will resist its charge of powder; when it is greater than this value, the sphere will burst.

The density of the gaseous products of the powder necessary to burst the sphere can be easily found by Rumford's formula:

$$\overset{\text{Atmos.}}{p = 1.841}\ (905d)^{1+0.362d};$$

but d, or the density of the gaseous products, is equal

to their weight, or the weight of the bursting charge, divided by the interior space of the sphere.

Or,
$$d=\frac{w}{\frac{4}{3}\pi r^3}.$$
$$w=\tfrac{4}{3}\pi r^3 d.$$

51. **Loss of gas by fuze-hole.** This loss of force by the fuze-hole may be ascertained with sufficient accuracy, provided we know from actual experiment the amount of the loss from the fuze-hole of any one shell.

Let R and r be the exterior and interior radii of a spherical projectile; T, the tenacity of the metal; i, the radius of the fuze-hole; w', the weight of powder necessary to burst it under the supposition that there is no loss of force at the fuze-hole; w, the weight of powder that is actually required to burst it. By the preceding formulas we obtain the value of w'; $w-w'$ is therefore the amount of loss by the fuze-hole. Take another projectile, and let $w'_{\prime}$ represent the charge which is necessary to burst it, under the supposition that there is no loss, and $w_{\prime}$ the weight that is found by experiment necessary to burst it; $w_{\prime}-w'_{\prime}$ will represent the loss. We are at liberty to suppose the loss from the two fuze-holes is proportional to the size of the holes, and the density of the gases at the moment of rupture; we shall, therefore, have this proportion,

$$w-w' : w_{\prime}-w'_{\prime} :: i^2 d : i_{\prime}^2 d_{\prime}.$$

$$\text{Or, } w=w'+(w_{\prime}-w'_{\prime})\frac{i^2 d}{i_{\prime}^2 d_{\prime}}.$$

From the experiments made at Metz in 1835, it was shown that this mode of estimating the loss of force by the fuze-hole, was sufficiently exact for practical purposes.

FABRICATION OF PROJECTILES.

52. **Materials.** Shot and shells should be made of gray or mottled iron, of good quality.

Spherical case-shot should be made of the best quality of iron, and with peculiar care, in order that they may not break in the gun.

All projectiles should be cast in sand and not in iron moulds, as those from the latter are generally not spherical in form, nor uniform in size; they are also full of cavities, and are cracked by being heated.

Sand. The sand used should be silicious, of an angular grain, and moderate degree of fineness. It should be mixed with a sufficient quantity of clay, so that, when slightly moistened, it will retain its shape when pressed in the hand.

Pattern. The pattern of a spherical projectile is composed of two hollow cast-iron hemispheres, which unite in such a manner as to form a perfect sphere; on the interior of each hemisphere is fastened a handle to enable the operator to draw it from the sand when the half-mould is completed. The *flasks* which contain the mould are made of cast iron, in two equal parts united at their larger bases.

Fig. 17.

Moulding. This operation is performed by placing the flat side of one of the hemispheres on the moulding-board, and covering it with a flask. Sand is then poured into the flask, filling up the entire space between it and the hemisphere, and well rammed. The flask is then turned over, the hemisphere is withdrawn, and the entire surface of the sand painted with coke-wash, and dried. The remaining half of the mould is formed in the same way, except that a channel for the introduction of the melted iron is made by inserting a round stick in the sand before it is rammed, and withdrawing it afterward.

To secure strong metal and a homogeneous casting for large size solid shot, they are cast in a string, one above the other, with a large sinking head and connecting branches. They are afterwards cut apart and turned down to the required shape and size. Three 20-inch and four 15-inch shot may be cast in this manner.

Hollow projectiles. Thus far, the operation of moulding and casting solid and hollow projectiles are the same. The cavity of a hollow projectile is made by inserting a *core* of sand, which is formed around a stem fastened into the lower half of the mould. The stem is hollow, and perforated with small holes to allow of the escape of steam and gas generated by the heat of the melted metal. It is also made of iron, but that part of it which comes in contact with the melted iron, and forms the fuze-hole, is coated with sand.

In pouring the melted iron into the mould with the ladle, care should be taken to prevent scoria and dirt from entering with it; and, for this purpose, the surface should be skimmed with a wooden stick.

Before the iron is fairly cooled, the flasks are opened, and the sand knocked from the castings. After this, the core is broken up and knocked out, and the interior surface cleaned by a scraper. The sinking head and other excrescences are knocked off, and the surface smoothed in a rolling-barrel, or with a file, or chisel, if necessary. The fuze-hole is then reamed out to the proper size, and the projectile is ready for inspection.

INSPECTION OF PROJECTILES.

53. **Object of inspection.** The principal points to be observed in inspecting shot and shells are, to see that they are of proper size in all their parts; that they are made of suitable metal; and that they have no defects, concealed or otherwise, which will endanger their use, or impair the accuracy of their fire.

As it would be impracticable to make all projectiles of exact dimensions, certain variations are allowed in the fabrication. See Ordnance Manual.

Inspection of shot. The instruments are one *large* and one *small* gauge, and one *cylinder gauge*; the cylinder gauge has the same diameter as the large gauge, it is made of cast iron, and is five calibres long. There are also, one *hammer* with a conical point, six *steel punches*, and one *searcher* made of wire.

The shot should be inspected before they become rusty; after being well cleaned, each shot is placed on a table and examined by the eye to see that its surface is smooth, and that the metal is sound and free from seams, flaws, and blisters. If cavities or small holes appear on the surface, strike the point of the hammer or punch into them, and ascertain their depth with the searcher; if the depth of the cavity exceed 0.2 inch, the shot is rejected; and also if it appear that an attempt has been made to conceal such defects by filling them up with nails, cement, &c.

The shot must pass in every direction through the large gauge, and not at all through the small one; the founder should endeavor to bring the shot up as near as possible to the *large gauge*, or to the true diameter.

After having been thus examined, the shot are passed through the *cylinder gauge*, which is placed in an inclined position, and turned from time to time, to prevent its being worn into furrows; *shot* which *slide* or *stick* in the cylinder are rejected.

Shot are proved by dropping them from a height of twenty feet on a block of iron, or rolling them down an inclined plane of that height, against another shot at the bottom of the plane.

The average weight of the shot is deduced from that of three parcels of twenty to fifty each, taken indiscriminately from the pile; some of those which appear to be the smallest should also be weighed, and they are rejected if they fall short of the weight expressed by their calibre, more than one *thirty-second* part. They almost invariably exceed that weight.

Inspection of grape and canister shot. The dimensions are verified by means of a large and small gauge, attached to the same handle. The surface of the shot should be smooth, and free from seams.

Inspection of hollow projectiles. The inspecting instruments are a *large* and *small gauge* for each calibre, and a *cylinder gauge* for shells of eight inches and under.

Calipers for measuring the thickness of shells at the sides.

Calipers to measure the thickness at the bottom.

Gauges to verify the dimensions of the fuze-hole, and the thickness of the metal at the fuze-hole.

A pair of hand-bellows; a wooden plug to fit the fuze-hole, and bored through to receive the nozzle of the bellows.

A hammer; a searcher; a cold chisel; steel punches.

Inspection. The surface of the shell and its exterior dimensions, are examined as in the case of shot. The shell is next struck with the hammer, to judge by the sound whether it is free from cracks; the position and dimensions of the ears are verified; the thickness of the metal is then measured at several points on the great circle perpendicular to the axis of the fuze-hole. The diameter of the fuze-hole, which should be accurately reamed, is then verified, and the soundness of the metal about the inside of the hole is ascertained by inserting the finger.

The shell is now placed on a trivet, in a tub containing water deep enough to cover it nearly to the fuze-hole; the bellows and plug are inserted into the fuze-hole, and the air forced into the shell; if there be any holes in the shell, the air will rise in bubbles through the water. This test gives another indication of the soundness of the metal, as the parts containing cavities will dry more slowly than other parts.

The mean weight of shells is ascertained in the same manner as that of shot. Shot and shells rejected in the inspection, are marked with an X made with a cold chisel—on shot near the gate, and on shells near the fuze-hole.

PRESERVATION AND PILING OF BALLS.

54. **Lackering.** Projectiles should be carefully lackered as soon as possible after they are received. When it is necessary to renew the lacker, the old lacker should be removed by rolling or scraping the balls, which should never be heated for that purpose.

Piling. Balls should be piled according to kind and calibre, under cover if practicable, in a place where there is a free circulation of air; to facilitate which, the piles should be narrow if the locality permits; the width of the bottom tier may be from twelve to fourteen balls, according to the calibre.

Prepare the ground for the base of the pile by raising it above the level of the surrounding ground, so as to throw off the water; level it, ram it well, and cover it with a layer of screened sand. Make the bottom of the pile with a tier of unserviceable balls, buried about two-thirds of their diameter in the sand; this base may be made permanent; clean the base well, and form the pile, putting the fuze-holes of the shells downward, in the *intervals*, and not resting on the shells below.

The base may be also made of bricks, concrete, stone, or with borders and braces of iron.

55. **To find the number of balls in a pile.** *Multiply the sum of the number of balls in the three parallel edges by one-third of the number in a triangular face.*

In a square pile, one of the parallel edges contains but one ball; in a triangular pile, two of the edges have but one ball in each.

The number of balls in a triangular face is $\frac{n(n+1)}{2}$, n being the number in the bottom row.

The sum of the three parallel edges in a triangular pile is $n+2$; in a square pile, $2n+1$; in an oblong pile, $3N+2n-2$; N being the length of the top row, and n the width of the bottom tier; or $3m-n+1$; m being the length, and n the width of the bottom tier.

If a pile consist of two joined at right angles, calculate the contents of one as a common pile, and the other as a pile of which three parallel edges are equal.

THEORY AND CONSTRUCTION OF ROCKETS.

56. **Structure.** A rocket is a projectile which is set in motion by a force residing within itself; it therefore performs the two-fold function of piece and projectile.

It is essentially composed of a strong case of paper or wrought iron, enclosing a composition of *nitre*, *charcoal*, and *sulphur*—the same as gunpowder, except that the ingredients are proportioned for a slower rate of combustion. If penetration and range be required, its head is surmounted by a *solid shot;* if explosion and incendiary effect, by a *shell* or *spherical case-shot*, to which is attached a fuze, which is set on fire when it is reached by the flame of the burning composition. The base is perforated by one or more *vents* for the escape of the gas generated within, and sometimes with a screw-hole to which a guide-stick is fastened.

The disposition of the different parts will be readily understood by reference to the subjoined figure, which represents a section through the long axis of a Congreve rocket.

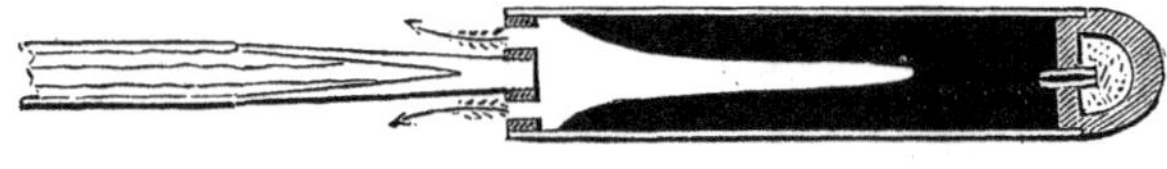

Fig. 18.

57. **Motion.** A rocket is set in motion by the reaction of a rapid stream of gas escaping through its vents. If it be surrounded by a resisting medium, the atmosphere for instance, the particles of gas, as they issue from the vent, will impinge against and set in motion certain particles of air, and the force expended on the inertia of these particles will react and increase the propelling force of the rocket. It follows, therefore, that, though a rocket will move *in vacuo*, its propelling force will be increased by the presence of a resisting medium. Whether the effect will be to accelerate the rocket depends upon the relation between the resistance which the medium offers to the motion of the gas, and that which it offers to the motion of the rocket.

Vent. As the rate of combustion of the composition is independent of the pressure of the gas in the bore, it follows, that if the size of the vent be contracted, the flow of gas through it will be accelerated. The strength of the case, and the friction of the gas, which increases as the vent diminishes, alone limit the reduction of the size of the vent.

For vents of the same size, but of different shapes, that one which allows the gas to escape most freely, will be most favorable to the flight of the rocket. A conical form of vent, with the larger orifice next to the bore, will allow the gas to escape more rapidly than one of cylindrical form. This may be shown by burning portfire composition in tubes with different-shaped vents. It will be found that the sparks from a conical vent will be thrown much higher than those from a cylindrical vent; the relative heights depending on the slope of the sides of the conical vent.

Bore. As the composition of a rocket burns in parallel layers of uniform thickness, the amount of gas generated in a given time, or the velocity of its exit from the case, depends on the extent of the inflamed surface.

Experience shows that to obtain the required surface of inflammation, it is necessary to form a long cavity in the mass of the composition. This cavity is called the *bore*. In small rockets, the bore is formed by driving the composition around a spindle which is afterward withdrawn; but in the large ones, the composition is driven into the case in a solid mass by a powerful hydrostatic press, and then bored out with a bit. In all rockets the bore should be concentric with the case; its shape should be made conical to facilitate the drawing out of the spindle, and to diminish the strain on the case near its head, by reducing the amount of surface where the pressure on the unit of surface is greatest.

Nature of movement. Suppose the rocket in a state of rest, and the composition ignited; the flame immediately spreads over the surface of the bore, forming gas, which issues from the vent. The escape is slow in the first moments, as the density of the gas is slight; but as the surface of the inflammation is large compared to the size of the vent, the gas accumulates rapidly, and its density is increased until the velocity of the escape is sufficient to overcome the resistances which the rocket offers to motion. These resistances are, inertia, friction, the component of weight in the direction of motion, and, after motion takes place, the resistance of the air.

The constant pressure on the head of the bore accelerates the motion of the rocket until the resistance of the air equals the propelling force; after this, it will remain constant until the burning surface is sensibly diminished. When the gas ceases to flow, the rocket loses its distinctive character, and becomes, so far as its movement is concerned, an ordinary projectile.

The increase in the surface of combustion whereby more gas is developed in the same time, and the diminution in the weight of the remaining composition, cause the point of maximum velocity to be reached with increased rapidity. If the weight of the rocket be increased, the instant of maximum velocity will be prolonged, but the amount will remain the same. A change in the form of the rocket which increases the resistance of the air, will have the effect to diminish the maximum velocity.

The maximum velocity of French rockets, and the distances at which they are attained, are given in the following table:—

CALIBRE.	DISTANCE.	MAXM. VELOCITY.
2½ inches,	134 yds.	278 yds.
3½ "	141 "	370 "

According to the calculations of Piobert, for small rockets it takes about ¾ second for the gas to attain its maximum velocity of 850 yds.

58. **Guiding principle.** The propelling force of a

rocket changes its direction with the axis along which it acts; it follows, therefore, that without some means of giving stability to this axis, the path described will be very irregular, so much so, at times, as to fold upon itself; and instances have been known where these projectiles have returned to the point whence they started.

An example of this irregular motion may be seen in "serpents," a species of small rockets without guide-sticks.

The two means now used to give steadiness to the flight of a rocket are, *rotation*, as in the case of a rifle-ball, and the *resistance of the air*, as in an arrow.

Hale's system. The first is exemplified in Hale's rocket, where rotation is produced around the long axis by the escape of the gas through five small vents situated obliquely to it. In his first arrangement, the inventor placed the small vents in the base, surrounding the large central vent, so that the resultant of the tangential forces acted around the posterior extremity of the axis of rotation. In 1855, this arrangement was changed by reducing the number of the small vents to three, and placing them at the base of the head of the rocket. The rocket thus modified, and shown in fig. 20, is the one now used by the United States government for war purposes.*

* A still later improvement in Hale's rocket consists in screwing a cast-iron piece (*a*) into the bottom of the case, which is perforated with three vents. A corresponding side of each vent is surrounded with a fence (*b.b.b.*), the opposite sides being open. The gas in its efforts to expand after issuing from the vents, presses against the fences and rotates the rocket around its long axis.

Fig. 19.

Fig. 20.

a. .Bore and vent.
b. .Recess in the base of the head.
c. .Tangential vent (three).
d. .Head (solid).

Congreve's system. A Congreve rocket is guided by a long wooden stick attached to its base. If any cause act to turn it from its proper direction, it will be opposed by resistances equal to its moment of inertia and the lateral action of the air against the stick.

The effect of these resistances will be increased by placing the centre of gravity near the head of the rocket, and by increasing the surface of the stick.

In *signal* rockets, where the case is made of paper, the stick is attached to the side by wrapping around twine; and there is but one large vent, which is in the centre of the case.

In *war*-rockets the stick is attached to the centre of the base, and the large central vent is replaced by several small ones near its circumference. See fig. 18. The former arrangement is not so favorable to accuracy as the latter, inasmuch as rotation will be produced if the force of propulsion and the resistance of the air do not act in the same line.

59. **How fired.** Rockets are generally fired from *tubes* or *gutters;* but should occasion require it, they may be fired directly from the ground, care being taken to raise the forward end by propping it up with a stick or stone. As the motion is slow in the first moments of its flight, it is more liable to be deviated from its

proper direction at this time than any other; for this reason the conducting tube should be as long as practicable, say from five to ten feet.*

60. **Form of trajectory.** Take that portion of the trajectory where the velocity is uniform. The weight of the rocket applied at its centre of gravity, and acting in a vertical direction, and the propelling force acting in the direction of its length, are two forces the oblique resultant of which moves the rocket parallel to itself; but the resistance of the air is oblique to this direction, and acting at the centre of figure, a point situated between the centre of gravity and extremity of the guide-stick, produces a rotation which raises the stick, and thereby changes the direction in which the gas acts. As these forces are constantly acting, it follows that each element of the trajectory has less inclination to the horizon than the element of an ordinary trajectory in which the velocity is equal.

When the velocity is not *uniform*, the position of the centre of gravity has a certain influence on the form of the trajectory. To understand this, it is necessary to consider that the component of the resistance of the air which acts on the head of the rocket is greater than that which acts on the side of the stick. It is also necessary to consider that the pressure of the inflamed gas acts in a direction opposite to the resistance of the air, that is to say, from the rear to the front, and that the centre of gravity is near the rear extremity of the case.

* Mr. Hale has suggested a means of using a short tube, by applying a pressure to the rocket to retain it in its place until the gas has acquired the requisite velocity.

At the beginning of the trajectory, when the motion of the rocket is accelerated, its inertia is opposed to motion, and being applied at the centre of gravity, which is in rear of the vent, the point of application of the moving force, it acts to prevent the rocket from turning over in its flight. But when the composition is consumed, the centre of gravity is thrown further to the rear, and the velocity of the rocket is retarded, the inertia acts in the opposite direction, and the effect will be, if the centre of gravity or inertia is sufficiently far to the rear, to cause it to turn over in the direction of its length.

If the rocket be directed toward the earth, this turning over will be counteracted by the acceleration of velocity due to the weight, and the form of the trajectory will be preserved.

Effect of wind. When the wind acts obliquely to the plane of fire, its component perpendicular to this plane, acting at the centre of figure, will cause the rocket to rotate around its centre of gravity. As the centre of figure is situated in rear of the centre of gravity, the point will be thrown toward the wind, and the propelling force acting always in the direction of the axis, the rocket will be urged toward the direction of the wind. To make an allowance for the wind, in firing rockets, they should be pointed toward the opposite side from which the wind comes, or with the wind instead of against it.

If the wind act in the plane of fire from front to rear, it will have the effect to depress the point, and with it the elements of the trajectory in the ascending branch, and elevate them in the descending branch; as the latter is shorter than the former, the effect of a front wind will be to diminish the range. The converse will be true for a rear wind.

61. **History.** Rockets were used in India and China for war purposes before the discovery of gunpowder; some writers fix the date of their invention about the close of the ninth century. Their inferior force and accuracy limited the sphere of their operations to incendiary purposes, until the year 1804, when Sir William Congreve turned his attention to their improvement. This officer substituted sheet-iron cases for those made of paper, which enabled him to use a more powerful composition; he made the guide-stick shorter and lighter, and removed a source of inaccuracy of flight by attaching the stick to the centre of the base instead of the side of the case. He states that he was enabled by his improvements to increase the range of 6-pdr. rockets from 600 to 2,000 yards. Under his direction they were prepared, and used successfully at the siege of Boulogne and the battle of Leipsic. At the latter place they were served by a special corps.

62. **Advantages.** The advantages claimed for rockets over cannon are, unlimited size of projectile; portability; freedom from recoil; rapidity of discharge; and the terror which their noise and fiery trail produce on mounted troops.

The numerous conditions to be fulfilled in their construction in order to obtain accuracy of flight, and the uncertainty of preserving the composition uninjured for a length of time, are difficulties not yet entirely overcome, and which have much restricted their usefulness for general military purposes.

63. **Kind used.** The two sizes of Hale's rockets in use in the American service are, the

2-inch (interior diameter of case), weighing 6 lbs.; and
3-inch " " " " 16 lbs.

Under an angle of from 4° to 5° the range of these rockets is from 500 to 600 yds. Under an angle of 47° the range of the former is 1,760 yds. and the latter 2,200.

CHAPTER III.

HISTORY OF CANNON.

64. The terms *cannon* and *ordnance* are applied to all heavy fire-arms which are discharged from carriages, in contradistinction to *small-arms*, which are discharged from the hand.

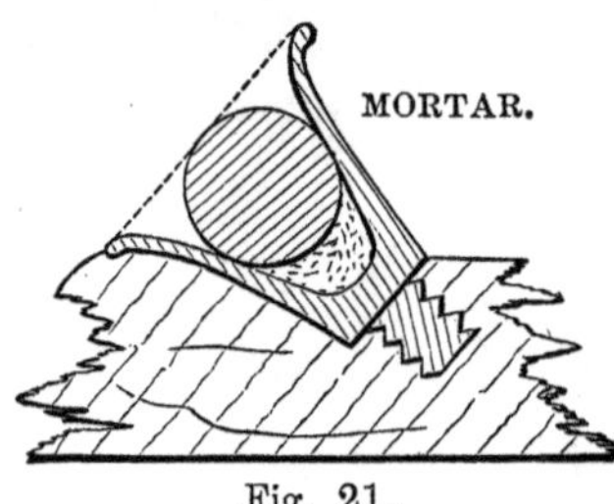

Fig. 21.

65. **Early cannon.*** The shape of the first cannon used, after the invention of gunpowder was conical, internally and externally resembling an apothecary's mortar. They were called *mortars*, *bombards* and *vases;* were fired at high angles; and, in consequence of the slow burning of the powder of that day, and the conical shape of the bore, the stone balls which they projected moved with very little velocity and accuracy.

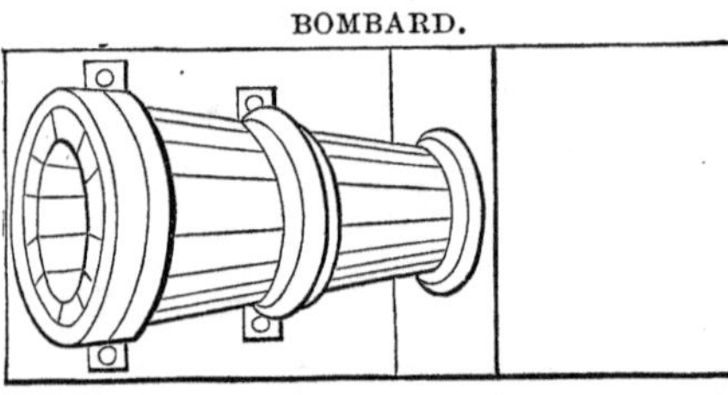

Fig. 22.

Perriere. To economize the action of the powder, and give a more accurate direction to the projectile, the interior space, or bore, was afterward made nearly cylindrical, from 4 to 8 calibres long; it was terminated at

* Vide Thiroux.

the bottom by a very narrow and deep chamber, the object of which was to increase the effect of the powder, by retarding the escape of the gas before it acted on the projectile. These cannon were further improved by making the bores perfectly cylindrical; and were called *perrieres* (fig. 23), from the fact that they fired stone balls. They were principally employed to breach stone walls, and for this purpose were fired horizontally.

PERRIERE.

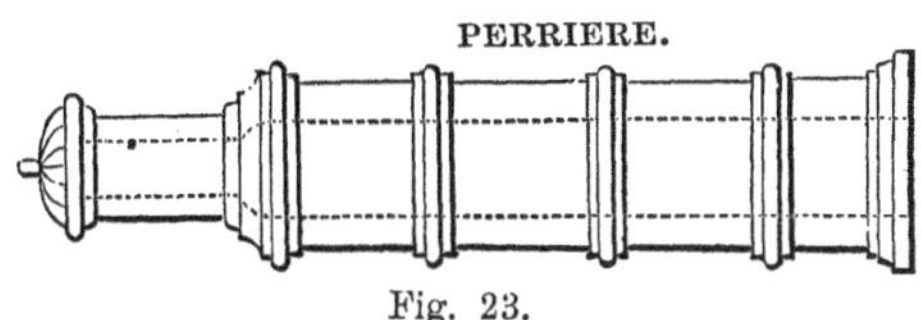

Fig. 23.

66. **Construction of early cannon.** *The first bombards* were made of bars of iron, bound together by hoops, after the manner of the staves of a barrel. Fig. 22. Afterward they were made of wrought iron, and finally of cast metal. Bronze guns were used in the time of King John of France.

Among the earliest cannon are found those which were loaded at the breech instead of the muzzle. One of the methods is shown in fig. 24, in which *A* represents a rectangular opening formed at the breech for the purpose of receiving a movable chamber, *C*, which contained the charge, and was held in its place by a key,

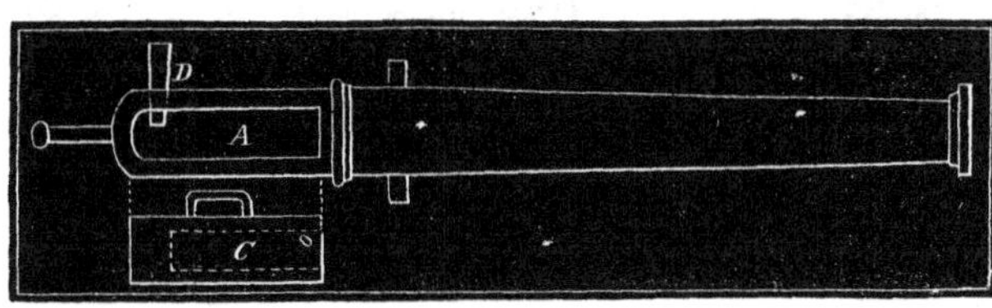

Fig 24.

D. Though these pieces possessed great facility in loading, they were abandoned for want of strength and solidity.

67. **Ancient guns.** The introduction of cast-iron projectiles, which are much stronger and denser than those of stone, led to the invention of a new species of cannon, called *culverins*, which very nearly correspond in construction and appearance to the *guns* of the present day. The great strength of these pieces and their projectiles, permitted the use of a large charge of powder; and their introduction proved an important step in the improvement of artillery.

CULVERIN.

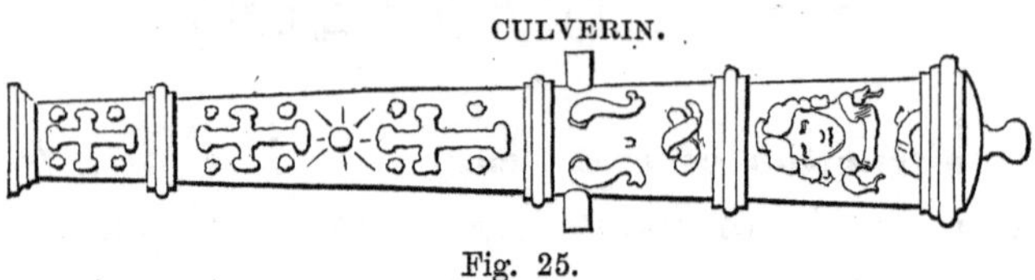

Fig. 25.

The idea was entertained by ancient artillerists—founded on the relation which cannon were erroneously supposed to bear to small arms—that the range increased with the length of the piece; and in consequence, many culverins were made of enormous length. A remarkable piece of this description still exists at Dover, England, familiarly known as Queen Anne's pocket-piece. While it carries a ball weighing only 18 lbs., it is more than 25 feet long.

68. **Ancient mortars.** From the earliest days of artillery there existed short, chambered pieces, which projected stone balls under great angles of elevation. In 1478, an attempt was made to use in these pieces, hollow projectiles filled with powder, to which was attached a burning match to set the powder on fire; but it is probable that the accidents which accompanied

their use caused them to be abandoned for the time. In 1634, however, means were devised to overcome this difficulty; and, thus perfected, these pieces were introduced into the French service as a class of cannon now known as *mortars*.

In the reign of Louis XIV., a great variety of mortars were used; and some of them, called Comminges, after their inventor, threw bombs weighing 550 lbs.

69. **Ancient howitzers.** Early attempts were also made to throw hollow projectiles from "perrieres," and "culverins," or guns; but great difficulties were experienced in loading them; and the accidents to which they were liable, as in the case of mortars, caused them to be abandoned. Subsequently, however, the Dutch artillerists conceived the idea of reducing their length, so that the projectile could be inserted in its place by hand; and, thus improved, these cannon rapidly came into use, under the name of howitzers, from the German, Haubitz.

70. **Calibre.** The calibre of a cannon is the diameter of its bore expressed in inches, or the weight of the shot corresponding to it.* Each nation early adopted a series of calibres, decreasing in a geometrical progression. The principal series were the French, or 32-pounder, 16-pounder, 8-pounder, 4-pounder, and

* In some services the calibre refers to the size of the bore, in others, to the size of the shot. In Germany, the projectile referred to is a stone ball.

TABLE OF CALIBRES IN AMERICAN SERVICE.

6-pd.	9-pd.	12-pd.	18-pd.	24-pd.	32-pd.	42-pd.
3.67″	4.2″	4.62″	5.2″	5 82″	6.4″	7.0″

2-pounder; and the German, or 48-pounder, 24-pounder, 12-pounder, 6-pounder, 3-pounder, and 1½-pounder.

71. **Devices.** Down to the time of the French revolution, bronze cannon were highly ornamented with carved figures representing some fanciful design, together with the national coat-of-arms and cypher of the reigning monarch. Each piece also bore a particular name borrowed from some animal or passion; and French cannon of the time of Louis XVI. bore the mottoes, "Nec pluribus impar," and "Ultima ratio regum." That this custom of naming cannon is not entirely obsolete, is shown by a piece of singular shape and construction, captured in the late war with Mexico, which bears the title of "El Terror del Norte Americano."

72. **"Materiel."—System.** The expression, "*matériel of artillery*," embraces all cannon, carriages, implements, ammunition, &c., necessary for artillery purposes, and is used in contradistinction to "*personnel of artillery*," which refers to the officers and men. The expression, "*system of artillery*," refers to the character and arrangement of the matériel of artillery, as adopted by a nation at any particular epoch.

In the United States' service, the term "ordnance and ordnance stores," embraces not only all the *matériel* of artillery, but the swords, small-arms, and accoutrements used by infantry and mounted troops.

The principal qualities to be observed in establishing a system of artillery are *simplicity*, *mobility*, and *power*; and the improvements which have been made in artillery in the last four hundred years have had these qualities steadily in view.

The American systems of field and siege artillery are

chiefly derived from those of France; it will, therefore, be useful to the pupil to study the history of the latter, and compare the successive steps of improvement which have brought them to their present state of perfection.*

73. **First system.** Toward the middle of the sixteenth century, the various guns of the French artillery were reduced to six. The weights of the balls corresponding to these calibres were $33\frac{1}{4}$, $15\frac{1}{4}$, $7\frac{1}{4}$, $2\frac{1}{2}$, $1\frac{1}{2}$, and $\frac{3}{4}$ lb. respectively. This range of calibres was thought to be necessary, for the reason that it required guns of large calibre to destroy resisting objects, while guns of small calibre were necessary to keep up with the movement of troops.

Each of the five principal calibres was mounted on a different carriage, and the ammunition, stores, and tools were carried on different store-carts.

Three kinds of powder were used, viz.: large-grain, small-grain, and priming, which were carried in barrels of three sizes.

The axle-trees, which were of wood, varied for the different wheels, as well as for the different guns. The gun-carriages were without limbers, and had only two wheels, the shafts being attached to the trails, which often dragged along the ground. No spare wheels were used, except for pieces of large calibre; and for facility of transportation these were put on an axle-tree, so as to form a carriage.

With the exception of replacing injured wheels, all repairs were made on the spot, from the resources of the country, and no spare articles were carried with the train. There was no established charge of powder for

* Vide New System of Field Artillery, by Captain Favé.

the guns; although a weight equal to that of the shot was generally used.

Such was the character of the artillery which accompanied the French armies up to the middle of the seventeenth century.

74. **Second system.** In the reign of Louis XIV., the calibres of cannon were gradually changed by the introduction of several foreign pieces. There were 48, 32, 24, 16, 12, 8, and 4 pdrs.; and those of the same calibre varied in weight, length, and shape.

Uniformity existed in general in each district commanded by a lieutenant-general of artillery, but the cannon of one district differed from another. Each district had (for the six kinds of cannon) six carriages, with different wheels, and three kinds of limbers, with different wheels, making nine patterns of wheels, without counting those for the platform wagons used to transport heavy guns, the ammunition carts, the trucks, and the wagons for small stores and tools.

Spare carriages were carried into the field, but those of one district would not fit the guns of another. There was but one kind of powder, and this was carried in barrels. The charge was usually two-thirds the weight of the projectile, roughly measured. Besides this, the powder often varied in strength according to the district from which it came.

75. **Valiere's system.** In 1732, General Valière abolished the 32-pdr., as being heavy and useless, and gave uniformity to the five remaining calibres. Toward the end of the 18th century, mortars, or Dutch howitzers, were sometimes attached to the field trains; for the latter, a small charge, and calibre of 8 inches, were adopted.

There were also light 4-pdr. guns attached to each regiment. Up to that time an army always carried with it heavy guns (24-pdrs.), and light guns (4-pdrs.), which were combined in the same park.

Valière established a system of uniformity for cannon throughout France; but such was not the case with the carriages and wagons used with them. Great exactness was not then sought for, and there existed as many plans for constructing gun-carriages as there were arsenals of construction. The axle-trees were made of wood, the limbers were very low, and the horses were attached in single file.

76. **Gribeauval's system.** In 1765, General Gribeauval founded a new system, by separating the field from the siege artillery. He diminished the charge of field-guns from a half to a third the weight of the shot, but as he diminished the windage of the projectile at the same time, he was enabled to shorten them and render them lighter, without sensibly diminishing their range.

Field artillery then consisted of 12, 8, and 4 pdr. guns, to which was added a 6-inch howitzer, still retaining a small charge, but larger in proportion than that before used. For draught, the horses were disposed in double files, which was much more favorable to rapid gaits. Iron axle-trees, higher limbers, and travelling trunnion-holes, rendered the draught easier. The adoption of cartridges, elevating screws, and tangent scales, increased the rapidity and regularity of the fire. Stronger carriages were made for the lighter guns, and the different parts of all were made with more care, and strengthened with iron work. Uniformity was established in all the new constructions, by compelling all the arsenals to make every part of the carriages, wagons, and limbers according to certain fixed dimensions. By this exact correspondence of all the parts of a carriage, spare parts could be carried into the field ready made, to refit. Thus an equipment was obtained which could be easily repaired, and could be moved with a facility hitherto unknown.

In order to reduce the number of spare articles necessary for repairs, Gribeauval gave, as far as practicable, the same dimensions to those things which were of the same nature.

The excellence of this system was tested in the wars of the French Republic and Empire, in which it played an important part.

77. **Stock trail system.** In 1827, the system of Gribeauval was changed by introducing the 24 and 32 pdr. howitzers, lengthened to correspond with the 8 and 12 pdr. guns, and abolishing the 4-pdr. gun and 6-inch howitzer. Afterward some important improvements were made in the carriages, chiefly copied from the English system; the number for all field cannon was reduced to two, the wheels of the carriage and limber were made of the same size; the weight of the limber was reduced, and an ammunition chest placed on it; the method of connecting the carriage and limber was simplified, and the operations of limbering and unlimbering greatly facilitated; and the two flasks which formed the trail were replaced by a single piece, called the *stock*, which arrangement allowed the new pieces to turn in a smaller space than that required by the old ones.

78. **Louis Napoleon's system.** In 1850, the present Emperor of the French caused a series of experiments to

be made, at the principal artillery schools of France, to test the merits of a new system of field artillery proposed by himself. The principal idea involved in this system was, to substitute a single gun of medium weight and calibre, capable of firing shot and shells, for the 8 and 12 pdr. guns and 24 and 32 pdr. howitzers, then in use. The calibre selected was the 12 pdr.

The favorable results of all these experiments, and the simplicity of the system, led to the adoption of this, the Napoleon gun, as it is sometimes called, into the French service; and others of similar principle were introduced into various European services, and also into our own. As this piece unites the properties of gun and howitzer, it is called *canon-obusier*, or gun-howitzer.

79. **Recent improvements.** At no time since the discovery of gunpowder, have such important improvements been made in fire-arms, as within the past few years. These improvements may be summed up as follows, viz.:—

1st. Improvement in the quality of cast iron, and the consequent increase in the calibre of sea-coast cannon. In 1820, the heaviest gun mounted on our sea-coast batteries, was the 24-pdr.; at present, the heaviest is a 20-inch gun, carrying a shell weighing 1,080 lbs. with from 100 to 200 lbs. of powder. 2d. The use of wrought-iron as a material for fortress carriages, and for covering ships of war. 3d. The extensive introduction of shells in sea-coast defences and naval warfare; and spherical case-shot into the field-service; and, 4th. The successful application of the rifle principle to small-arms and cannon.

CONSTRUCTION, &c., OF CANNON.

80. **Definition.** A cannon is a heavy machine, used to set projectiles in motion by means of gunpowder. Its general form is that of a tube closed at one end.

81. **Classification.** All cannon may be classified, according to their nature, as *guns*, *howitzers*, and *mortars;* and, according to the uses to which they are applied, as *field*, *mountain*, *prairie*, *siege*, and *sea-coast* cannon. The recent introduction of rifle-cannon into the military service, requires that a further distinction should be made, between *rifled* and *smooth-bored* cannon. How far this change will affect the distinction now made between guns and howitzers, remains to be determined by future experience.

In treating of cannon, it is proposed, in the first place to discuss those parts and principles common to all; and, in the second place, to consider the peculiar characteristics of each class and calibre.

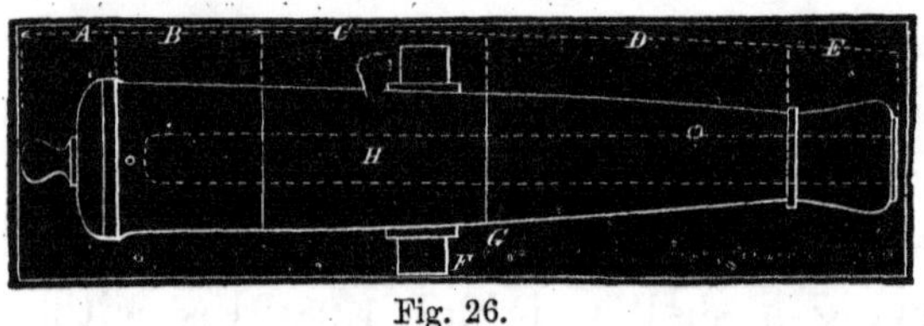

A. .Cascable.
B. .First reinforce.
C. .Sec'd reinforce.
D. .Chase.
E. .Swell of the muzzle.
F. .Trunnions.
G. .Rimbases.
H. .Bore.

Fig. 26.

82. **Nomenclature.*** The *cascable* is that part of the gun in rear of the base of the breech; it is composed generally of the following parts: the *knob*, the *neck*, the *fillet*.

* This nomenclature refers more particularly to guns of the old pattern, large numbers of which will probably remain in service for some time to come. The most recent models are characterized by an entire absence of mouldings and ornaments, and the elements, in most cases, are curved instead of right lines. The modifications which it is necessary to make to suit the present nomenclature to the new system, will be seen by reference to Fig. 46.

The *base of the breech* is a frustum of a cone, or a spherical segment, in rear of the breech.

The *base-ring* is a projecting band of metal adjoining the base of the breech, and connected with the body of the gun by a concave moulding. It serves as a point of support for the breech sight, and rests upon the head of the elevating screw. The ring is omitted in guns of recent model.

The *breech* is the mass of solid metal behind the bottom of the bore, extending to the rear of the base-ring.

The *reinforce* is the thickest part of the body of the piece, in front of the base-ring. If there be more than one reinforce, that which is next to the base-ring is called the *first reinforce;* the other, the *second reinforce.*

The *chase* is the conical part of the piece in front of the reinforce.

The *astragal* and *fillets*, in field guns, and the *chase-ring* in other pieces, are the mouldings at the front end of the chase.

The *neck* is the smallest part of the piece, in front of the astragal or chase-ring.

The *swell of the muzzle* is the largest part of the piece in front of the neck. It is terminated by the muzzle mouldings, which, in field and siege guns, consist of the *lip* and *fillet.* In sea-coast guns, and heavy howitzers and columbiads, there is no fillet. In field and siege howitzers, and in mortars, a muzzle *band* takes the place of the swell of the muzzle.

The *face* of the piece is the terminating plane perpendicular to the axis of the bore.

The *trunnions* are cylinders, the axes of which are

in a plane perpendicular to the axis of the bore, both axes being in the same plane.

The *rimbases* are short cylinders, uniting the trunnions with the body of the gun. The ends of the rimbases, or the *shoulders of the trunnions*, are planes perpendicular to the axis of the trunnions.

The *bore* of the piece includes all that part bored out, viz.: the *cylinder*, the *chamber* (if there be one), and the conical or spherical surface connecting them.

The *muzzle*, or mouth of the bore is chamfered, in order to prevent abrasion and facilitate loading.

The *lock-piece* is a block of metal at the outer opening of the vent for the attachment of the lock. As friction-tubes are now used for firing cannon in the land service, this part is omitted.

The *natural line of sight* is a line drawn, in a vertical plane through the axis of the piece, from the highest point of the base-ring to the highest point of the swell of the muzzle, or to the top of the *sight* if there be one.

The *natural angle of sight* is the angle which the natural line of sight makes with the axis of the piece.

The *dispart* is the difference of the semi-diameters of the base-ring and the swell of the muzzle, or muzzle-band. It is, therefore, the tangent of the natural angle of sight, to a radius equal to the distance from the rear of the base-ring to the highest point of the swell of the muzzle, the sight, or the front of the muzzle-band, as the case may be.

INTERIOR FORM.

83. **Division of parts.** The *interior* of cannon may be divided into three distinct parts; 1st, the *vent*, or

channel which communicates fire to the charge; 2d, *the seat of the charge*, or chamber, if its diameter be different from the rest of the bore; 3d, the *cylinder*, or that portion of the bore passed over by the projectile.

84. **The vent.** The size of the vent should be as small as possible, in order to diminish the escape of the gas, and the erosion* of the metal which results from it. All vents in the United States' service are 0.2 inch in diameter.

In bronze pieces which fire large charges of powder, the heat of the inflamed gases would be sufficient to melt the tin, and rapidly enlarge its diameter. For this reason, they are bushed by screwing in a perforated piece of pure wrought copper, called the *vent-piece*. See fig. 27. This arrangement allows the vent to be renewed when too much enlarged by continued use, or when closed with a spike. Copper vent pieces are especially necessary in rifle-guns, in consequence of the prolonged action of the gas arising from the heavy weight of projectile to be moved.

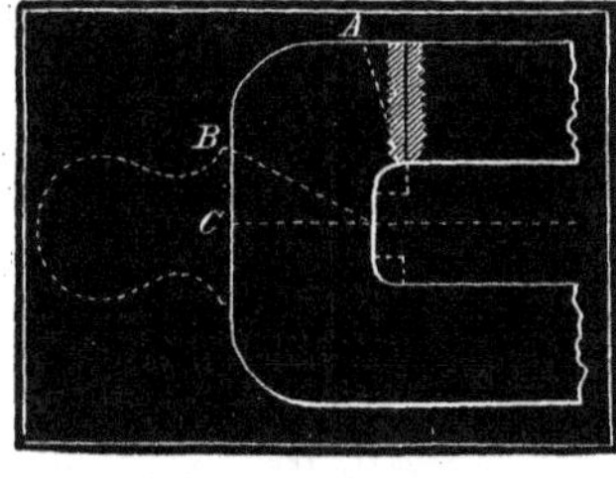

Fig. 27.

Position. Columbiads and Mortars of the model of 1861, have each two unbushed vents, which are situated in two vertical planes on opposite sides of and parallel to the axis of the bore, and distant from it one half the radius of the bore. The one on the left is bored entirely through; the other stops one inch short of the surface of the bore. When the open vent is too much enlarged by wear for further use, it is closed with melted zinc and the other is bored out. Each vent should endure at least five hundred service rounds.

The interior orifice of the vent is placed at a distance from the bottom of the chamber equal to a fourth of its diameter, or at the junction of the sides of the chamber with the curve of the bottom. Experiment shows that this position of the vent is more favorable to the full development of the force of the charge than any other along its length.

Many authors have attributed the injuries which are observed to take place about the lodgment of the projectile, to the position of the vent at the bottom of the bore, supposing that the evolution of the elastic gases begins at the upper portion of the charge, and that the projectile is consequently pressed down upon the lower side of the bore before it is set in motion. To remedy this, it was proposed to place the orifice at the centre of the bottom of the bore; and to determine the merits of this proposition, special experiments were made at the artillery schools of Douai, Toulouse, and Strasbourg, on new guns of 24 and 16 lbs. calibre.

The first gun had the ordinary, old-fashioned vent; see fig. 27 (*A*); in the second the orifice of the vent was placed at the centre of the bottom, with its axis making an angle of 30° with that of the gun (*B*); and the third had its orifice at the centre of the bottom, with its axis coincident with that of the gun (*C*).

The several pieces were fired under the same circumstances, and the injuries noted with great care. It was found that the gun with the ordinary vent had only experienced slight injuries, while the others became unserviceable in a few rounds; as will be seen by an examination of the following table:

Position of the vent.	Depth of Lodgement.		
	Strasbourg 24-pdr. Gun.	Toulouse 24-pdr. Guns.	Douai 16-pdr. Gun.
Vent in the axis.	37 points after 40 shots.	23 points after 6 sh. 25 " " 30 "	8 points aft. 6 sh. 17 " " 30 " 24 " " 60 "
Vent inclined 30°	34 points after 60 shots.	14½ " " 6 " 33 " " 30 "	14 " " 30 " 25 " " 90 "
Ordinary vent		3 " " 30 "	3 " " 60 " 4 " " 90 "

The most probable explanation of these results is this: In guns with the ordinary vent, the gas which is developed in the first moments of combustion, expands freely into the space between the top of the cartridge and bore; it has therefore less tension when it passes over the ball, which will have been moved before all the charge is inflamed. In the two cases in which the orifice is situated at the centre of the bottom, the gas formed cannot develop itself in the space over the charge, but it expands into the interstices of the charge with a greater tension than it had in the first case, and thereby accelerates the inflammation of the charge. From this it follows, that the ball is not moved from its place quite so soon as in the first case, but it begins to move at an epoch more nearly approximating that of the maximum tension of the gas of the charge; and the pressure, therefore, of the gas as it passes over the ball, will be greater; which will account for the greater depth of the lodgment.

85. **Loss.** Experiment also shows that the actual loss of force by the escape of gas through the vent, as compared to that of the entire charge, is inconsiderable, and may be neglected in practice.

SEAT OF THE CHARGE.

86. **Seat of the Charge.** The *form* of that part of the bore of a fire-arm which contains the powder, will have an effect on the force of the charge, and the strength of the piece to resist it.

The points to be considered as most likely to affect the force of the powder, are, the form of the surface, and its extent compared to the enclosed volume. In the first place, to obtain the full force of a charge, its form should be such that its inflammation will be nearly completed before the gas begins to escape through the windage, and the projectile is sensibly moved from its place—in other words, the length of the space occupied by the charge should be nearly equal to its diameter; in the second place, as the tension depends much upon the heat evolved by the combustion, the absorbing surface should be a minimum compared to the volume.

87. **Heavy charges.** The charges with which solid projectiles are generally fired being greater than $\frac{1}{6}$ of their weight, the cartridge occupies a space, the length of which is greater than the diameter; in cannon, therefore, which fire solid projectiles, the form of the seat of the charge is simply the bore prolonged; this arrangement, when compared with the chamber, makes the absorbing surface of the metal a minimum, and reduces the length of the charge so that its inflammation will be as complete as possible, before the gas escapes and the projectile is moved.

To give additional strength to the breech, and to prevent the angle formed by the plane of the bottom and sides of the bore from becoming a receptacle for dirt, and burning fragments of the cartridge-bag, it is rounded with the arc of a circle whose radius is one-fourth the diameter of the bore at this point. See fig. 27. Instead of being a plane bottom, it is sometimes made hemispherical, tangent to the surface of the bore. In all cannon of the most recent model, the bottom of the bore is a semi-ellipsoid. This is thought to fulfil the condition of strength more fully than the hemisphere.

88. **Light charges.** When a light piece is used to throw a projectile of large diameter and great weight, the effect of the recoil can only be diminished by employing a small charge of powder.

If such a charge were made into a cartridge of a form to fit the bore, its length would be less than its diameter, and being ignited at the top, a considerable portion of the gas generated in the first instants of inflammation, would pass through the windage, and a part of the force of the charge would be lost.

To obviate this defect, to give the cartridge a more manageable form in loading, and to make the surface a minimum, as regards the volume, the diameter of this part of the bore is reduced so as to form a *chamber*.

The shape of the chambers of fire-arms is either *cylindrical*, *conical*, or *spherical*.

The effect of these different forms of chambers on the velocity of the projectile will be modified by the size of the charge and the length of the bore. Up to a charge of powder equal to $\frac{1}{7}$ of the weight of the projectile, and a length of bore equal to 9 or 10 calibres, experience shows that the presence of a chamber is advantageous; but beyond these, it possesses no advantages to compensate for its inconvenience.

Cylindrical chamber. For very small charges of powder, and short lengths of bore, the cylindrical chamber gives better results than the conical chamber. This may be explained by the fact, that in this chamber the charge acts a longer time on the projectile, inasmuch as it acts on a smaller portion of its surface, and the grains of powder are therefore more completely consumed when the projectile leaves the piece. But for larger charges the conical chamber is found to answer best; which may be seen from the following table taken from actual firing:

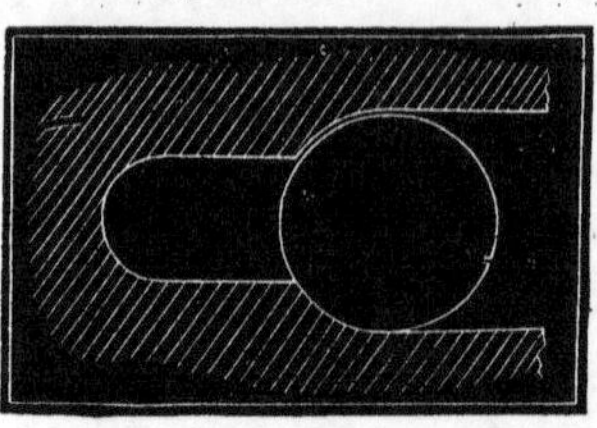
Fig. 28.

MORTARS.	CHARGE.	CYLINDRICAL CHAMBER.	CONICAL CHAMBER.
10-inch.	1.10 lbs.	456 meters.	390 meters.
"	1.65 "	790 "	695 "
"	2.20 "	1060 "	969 "
"	2.75 "	1290 "	1297 "
"	7.00 "		2530 "
"	7.90 "	2530 "	2750 "
8-inch.	0.50 "	325 "	210 "
"	0.60 "	775 "	540 "
"	1.30 "	1250 "	1308 "

NOTE.—Supposed to have been fired at 45° elevation.

Conical chambers. For the same capacity, the conical chamber gives a shorter cartridge, and is therefore better suited to the rapid inflammation of a large charge of powder than the cylindrical chamber. It

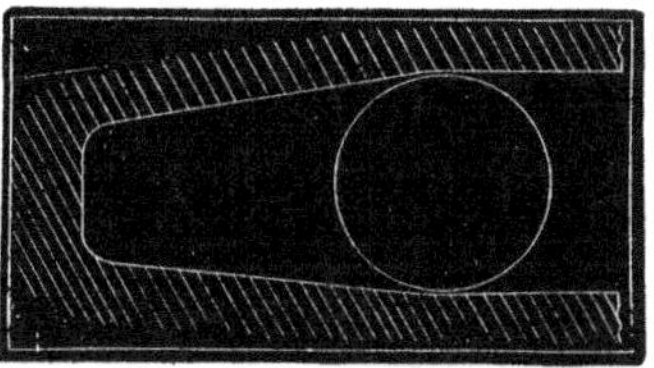
Fig. 29.

also presents less surface of metal for the absorption of heat. The particular kind of chamber represented in the diagram is called a Gomer chamber, after its inventor. Its principal advantages are, that of distributing the force of the charge over a large portion of the surface of the projectile, thereby rendering it less liable to break, if it be hollow; and that of destroying the windage when the projectile is driven down to its proper place.

Spherical chamber. This chamber was formerly used in mortars, but, owing to the inconveniences which attend its construction and use, and its liability to deterioration, it is now entirely abandoned. Experiment shows that when a chamber of this kind is entirely filled with powder, it gives a greater initial velocity to the projectile than any other; and this, probably, for the reasons that its form is better suited to the rapid inflammation of the charge; that it has the least surface compared to its capacity; that sensible motion does not take place so soon; and that the escape of gas by windage is comparatively small.

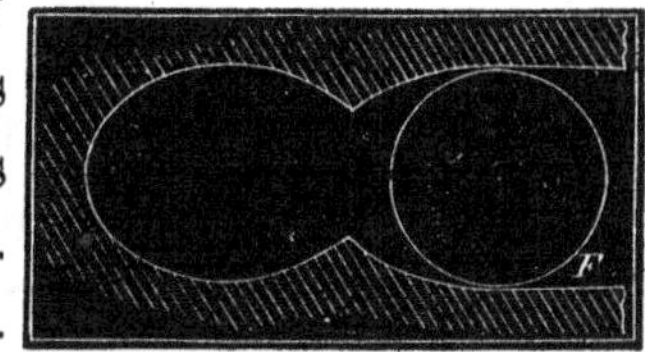
Fig. 30.

In all the regulation guns of the U. S. land service the bottom of the bore is *semi-ellipsoid.* The adoption of this form simplifies the whole subject of chambers, and it is found to give increased ranges for small charges.

89. **Effect on strength.** No very careful experiments have been made to determine, in a general way, the

effect of the chamber on the strength of cannon; but late experience indicates that *cylindrical* chambers in heavy iron guns, have an injurious effect on their endurance, and they have consequently been abandoned in these pieces.

WINDAGE.

90. **Definition.** Windage is the space left between the bore of a piece and its projectile, and is measured by the difference of their diameters. The objects of windage are to facilitate loading, and to diminish the danger of bursting the piece; it is rendered necessary by the mechanical impossibility of making every projectile of the proper size and shape, by the unyielding nature of the material of which large projectiles are made, by the foulness which collects in the bore after each discharge, and by the use of hot and strapped shot.

The *true windage*, which is the difference between the true diameters of the bore and projectile, increases slightly with the size of the bore, and is greater for solid shot, which are sometimes fired hot, than for hollow projectiles, which are never heated.

91. **Loss of force.** The ordinary windage of smooth-bored cannon, used in the United States' service, is about $\frac{1}{40}$ of the diameter of the bore, and the loss of force arising from the escape of gas through this windage amounts to a very considerable portion of the entire charge.

The amount of loss in any case depends on

1. The degree of windage;
2. The calibre of the gun;
3. The length of the bore;
4. The kind of powder;
5. The charge of powder;
6. The weight, or density of the ball.

It is probable that the influence which some of these causes exert on the force of the charge is very slight; and that to determine the exact influence of each of the others would be a very difficult if not an impracticable problem.

The most important question is, to determine what allowance must be made for a given difference of ordinary windage.

The pressure, or force, exerted by a charge of powder on different balls at the same point of the bore of a piece, will be proportioned to the surfaces, or squares of their diameters. If the weight of the balls be the same, the pressure will be proportional to the square of the velocity communicated to the balls in a given time. We, therefore, have the proportion—

$$D^2 : d^2 :: V^2 : v^2$$

$$\text{and } D^2 : d'^2 :: V^2 : v'^2$$

$$\text{or } D : d :: V : v$$

$$\text{and } D : d' :: V : v'$$

From these last two proportions we have

$$D : D-d :: V : V-v$$

$$D : D-d' :: V : V-v'$$

$$\text{or } D-d : D-d' :: V-v : V-v'$$

In which, D, d, and d' represent the diameters of three balls, and V, v, v' their initial velocities, respectively. If D equal the diameter of the bore, $D-d$ is the windage of the ball whose diameter is d, and $D-d'$ is the windage of the ball whose diameter is d'. If we multiply the extremes and means of the last proportion, and divide the resulting equation by $V(D-d)$, we shall have the expression

$$\frac{V-v'}{V}=(D-d')\frac{V-v}{V(D-d)}.$$

By making $m=\frac{V=v}{V(D-d)}$, the equation becomes

$$V-v'=V\times m(D-d').$$

This equation expresses the relation between a certain windage, $D-d'$ and the loss of velocity due to that windage, or $V-v'$.

In a series of experiments made by Major Mordecai, with the ballistic and gun pendulums, it was found that m was constant for all values of $D-d'$ that would be likely to arise in service. From this it follows that $V-v'$ is proportional to $D-d'$; or, in other words, *that the loss of velocity by windage is proportional to the windage.*

When the charge of powder was varied, it was found that the *absolute* loss of velocity by a given increase of windage, was very nearly the same for all the charges used. *It follows from this that the proportional loss is less for the higher charges.*

Both the absolute and relative loss of velocity by a given difference of windage (say one-tenth of an inch) increase as the calibre of the piece decreases.

From the foregoing, it may be stated, *that the loss of velocity by a given windage, is directly as the windage, and inversely as the diameter of the bore, very nearly.*

The loss of velocity of a 24-pdr. ball by a windage of $\frac{1}{40}$, and a charge of 6 lbs. of powder, is 9 per cent.

LENGTH OF BORE.

92. **Ancient theory.** The *slow rate* of burning of mealed powder, which was originally used in cannon, led to the belief that the longest pieces gave the greatest ranges. In spite of much experience to the contrary, this belief was entertained, even after gunpowder received its granular form; and several pieces were made of enormous length, with the expectation of realizing corresponding ranges.

A culverin was cast during the reign of Charles V. which was 58 calibres long, and fired a ball weighing 36 lbs.; but on trial, this piece was found to have actually less range than an ordinary 12-pdr. gun.

The experiment of reducing its length, by successively cutting it off to 50, 44, and 43 calibres, was tried, and it was found that the range increased at each reduction until it gained 2,000 paces.

93. **What governs the length.** That the length of the bore has an important effect on the velocity of the projectile, will be readily seen by a consideration of the forces which accelerate and retard its movement in the piece.

The *accelerating* force is due to the expansive effort of the inflamed powder, which reaches its maximum when the grains of the charge are completely converted into vapor and gas. This event depends on the size of the charge, and the size and velocity of combustion of the grains. With the same accelerating force, the point at which a projectile reaches its maximum velocity depends on its density, or the time necessary to overcome its inertia.

The *retarding* forces are—1st. The friction of the projectile against the sides of the bore: this is the same for all velocities, but different for different metals; 2d. The shocks of the projectile striking against the sides of the bore: these will vary with the angle of incidence, which depends on the windage, and the extent of the injury due to the lodgment and balloting of the projectile; 3d. The resistance offered by the column of air in front of the projectile: this force will increase in a certain ratio to the velocity of the projectile and length of the bore.

As the accelerating force of the charge increases up to a certain point, after which it rapidly diminishes, as the space in rear of the projectile increases, and as the retarding forces are constantly opposed to its motion, it follows, that there is a point where these forces are equal and the projectile moves with its greatest velocity; it also follows that after the projectile passes this point, its velocity decreases until it is finally brought to a state of rest, which would be the case in a gun of great length.

94. **Experiments to determine it.** Elaborate experiments have been made in this country and abroad, to determine accurately the influence which the length of the piece exercises on the velocity of its projectile.

The curves in the accompanying figure show to the eye the relation existing between the different lengths of the bore of a 12-pdr. gun and the corresponding velocities, for charges of 2.2, 3.3, and 4.4 lbs.

The ordinates represent the lengths of the bore in

calibres, and the abscissas represent the velocities, as determined by the electro-ballistic pendulum.

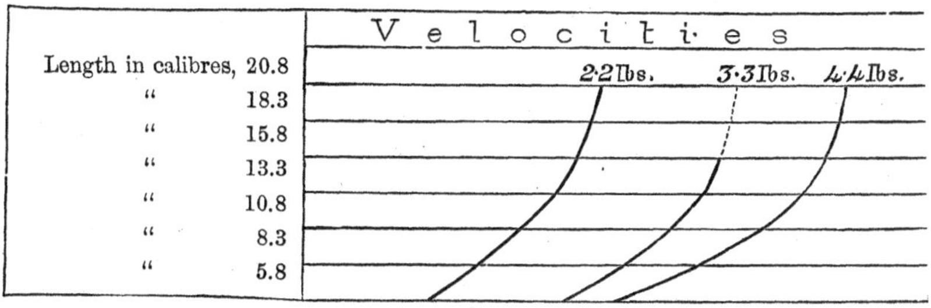

Velocities

2·2 lbs. 3·3 lbs. 4·4 lbs.

Length in calibres, 20.8
" 18.3
" 15.8
" 13.3
" 10.8
" 8.3
" 5.8

Fig. 31.

An inspection of the figure shows that the velocity increases with the length of the bore in a variable ratio, the increase of velocity for the short lengths being much greater than for the long lengths.

The experiments made by Major Mordecai, some years before these, on a gun of the same calibre, show that the velocity increases with the length of the bore up to 25 calibres; but that the entire gain beyond 16 calibres, or an addition of more than one half to the length of the gun, gives an increase of only *one-eighteenth* to the effect of a charge of 4 lbs.

95. **Conclusions.** It follows from the foregoing, that the length of bore which corresponds to a maximum velocity, depends upon the projectile, charge of powder, and material of which the piece is made; and taking the calibre as the unit of measure, it is found that this length is greater for small arms, which fire leaden projectiles, than for guns which fire solid iron shot, and greater for guns than for howitzers and mortars, which fire hollow projectiles.

For the same charge of powder, it may be said that *the initial velocity of a projectile varies, nearly, with the fourth root of the length of the bore, provided the variation in length be small.*

CHARGE.

96. **Maximum charge.** By increasing the charge of powder of a fire-arm, the greater and (in consequence of the wedging of the unburned grains among each other) the more difficult will be the mass to be set in motion; the space between the front of the charge and the muzzle will be diminished; and a larger number of grains will be thrown out unconsumed. It is evident, therefore, that the effect of a charge of powder on a projectile should not increase with the size of the charge; and experiment shows that beyond a certain point, an increase of charge is actually accompanied with a loss of velocity. The charge corresponding to this point is called the maximum charge.

The following are the results of experiments made in France on a 36-pounder gun, of 16 calibres in length:

Charge, lbs., . .	36,	42,	49,	56,	70,	77.
Initial velocity, feet,	1,320,	1,170,	950,	493,	454,	191.

It will therefore be seen that an excess of charge is almost as injurious to the velocity of a projectile, as an excess of length of bore.

97. **Effects on recoil.** Trials made at Turin show that the recoil, and consequently the strain on the gun and carriage, increase in a more rapid ratio than the charges,

viz.: 14 lbs. of powder gave a recoil of 70 inches; 15 lbs., 72 inches; 16 lbs., 74 inches; 18 lbs., 100 inches.

98. **Effect of length of bore on maximum charge.** All experience proves that the longer a piece is, in terms of its calibre, the greater will be the maximum charge in proportion to the weight of the projectile. For heavy cannon, 19 to 20 calibres long, the maximum charge may be stated to be $\frac{1}{2}$ the weight of the projectile; and for light cannon of the same length, $\frac{1}{2}$ to $\frac{2}{3}$ of this weight; the increase of range for charges above $\frac{1}{2}$ the weight of the projectile, being very small.

99. **Most suitable charge.** A charge of $\frac{1}{4}$ the weight of the projectile, and a bore of 18 calibres, is the most favorable combination that can be made in smooth-bored cannon, to obtain the greatest range with the least strain to the carriage.

In the early days of artillery, when *dust* instead of grained powder was used in cannon, the weight of the charge was equal to that of the projectile; after the introduction of grained powder, it was reduced to $\frac{2}{3}$, and in 1740 to $\frac{1}{2}$, this weight.

MATERIALS.

100. **Requirements.** Before discussing the exterior form of cannon, it is necessary to study the nature of the materials of which they are composed. The selection of a suitable material is a very important consideration in the construction of cannon, in consequence of the great difficulty of obtaining any one that possesses all the qualities required of it.

The qualities necessary in cannon-metals are, *strength* to resist the explosion of the charge, *weight* to overcome severe recoil, and *hardness* to endure the bounding of the projectile along the bore.

101. **Strength.** The term *strength*, as applied to a cannon-metal, should not be confined to *tensile* strength alone, which expresses the ability of a substance to resist rupture from extension produced by a simple pressure, as a weight, but should embrace a knowledge of its *elasticity*, *ductility*, and *crystalline structure*, which affect its power to resist the enormous and oft-repeated force of gunpowder—a force which resembles a blow, in the rapidity of its application.

Elasticity. It has been shown by experiment, that the feeblest strains produce permanent elongation or compression in iron; and the same is probably true of all other materials. Perfect elasticity cannot, therefore, be found in solids, although different substances possess it in different degrees. It follows that each discharge, however small, must impair the strength of a cannon, and an ordinary discharge, repeated a sufficient number of times, will burst it.

In the selection of a durable cannon-metal, it is necessary to know, not only the ultimate rupturing force, but also the relation between lesser forces, and the extension and compression produced by them, and the permanent extension, or compression, which remains after these forces are withdrawn, or what is technically known as the "permanent set." This knowledge will be useful in regulating the charge of a cannon to suit the required endurance.

Ductility. Ductility is the property which a metal

possesses of changing its form, without rupture, after it has passed its elastic limit, under the operation of extraneous forces, and, for present purposes, may be considered as opposed to brittleness.

Of two metals that possess the same tensile strength and elasticity, it is evident that it will require more "work" to rupture the one which possesses the greatest amount of ductility.

Crystalline structure. The size and arrangement of the crystals of a metal, have an important influence on its strength to resist a particular force. This arises from the fact that the adhesion of the crystals, by the contact of their faces, is less than the cohesion of the particles of the crystals themselves, and that, consequently, rupture takes place along the larger, or principal crystalline faces.

A metal will be strongest, therefore, when its crystals are small, and the principal faces are parallel to the straining force, if it be one of extension, and perpendicular to it, if it be one of compression.

The size of the crystals of a particular metal depends on the rate of cooling of the heated mass: the most rapid cooling gives the smallest crystals. Practically, there is a limit to the rate of cooling of certain metals; cast iron, for instance, is supposed to change its nature by losing a portion of its uncombined carbon, when suddenly cooled, as in iron moulds.

The position of the principal crystalline faces of a cooling solid, is found to be parallel to the direction in which the heat leaves it, or in a direction perpendicular to the cooling surface.

The result of this arrangement of crystals is to create

planes of weakness where the different systems of crystals intersect. Figure 32 represents sections of the cylinders of two hydraulic presses, used in the construction of the Britannia Bridge. The bottom of No. 2, which was flat, gave way along the lines of weakness A B and C D, while No 1, which was hemispherical, and presented no lines of weakness, resisted all the pressure applied to it.

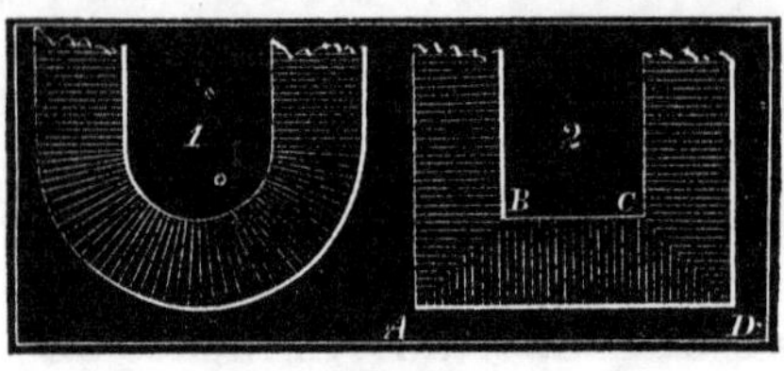

Fig. 32.

The effect of this law on the strength of cannon seems to have been first noticed by Mr. Mallet; and its truth has been confirmed in several instances by Captain Rodman, of the ordnance department, who finds that *radial* specimens are more tenacious than those cut *tangentially* from the same gun.

102. **Effect of cooling.** All solid bodies contract their size in the operation of cooling. It follows, therefore, that if the different parts of a body cool unequally, they will contract unequally, and the body will change its form, provided it be not restrained by the presence of a superior force; if it be so restrained, the contractile force will diminish the adhesion of the parts by an amount which depends on the rate of cooling of the different parts, and the contractibility of the metal.

This is an important consideration in estimating the strength and endurance of cannon, particularly those made of cast iron, as will be seen by examining the form of the casting and the method of cooling it.

The general form of the casting is that of a solid frustum of a cone; it is, therefore, cooled from the exterior, which causes the thin outer layer to contract first, and force the hotter and more yielding metal within, toward the opening of the mould. Following this, the adjacent layer cools, and tends to contract; but the exterior layer, to which it coheres, has become partially rigid, and does not fully yield to the contraction of the inner layer. The result is, the cohesion of the particles of the inner layer is diminished by a force of extension, and that of the outer layer increased by a force of compression.

As the cooling continues, this operation is repeated until the whole mass is brought to a uniform temperature; and the straining force is increased to an extent which depends on the size and form of the mass, the rapidity with which it is cooled, and the contractibility of the particular metal used.

All cannon, therefore, that are cooled from the exterior are affected by two straining forces—the outer portion of the metal being compressed, and the interior extended, in proportion to their distances from the *neutral axis*, or *line* composed of particles which are neither extended nor compressed by the cooling process.

The effect of this unequal contraction may be so great as to crack the interior metal of cast-iron cannon, even before it has been subjected to the force of gunpowder; and chilled rollers, which are cooled very rapidly by casting them in iron moulds, have been known to split open longitudinally, from no other cause than the enormous strains to which they are thus subjected.

The strain produced by the action of a central force, as gunpowder acting in a cannon, is not distributed equally over the thickness of metal. Barlow shows *that it diminishes as the square of the distance from the* *centre increases.** It follows from this, that the sides of a cannon are not rent asunder as by a simple tensile force, but they are torn apart like a piece of cloth, commencing at the surface of the bore. This is confirmed by experience; for the inner portion of the fractured surface of a ruptured gun, is found to be stained with the smoke of the powder, while the outer portion is untouched by it.

It will thus be seen that the effect of ordinary cooling is, to diminish the strength and hardness of the metal of cannon at, or near, a point where the greatest strength and hardness are required, *i. e.*, at the surface of the bore.

Circumstances affecting it. The strains produced by unequal cooling increase with the diameter of the casting, and the irregularity of its form. This explains the great difficulty which is found in making large cast-iron cannon proportionally as strong as small ones; and also, how it is that projections, like bands, mouldings, &c., injure the strength of cannon. It also explains why cannon made of "*high*" cast iron, or cast iron made more tenacious by partial decarbonization, are not so strong as cannon made of weaker iron; for it is well known that such iron contracts more than the latter in cooling, and therefore produces a greater strain of extension on the surface of the bore.

Rodman's plan. The foregoing considerations led Captain Rodman to propose a plan for cooling cannon from the interior, hoping thereby to reverse the strains

* From this law it can be shown that a piece with a thickness of metal equal to one calibre, experiences nine times greater strain on the surface of the bore than on the exterior.

produced by external cooling, and make them contribute to the endurance rather than to the injury of the piece.

The method employed is, to carry off the internal heat by passing a stream of water through a hollow core, inserted in the centre of the mould-cavity before casting, and to surround the flask with a mass of burning coals to prevent too rapid radiation from the exterior.

Extensive trials have been made to test the merits of this plan; and the results show that cast-iron cannon made by it are not only stronger but are less liable to enlargement of the bore from continued firing.

Indications were shown, however, in these and in other trials, that the strains produced by unequal cooling are modified by time, which probably allows the particles to accommodate themselves, to a certain extent, to their constrained position, as in the case of a bent spring or hoop.

103. **Weight.** When a material possesses great strength, but cannot be easily wrought into a heavy mass, it is customary to diminish the recoil by applying an extraneous weight to the piece, or by some contrivance for increasing the weight or friction of the carriage. It is evident that these methods want that unity and solidity which are necessary to great endurance in cannon.

104. **Hardness.** Without a certain degree of hardness, the shape of the bore will be rapidly altered by the compressive force of the powder and projectile, and the accuracy and safety of the piece will be destroyed.

In rifle cannon, hardness is particularly necessary, to

enable the spiral grooves to resist this action; at least, the surface of the bore should be relatively harder than the projectile.

105. **Corrosion, &c.** Cannon metals should be able to resist the corroding action of the atmosphere, and the heat and the products of combustion of the powder; should be susceptible of being easily bored and turned; and should not be too costly, on account of the great number and weight of cannon required for the military service.

106. **Kind of Metals.** The principal materials heretofore used, in the fabrication of cannon, are *bronze*, *steel*, *wrought iron*, and *cast iron*, each of which possesses its peculiar advantages and disadvantages.

107. **Bronze.** Bronze for cannon (commonly called brass), consists of 90 parts of copper and 10 of tin, allowing a variation of one part of tin, more or less. By increasing the proportion of tin, bronze becomes harder, but more brittle and fusible; by diminishing it, it becomes too soft for cannon, and at the same time loses a part of its elasticity. Bronze is more fusible than copper, much less so than tin, more sonorous, harder, and less susceptible of oxidation, and much less ductile than either of its constituents. Its fracture is of a yellowish color, with little lustre, a coarse grain, irregular, and often exhibiting spots of tin, which are of a whitish color. These spots indicate defects of metal, which are supposed to arise from a disposition of the ingredients to separate, in the melted state, into two distinct alloys, or chemical compounds, possessing different degrees of fusibility. The amount of tin which the lighter-colored alloy contains, never exceeds 25 per cent.

Properties. The *density* and *tenacity* of bronze, when cast into the form of cannon, are found to depend upon the pressure and mode of cooling. This is exhibited by the mean of observations made on five guns cast at the Chicopee Foundry, in a vertical position, with the breech down, viz.:

Density.			Tenacity per square inch.	
Breech-square.	Gun-head.	Finished Gun.	Breech-square.	Gun-head.
8.765	8.444	8.740	46.509 lbs.	27.415

In consequence of the difference of fusibility of tin and copper, the perfection of the alloy depends much on the nature of the furnace, the treatment of the melted metal, and the material of which the moulds are made. Cast-iron moulds, by the more sudden cooling of the mass, give greater density and tenacity to bronze than those made of sand or clay.

Bronze is but slightly corroded by the action of the gases evolved from gunpowder, or by atmospheric causes; but its tin is liable to be melted away at the sharp corners by the great heat generated by rapid firing. It is soft, and therefore liable to serious injury by the bounding of the projectile in the bore; this injury is augmented, as the force of the rebound is increased by the elasticity of the metal. When bronze cannon are worn out they can be easily recast.

108. **Steel.** Steel is a compound of iron and carbon, in which the proportion of the latter seldom exceeds 1.7 per cent. It may be distinguished from iron by its

fine grain; its susceptibility of hardening by immersing it, when hot, in cold water; and with certainty by the action of diluted nitric acid, which leaves a black spot on steel, and on iron a spot which is lighter colored in proportion as the iron contains less carbon.

There are many varieties of steel, the principal of which are:

Natural steel, which is obtained by reducing the rich and pure kinds of iron ore with charcoal, and refining the cast-iron, so as to bring it to a malleable state. It is made principally in Germany, and is used for making files and other tools. The India steel, called *wootz*, is said to be a natural steel, containing a small portion of other metals.

Blistered steel is prepared by exposing alternate layers of bar iron and charcoal in a close furnace for several days. When taken out, the bars are covered with blisters, and are brittle in quality and crystalline in appearance. The purpose for which this steel is to be used determines the degree of carbonization. The best qualities of iron (Russian and Swedish) are used for the finest kinds of steel.

Tilted steel is blistered steel moderately heated and subjected to the action of a tilt-hammer, by which means its density and tenacity are increased.

Shear steel is blistered or natural steel refined by piling thin bars into faggots, and then rolling or hammering them into bars after they have been brought to a welding heat in a reverberatory furnace. The quality is improved by a repetition of this process, and the steel is accordingly known by the names *half shear*, *single shear*, and *double shear;* or 1 *mark*, 2 *marks*, and 3 *marks*.

Cast steel is made by breaking blistered steel into small pieces and melting it in close crucibles, from which it is poured into iron moulds. The *ingot* is then reduced to a bar by hammering, or rolling with great care. Cast steel is the finest kind of steel, and is best adapted for most purposes; it is known by a very fine, even, and close grain, and a silvery homogeneous fracture.

High steel. For the construction of cannon, steel may be divided into *high* steel and *low* steel, the difference being that the former contains more carbon than the latter. High steel is very hard and has great ultimate tenacity. It has but little extensibility within or without its elastic limit; it is therefore too brittle for use in cannon, unless used in such large masses that the elastic limit will not be exceeded by the explosive force of the powder. It melts at a lower temperature than wrought iron, and is difficult to weld as its welding temperature is but little less than that at which it melts.

Low steel is often known as "mild steel," "soft steel," "homogeneous metal," and "homogeneous iron," and is made by fusing wrought-iron with carbon in a crucible, after which it is cast into an ingot and worked under a hammer. As it contains less carbon than high steel, it has a greater specific gravity. It can be welded without difficulty, although overheating injures it. It more nearly resembles wrought-iron in all its properties, although it has much greater hardness and ultimate tenacity, and a lower range of ductility depending on its proportion of carbon. It has less extensibility within the elastic limit than high steel, but greater beyond it; or, in other words, greater ductility. Its great advantage over wrought-iron for general purposes is that it can be melted at a practicable heat and run into large masses, possessing soundness and tenacity. Its advantages for cannon are greater* elasticity, tenacity and hardness. Its tenacity when suitable for cannon is about 90,000 lbs., or three times as much as cast gun-iron, and 50 per cent. more than the best wrought-iron.

The difficulty in the use of steel for large, homogeneous guns is the great size of the hammers required to work the blocks into which it is cast.

Bessemer Steel. Bessemer steel is formed by forcing air into a mass of melted pig-iron, which burns out the silicon and carbon. If the process of combustion be stopped when a certain portion of the carbon remains, steel is the result; if it be all consumed, pure iron is the result. The melted mass is then cast into an ingot, in iron moulds, and while its interior is still in a pasty state, the ingot is forged or rolled into a gun, of any size consistent with the power of the machinery used. No large cannon have yet been made wholly of this metal, but several small ones have, which have shown great endurance. Trials at the Woolwich Arsenal show that the tenacity of Bessemer steel is increased by hammering from 60,000 to 145,000 lbs.

109. **Semi-steel.** If, in the operation of puddling, or decarbonizing cast-iron, the process be stopped at a particular time, determined by indications given by the metal to an experienced eye, an iron is obtained of

* Mr. Anderson, assistant superintendent of the Royal Gun Factory at Woolwich, states: "In the qualities of steel employed for guns, the limit of elasticity is about 13 tons; when tempered in oil, it rises to about 32 tons. The ultimate strength is, in the untempered state, about 33 tons; and tempered in oil, about 52 tons, but sometimes a little over and sometimes a little under, according to quality."

greater hardness and strength than ordinary iron, to which the name of semi-steel, or puddled steel, has been applied. The principal difficulty in its manufacture is that of obtaining uniformity in the product, homogeneity and solidity throughout the entire mass. It is much improved by reheating and hammering under a heavy hammer, but it has not been found a reliable material for even cannon of small calibre.

11. **Wrought-iron.** Wrought-iron is made by subjecting melted pig-iron to the decarbonizing action of the flame of bituminous coal in a reverberatory furnace, and then expressing the cinder by the squeezer and rolls. It is afterwards refined by subjecting a pile of short bars to a welding heat and then hammering or rolling it. The quality of wrought-iron may be judged by the appearance of the grain, which should have an even fibrous look in a fresh fracture, and by bending a small bar back and forth when cold. Tough iron will not crack or break.

Strength. The tenacity of good wrought-iron is about 60,000 lbs.; about double that of cast gun-iron. Drawing into wire increases its tenacity, and particularly so if drawn at a little less than a dull red heat. The tenacity of wrought-iron depends on its crystalline structure and the manner of applying the tensile force; i. e., it offers the greatest resistance to a force of extension when the structure is fibrous and the force acts in the direction of the fibre. Mr. Mallet states as the result of experiment, that for artillery purposes, the ultimate strength of a fire-arm in which the explosive strains are all resisted by wrought-iron in the direction of the fibre, is to the resistance in a transverse direction as 7 1-3 to 1.

Elasticity. Under a force of about $\frac{5}{9}$ths of its breaking weight, it is said that wrought-iron will elongate 0.0015 of its length without injury to its elasticity.

Mr. Anderson, however, says that trials made with the wrought-iron used in the Armstrong guns show that it begins to yield permanently under a force of 28,000 lbs. The ductility of wrought-iron is very great; a bar of the Princeton's gun 0.6 in. diameter was stretched so much that it was reduced to 0.5 in. before rupture. It is in consequence of this quality that wrought-iron guns show premonitory symptoms of rupture. These symptoms are undue enlargement of the bore or exterior of the piece, arising from the stretching of the metal. It is claimed that out of 3000 Armstrong guns in the British service not one has burst explosively, as the indications of wear were such as to lead to their timely removal.

Hardness. On account of its softness, the bore of a cannon made of wrought-iron is liable to be enlarged under heavy charges of powder. Wrought-iron being pure iron is easily corroded by the action of the atmosphere, and of the powder. It is in consequence of the latter action that the vent and small cavities of a wrought-iron gun become rapidly enlarged by use.

Difficulties of Manufacture. The defects of large masses of wrought-iron are false welds, cracks, and a spongy and irregular crystalline structure. These are chiefly internal, and arise from the rapid cooling of the exterior, and the difficulty of rolling, or hammering the mass thoroughly. For these reasons it has not been found practicable to make reliable solid wrought-iron cannon of large calibre.

111. **Cast-iron.** Cast-iron for cannon is made by re-melting certain kinds of pig-iron. Its qualities are *hardness*, *tenacity*, and *cheapness*; it has, however, very little elasticity and ductility compared to wrought-iron and other cannon metals. Its structure is generally uniform and homogeneous even in large masses. It is used for making smooth-bored guns of large calibre, but it is not generally considered reliable for heavy rifle-cannon.

Causes which affect its quality. Cast-iron is a well known compound of iron and carbon. The amount of carbon, the state of its combination, together with the ore, fuel, and fluxes, and the process of manufacture, materially affect the quality of cast-iron for artillery purposes.

Many experiments have been made by the ordnance department in the last few years, to ascertain, by chemical and mechanical means, the precise causes which improve or injure the quality of cast iron; but with little success. The utmost that has been accomplished, is the knowledge that certain ores, treated in a certain way, make cast iron suitable for cannon; but the reasons for such results are but little understood.

A slight variation in the ore—even when taken from the same deposit—in the fuel, in the model of the piece, or in the character of the powder, has been known to produce the most disastrous results.

How tested. The fitness of a particular kind of cast iron for cannon-metal, can only be determined by submitting it to the tests of the service; after this is known, a knowledge of certain physical properties, such as tenacity, hardness, density, and color, form and size of crystals, presented in a freshly-fractured surface, will be useful in keeping the metal up to the required standard, and securing its presence in the finished piece.

The course adopted in the land service is to leave the selection of the metal to the private founders. A trial gun is then made of it, generally of 10-inch calibre, which is fired one thousand service-rounds (two hundred being with solid shot and eight hundred with shells.) The tenacity to be not less than 30,000 lbs. to the square inch, to be determined by testing specimens from the sinking head of the gun, and from a cylinder cast from the same heat, and from metal of the same quality as that in the gun. The cylinder to be cast on end in dry sand moulds, and to be 72 inches high, with an elliptical base of 24 in. greater and 16 in. lesser axis. The specimens are to be cut from the gun head, and a slab 4 1-2 in. from the cylinder by planes parallel to and equi-distant from the axis of the base. Among the principal ore beds are the Bloomfield in Penn., the Greenwood in N. Y., and the Richmond in Mass.

Character of pig-iron. Ores suitable for "*gun-metal*" should be reduced in the smelting furnace, with charcoal and the *warm blast.** Iron thus made, or pig-iron should be soft, yielding easily to the file and chisel; the appearance of the fracture should be uniform, with a brilliant aspect, dark gray color, and medium-sized crystals.

Character of gun metal. When remelted and cast into cannon, it should approach that degree of hardness which resists the file and chisel, but not so hard as to be bored and turned with much difficulty. Its color should be a bright, lively gray; crystals small, with acute an-

* Varying from 125° to 300°, Fahr., depending upon the ore used.

gles, and sharp to the touch; structure uniform, close, and compact.

If pig-iron be too soft, coarse, and loose, its strength and density may be increased by remelting it once or twice, and by allowing it to remain in a state of fusion, subjected to a high degree of heat.

The density of pig-iron is about 7.00 and its tenacity about 16.000 pounds to the square inch. The density of gun-metal, or remelted pig, is about 7.250, and its tenacity about 30.000—nearly double the former.

General properties of cast iron. There are several varieties of cast iron, differing from each other by almost insensible shades; the principal divisions are *gray* and *white*, called so from the color of the fracture when recent.

Gray iron is softer and less brittle than white iron; it is in a slight degree malleable and flexible; and is not sonorous; it can be easily drilled and turned in the lathe, and does not resist the file. It has a brilliant fracture, of a gray, or sometimes bluish-gray color; the color is lighter as the grain becomes closer, and its hardness increases at the same time. A medium-sized grain, bright-gray color, lively aspect, fracture sharp to the touch, and close, compact texture, indicate a good quality of iron. A grain either very large or very small, a dull, earthy aspect, loose texture, dissimilar crystals mixed together, indicate an inferior quality.

Gray iron melts at a lower temperature than white iron, becomes more fluid, and preserves its fluidity longer; it runs smoothly; the color of the metal is red, and deeper in proportion as the heat is lower; it does not stick to the ladle; it fills the mould well; contracts less; and contains fewer cavities than white iron; the edges of a casting are sharp, and the surface smooth, convex, and covered with carburet of iron. Gray iron is the only kind suitable for making castings which require great strength, such as cannon. Its tenacity and specific gravity are *diminished* by slow cooling or annealing.

White iron is very brittle and sonorous; it resists the file and the chisel, and is susceptible of high polish; the surface of the casting is concave; the fracture presents a silvery appearance, generally fine-grained and compact, sometimes radiating, or lamellar. Its qualities are the reverse of those of gray iron; it is therefore unsuitable for ordnance purposes. Its tenacity is *increased,* and its specific gravity diminished by annealing. Its mean specific gravity is 7.500, while that of gray iron is only 7.200.

Mottled iron is a mixture of white and gray; it has a spotted appearance—hence its name; it flows well, and with few sparks; the casting has a plane surface, with edges slightly rounded. It is suitable for making shot and shells.

Besides these general divisions, the manufacturers distinguish more particularly the different varieties of pig-metal by numbers, from 1 to 6, according to their relative hardness.

The qualities of these various kinds of iron would seem to depend on the proportion of carbon and the state in which it is found in the metal. In the darker kinds of iron, where the proportion is sometimes 7 per cent. of carbon, it exists partly in the state of graphite, or plumbago, which makes the iron soft. In white iron, the carbon is thoroughly combined with the metal, as in steel.

Cast iron frequently retains a portion of foreign ingredients from the ore, such as earths, or oxides of other metals, and sometimes sulphur and phosphorus, which are all injurious to its quality. Sulphur hardens the iron, and unless in very small proportions, destroys its tenacity. These foreign substances, and also a portion of the carbon, are separated by melting the iron in contact with air; and soft iron is thus rendered harder and stronger.

All cast iron expands forcibly at the moment of becoming solid, and again contracts in cooling; gray iron, as before remarked, expands more, and contracts less, than other iron—an important fact in considering the effect of unequal cooling.

The color and texture of cast iron depend greatly on the size of the casting and the rapidity of cooling: a small casting, which cools quickly, is almost always white; and the surface of large castings partakes more of the qualities of white metal than the interior.

112. **Combined metals.** The term "built up" is applied to those cannon in which the principal parts are formed separately and then united together in a peculiar manner. The object of this mode of manufacture is to correct the defects of one material by introducing another of opposite qualities. As for instance: trials have been made to increase the hardness, and therefore endurance of bronze cannon, by casting them around a core of steel, which formed the surface of the bore.

Built up cannon are not necessarily composed of more than one kind of metal; some of the most noted are made of steel, or wrought-iron alone. In this case, the defects which we have seen to accompany the working of large masses of wrought-iron, viz., crystalline structure, false welds, cracks, etc., are obviated, by first forming them in small masses, as rings, tubes, etc., of good quality, and then uniting them separately. The mode of uniting a built gun may be by welding the parts, by shrinking or forcing one over the other, or by screwing them together.

It has been shown by Barlow's law (page 135) that all parts of the sides of a cannon are not strained equally and are therefore not brought to the breaking point at the same time. Any arrangement of the parts by which the explosive strain is distributed equally over the entire thickness of the piece necessarily brings a greater amount of resistance into play to prevent rupture.

There are two general plans for accomplishing this, viz.: 1st, by producing a strain of compression on the metal nearest the surface of the bore. This is termed an "initial strain" and is brought about by shrinking heated bands or tubes around the part to be compressed, or by slipping a tube into the bore which has been slightly enlarged by heat. In either case it is apparent that the extent of the strain depends on the relative size of the fitting surfaces and the amount of heat used to produce expansion. Owing to the difficulty of regulating the heat, it is preferred by some to force the parts together by hydraulic pressure after they have been carefully bored and turned to the proper size.

The 2d plan is based on "varying elasticity," and is accomplished by placing that metal which stretches most within its elastic limit, around the surface of the bore, so that by its enlargement the explosive strain is transmitted to the outer parts.

By the selection of suitable materials and their proper management, both of these plans may be combined in the same gun, and thereby give it increased strength. Detailed descriptions of the principal built up cannon will be found in the appendix.

EXTERIOR FORM.

113. **Thickness of metal.** The exterior form of cannon is determined by the variable thickness of the metal which surrounds the bore at different points of its length. In general terms, the thickness is greatest at the seat of the charge, and least at or near the muzzle This arrangement is made on account of the variable action of the powder and projectile along the bore, and the necessity of disposing the metal in the safest and most economical manner, or, in other words, to proportion it according to the strain it is required to bear.

114. **To determine the pressure by calculation.** It has been proposed to determine the pressure of the powder at the different points of the bore, by supposing all the gases evolved in the first moment of combustion, and, as the space behind the projectile increased, applying Mariotte's law, that the tension of gas is proportional to its density, which, in turn, is inversely proportional to the space passed over by the projectile. This method of determining the pressure gives a very rapid taper to the exterior; and however well it may answer for cast-iron cannon, is unsuitable for those made of bronze;

which are found, in practice, to burst in the chase, in consequence of the enlargement of the bore from the striking of the projectile against its sides.

A more correct method of calculating the pressure is to apply the principles laid down in Chap. I.

In the case of a French 12-pdr. field gun, which is fired with a charge of $\frac{1}{3}$ the weight of the projectile, the formulas show that the projectile is moved 1.9 in. when the gases have reached their maximum density of 0.38, water being taken as unity.

Substituting this value in Rumford's formula, it will be found that the mean pressure of the charge, on the surrounding surface, is 1,500 atmospheres; and with powder made by the English, or "rolling-mill" process, the mean pressure is found to be as high as 2,400 atmospheres. The pressure at other points may be determined in a like manner.

115. **Determination of pressure by experiment.** About the year 1841, Colonel Bomford devised a plan for determining the pressure at various points of the bore by direct experiment. It essentially consists in boring a series of small holes through the side of a gun, at right angles to its axis (see fig. 36), the first hole being placed at the seat of the charge, and the others at intervals of one calibre. A steel ball was projected from each hole, in succession, into a target or ballistic pendulum, by the force of the charge acting through it; and the pressures at the various points were deduced from the velocities communicated to the balls; it was by this means that the form of the columbiads was determined. This plan has been lately tried in Prussia with great care and success.

Instead of the projectile, Captain Rodman uses a steel

punch which is pressed by the force of the charge into a piece of soft copper. The weight necessary to make an equal indentation in the same piece, is then ascertained by the "testing machine," or machine employed to determine the strength of cannon materials. This instrument is known as the "pressure piston," and is used in proving powder to measure the strain which is exerted on the bore of the eprouvette, or gun.

The following diagram shows the results obtained by this apparatus applied to a 42-pdr. gun.*

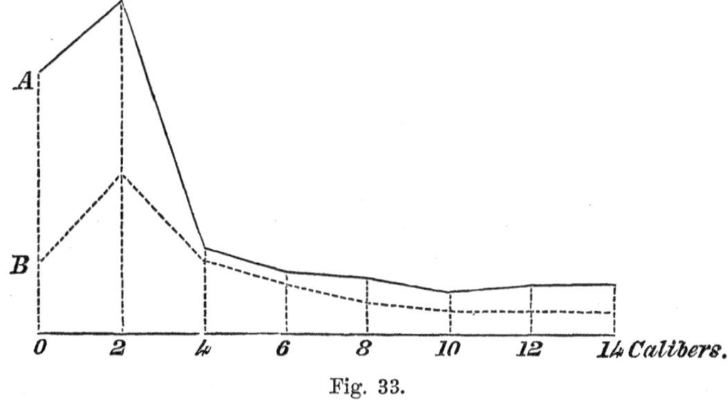

Fig. 33.

116. **Nature of force to be restrained.** In estimating the effect of any force upon a yielding material to which it may be applied, the rate of application, or the time which elapses from the instant when the force begins to act until it attains its maximum, should not be neglected; for, with equal ultimate pressures per square inch of surface, that powder will be most severe

* The data from which the above diagram was constructed, were taken from experiments made by Captain Rodman, to compare the effect of hollow cake-powder, with the ordinary grained powder.

The ordinates of the curve *A*, show the pressures on the bore at intervals of two calibres, commencing at the bottom of the bore, for the grain-powder; and those of the curve *B* the same for "cake-powder." The gun being suspended as a pendulum, and the recoil being equal, or nearly so, it follows that nearly the same velocity upon the gun which attains this pressure in the shortest period of time after ignition. The smaller the grain the more rapid will be the combustion of any charge of powder (other things being equal), and hence the greater will be the strain on the gun in which it is burned.*

117. **Various kinds of strain.** The strains to which all fire-arms are subjected, are four in number, viz.:

1st. The *tangential* strain, which acts to split the piece open longitudinally, and is similar in its action to the force which bursts the hoops of a barrel.

2d. The *longitudinal* strain, which acts to pull the piece apart in the direction of its length. Its action is the greatest at or near the bottom of the bore, and least at the muzzle, where it is nothing.

These two strains increase the volume of the metal to which they are applied.

3d. A strain of *compression*, which acts from the axis

was communicated to the projectile by the "cake" as by the grain powder, with only about one-half the mean pressure on the length of the bore.

No. of fires.	Distance from bottom of the bore.	Caked Powder.		Grained Powder.	
		Pressure, per square inch.	Recoil.	Recoil.	Pressure, per square inch.
3	0	10.989	22° 24′	22° 58′	41.289
3	2 Calibres.	26.001	24° 21′	22° 56′	57.512
3	4 "	12.457	22° 57′	22° 59′	14.103
3	6 "	8.620	25° 43′	23° 05′	10.878
3	8 "	5.801	24° 12′	22° 53′	10.417
3	10 "	4.870	21° 43′	22° 49′	7.127
3	12 "	4.071	22° 07′	22° 55′	8.932
3	14 "	4.071	21° 42′	22° 49′	9.007

* Report of Experiments, by Captain Rodman, to Colonel of Ordnance.

outward, to crush the truncated wedges of which a unit of length of the piece may be supposed to consist, and to give a cross-section the shape shown in the diagram (fig. *A*). This strain compresses the metal and enlarges the bore. If the metal were incompressible, the appearance of a cross-section of the rupture would be that of fig. *B*; and no enlargement of the bore would result from the crushing of the metal; and any enlargement of the bore caused by the action of the tangential force, would be accompanied by an equal enlargement of the exterior diameter of the piece; and hence the strain upon the metal, at the inner and outer surfaces of the gun, would be inversely as the radii of these surfaces, instead of inversely as their squares (as in the case of a compressible material). This quality would bring into action one-third more tangential resistance than the same metal would be capable of offering to a central force.

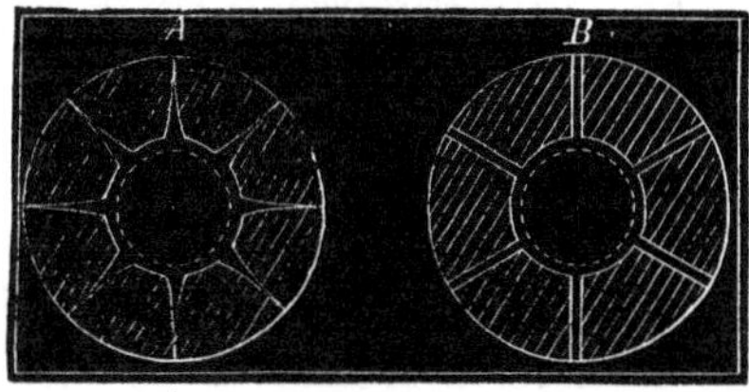

Fig. 34.

4th. A *transverse* strain, which acts to break transversely, by bending outward the *staves* of which the piece may be supposed to consist. This strain compresses the metal on the inner, and extends it on the outer surface. The resistance which a bar of iron, supported at its extremities, will offer to a pressure uniformly distributed over it, is directly as the square of its depth, and inversely as the square of its length; and this is the same as the

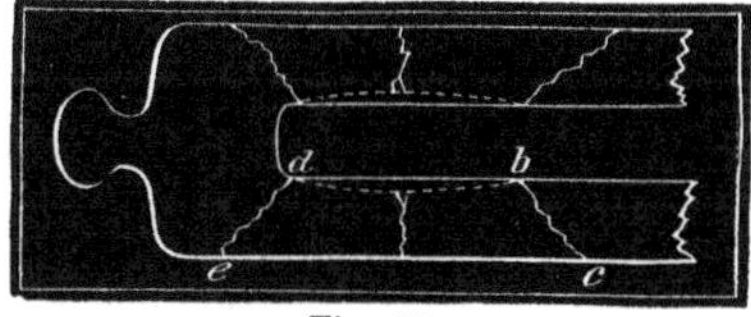

Fig. 35.

resistance offered by a stave of the piece when supported at the points d and b.

If p be the pressure on a unit of surface of the bore, and s the tensile strength of the metal, it can be shown by analysis, that the tendency to rupture, or the pressure on a unit of length of the bore, divided by the resistance which the sides are capable of offering to rupture, for a piece of one calibre thickness of metal, will be as follows, viz.:

$$\text{Tangential, } \frac{3p}{2s};$$

or, rupture will take place when three times the pressure is greater than twice the tensile strength.

$$\text{Longitudinal, } \frac{p}{2s};$$

or, rupture will take place in the direction of the length when the pressure is greater than twice the tensile strength.

$$\text{Transverse, } \frac{2p}{3s};$$

or, rupture will take place when twice the pressure is greater than three times the tensile strength.

From the above it appears, that the tendency to rupture is greater from the action of the tangential force than for any other; and for lengths above two, or perhaps three, calibres, the tangential resistance may be said to act alone, as the aid derived from the transverse resistance will be but trifling, for greater lengths of bore or stave; but for lengths of bore less than two calibres, this resistance will be aided by both the transverse and the longitudinal resistance. Every piece should, therefore, have sufficient thickness of breech to cause rupture to take place (if at all) along the lines, *b c* and *d e* (fig. 35), instead of splitting through the breech; and after this point has been attained, any additional thickness of breech adds nothing to the strength of the piece.

From the foregoing, we conclude that a fire-arm is strongest at or near the bottom of the bore, and that its strength is diminished rapidly as the length of the bore increases, to a certain point (probably not more than three calibres from the bottom); after which, for equal thicknesses of metal, its strength becomes sensibly uniform.

118. **Division of the exterior.*** The exterior of a cannon is generally divided into *five* principal parts, viz.: the *breech*, the 1*st reinforce*, the 2*d reinforce*, the *chase*, and the *swell of the muzzle*. See fig. 26.

The breech. The breech, or thickness of metal in the prolongation of the axis of the bore, should be at least equal to $1\frac{1}{4}$ times the diameter of the bore. A thickness of one diameter has been found insufficient for heavy iron guns.

First reinforce. This part extends from the base-ring

* The following formula, which is used for calculating the exterior form of cannon of large calibre for the land service, was deduced by Captain Rodman from a series of original experiments on the strength of hollow cylinders, &c.:

$$C = \frac{2 p r \sqrt{l'}}{S} \times \frac{R \sqrt{L}}{(R - r)\left(2 r L + R (R + r)(R - r) \frac{l^4}{L^5}\right)},$$

in which C is a constant quantity, r = interior radius, and R = exterior radius, p = pressure of gas, l = length of bore pressed, required to fully develop transverse resistance, L = length of bore corresponding to assumed values of R, S = tensile strength of metal, and l' = length of bore subjected to maximum pressure. The pressure of the gas is supposed to vary inversely as the square root of the length of the bore behind the projectile. The exterior forms thus obtained are entirely made up of curved lines, and a specimen of them is seen in fig. 46, which represents that of the new columbiads.

to the seat of the ball, and is the thickest part of the piece, for the reason that the pressure of the powder is found, both by experiment and calculation, to be greatest before the projectile is moved far from its place. The shape of this reinforce was formerly made slightly conical, under the impression that the pressure was greater at the vent than at the seat of the projectile; but it is now made cylindrical throughout. For bronze cannon, the thickness of this part is approximately given by the empirical formula $E=D\sqrt{\frac{C}{\frac{1}{2}P}}$, in which D represents the diameter of a solid cast-iron shot suited to the bore; C the proof charge; and P the real weight of the projectile. For cast-iron cannon, E should be multiplied by the coefficient 1.17. In general terms, the thickness of a bronze gun, at the seat of the charge, is a little less, and of a cast-iron gun a little greater, than the diameter of the bore. These dimensions exceed those determined by calculation, but are necessary to enable the piece to resist the shocks of the projectile, &c.

Second reinforce. This portion of the piece connects the first reinforce with the chase. It is made considerably thicker than necessary to resist the pressure of the powder, in order to serve as a proper point of support for the trunnions, and to compensate for certain defects of metal liable to occur in the vicinity of the trunnions of all cast cannon, arising from the crystalline arrangement, and unequal cooling of the different parts.

Chase. From the extremity of the second reinforce, cannon taper more or less rapidly to the vicinity of the muzzle. This part is called the *chase*, and constitutes the largest portion of the piece in front of the trunnions. The principal injury to which the chase is liable, arises from the striking of the ball against the side of the bore; and the thickness of metal should be sufficient to resist it. In pieces of soft iron, or bronze, the indentations thus made by the projectile may increase to the extent of bursting the piece; but in cast-iron cannon, where they never exceed 0.02 inch, the taper of the chase can be made more rapid, or, with the same weight of metal, longer than in bronze guns. An example of this is seen in the Dahlgren guns of the naval service.

For many years, cast-iron cannon have been made in Sweden of a form nearly approaching that called for by the actual pressure of the powder at different points of the bore. See fig. 36.

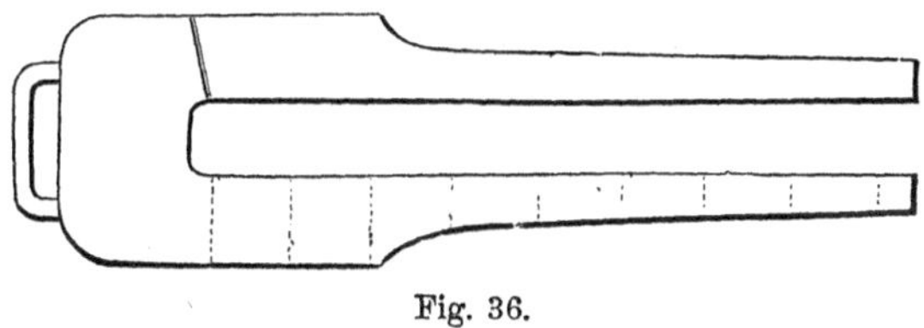

Fig. 36.

In the construction of bronze guns, the thickness of the metal at the neck, or thinnest part, is about equal to $\frac{5}{11}$ of that at the first reinforce, or $\frac{5}{11}E$, given in the empirical formula on p. 158.

Swell of the muzzle. Inasmuch as the metal situated immediately at the muzzle, is supported only in rear, it has been usually considered necessary to increase its thickness, to enable it to resist the action of the projectile at this point. This enlargement is called the *swell of the muzzle.* At present, the practice is to reduce the diameter of the swell of the muzzle of all cannon, and particularly that of heavy iron cannon, designed to be fired through embrasures. By a late order of the war department, the swell is to be hereafter omitted from all sea-coast cannon.

In field and siege howitzers, the muzzle band takes the place of the swell.

All projections on the surface of cannon, not absolutely necessary for the service of the piece, are omitted in cannon of late models. This omission simplifies their construction, renders them easier to clean, and obviates certain injurious strains that would otherwise arise from unequal cooling in fabrication.

119. **Trunnions.** The trunnions are two cylindrical arms attached to the sides of a cannon, for the purpose of supporting it on its carriage. They are placed on opposite sides of the piece, with their axes in the same line, and at right angles to its axis.

Size. The size of the trunnions depends on the recoil of the piece, and the material of which they are made. The resistance which a cylinder opposes to rupture, is proportional to the cube of its diameter, or the weight of a sphere of the same diameter. On the supposition that the strain is proportional to the weight of the charge, it is laid down as a rule that the diameter of the trunnions of guns shall be equal to the diameter of the bore, and the diameter of the trunnions of howitzers shall be equal to the diameter of their chambers—the recoil being less than for guns of the same calibre.

Position. The position of the trunnions, with reference to the axis of the bore, influences the amount of recoil, the endurance of the carriage, and the extent of the vertical field of fire.

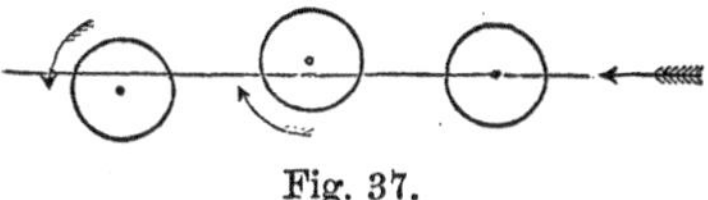

Fig. 37.

By reference to the figure, it will be seen that if the axis of the trunnions be placed below the axis of the piece, the resultant of the force of the charge, which acts against the bottom of the bore, will act to turn the piece around its trunnion, and cause the breech to press upon the head of the elevating screw, with a force proportioned to the length of the lever arm, or distance between the axes. The effect will be to press the trail on the ground and check the recoil—thereby throwing an additional strain on the carriage.

If the trunnions be placed above the axis of the piece, rotation will take place in the opposite direction, and the effects of the discharge on the carriage and recoil will be reversed. By placing the two axes in the same plane, the force of the charge will be communicated directly to the trunnions, without increasing or diminishing its effect on the carriage, or recoil; this position is given to them in all the cannon of the United States service.

It is evident that the space between the lower side of the piece and the carriage, limits the amount of elevation or depression that can be given to the piece, and that the greatest angle of fire can be attained when the axis of the trunnions is situated below the axis of the piece.

120. **Preponderance.** The unequal distribution of

the weight of a piece of artillery, with reference to the axis of the trunnions, is called the preponderance, the object of which is to give it stability in transportation and firing, by producing a pressure on a third point of support, generally the head of the elevating screw. In all guns and howitzers, the centre of gravity is situated in rear of the trunnions, and in all mortars it is situated in front of them.*

Formerly it was measured by the weight which it was necessary to apply to the plane of the muzzle to balance the piece, when suspended freely at the axis of the trunnions; but in pieces of late model it is considered equal to the pressure exerted on the head of the elevating screw, or a third point of support.

The position of the trunnions, with reference to the length of the piece, is an important consideration in siege and sea-coast cannon, for by placing them further to the rear, the piece can be elevated and depressed more rapidly, and its penetration into the embrasure is increased.

121. **Rimbases.** Rimbases are two larger cylinders (*b b*), placed concentrically around the trunnions, for the purpose of strengthening them at their junction with the piece, and by forming shoulders, to prevent the piece from moving sideways in the trunnion-beds.

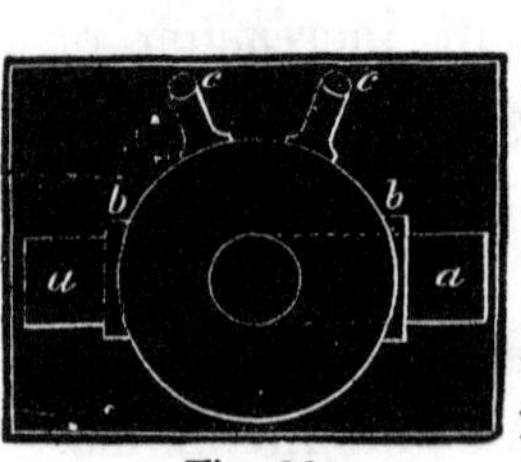

Fig. 38.

* The mortars and columbiads modelled in 1861 have no preponderance, as the axis of the trunnions intersects the axis of the piece at the centre of gravity. Captain Rodman has shown that, contrary to the generally-received opinion, cannon constructed in this way are found not to sensibly change their position before the projectile leaves the bore—and that the accuracy of the fire is not affected.

122. **Knob of the cascable.** This is a projection affixed to the breech of a cannon, for the purpose of attaching the sling in mounting and dismounting it from its carriage. Its axis is that of the bore prolonged.

123. **Handles.** These are two projections (*c c*, fig. 38,) placed over the centre of gravity of certain bronze field-pieces, for manœuvring purposes.

In the heavy sea-coast mortars, the handle is replaced by a *clevis* attached to a projection cast on the piece.

124. **Position of the centre of gravity.** Having determined the precise form of each part of a cannon, it becomes necessary to place the trunnions so that the breech shall exert a given pressure on the head of the elevating screw; or in other words, so that the piece shall have a certain preponderance.

In making the computation, it will be necessary to know the position of the centre of gravity of the piece; and this may be determined from the principle, that the sum of the moments of the weight of the several parts is equal to the moment of the weight of the entire piece.

For convenience, let the plane of reference be taken tangent to the knob of the cascable, and at right angles to its axis.

Let a be the volume of the breech and cascable,
b " " 1st reinforce,
c " " 2d "
d " " trunnions and rimbases,
e " " chase,
f " " swell of muzzle and mouldings,
g " " bore, including the chamber,

and a', b', &c., the distances of the centre of gravity of each part from the plane of reference, respectively. Let w be the weight of a unit of volume of the piece, W the weight of the piece, and x the distance of its centre of gravity from the plane of reference.

Taking the sum of the moments of all the parts, diminishing it by the moment of the bore, and placing the remainder equal to the moment of the piece, we have the relation:

$$x=\frac{(aa'+bb'+cc'+dd'+ee'+ff'-gg')w}{W}.$$

Cancelling the unit of weight in the numerator and denominator of the second member of the above equation, the distance of the centre of gravity of any cannon, from either extremity, *is equal to the algebraic sum of the products of the volumes of the parts by the distances of their centres of gravity from that extremity, divided by the volume of the metal.*

The weight of the piece is supported by the elevating screw and trunnions. The pressure on the screw, and its distance from the centre of gravity, are known; and the distance which the trunnions should be placed in front of the centre of gravity, to support the remainder of the weight, will become known from the proportion

$$p:(W-p)::y:l$$

$$\text{or, } y=\frac{p}{W-p}l;$$

in which p represents the preponderance, l the distance of the head of the elevating screw from the centre of gravity, $(W-p)$ the weight to be sustained by the trunnions, and y the distance of their axis from the centre of gravity.

125. **Weight of cannon.** The weight of a cannon is determined by the weight of the projectile, the maximum velocity it may be necessary to communicate to it, and the extent of the recoil.

The extent of the recoil being limited by the conditions of the service, the weight of the piece may be deduced from the principle, that action and reaction are equal and opposite; or, that the quantity of motion expended on the piece, carriage, and friction, is equal to that expended on the projectile, and the air set in motion by the charge.

Let w be the weight of the projectile,

v its maximum velocity,

c the weight of the charge of powder,

N a constant linear quantity, representing the velocity communicated to the piece by a unit of the charge, arising from its action on the air, independent of the projectile. For American powder, this has been found by experiment with the gun and ballistic pendulums, to be equal to 1,600 feet.

f the velocity lost by a unit of mass, from the friction of the carriage on ordinary ground,

W the weight of the piece,

V velocity of recoil,

C the weight of the carriage,

R the pressure of the trail on the ground, arising from the recoil,

g the force of gravity.

From the principle before enunciated, we have

$$\frac{w}{g}v+\frac{c}{g}N=\frac{(W+C)}{g}V+\frac{(W+C+R)}{g}f;$$

or, by reduction,

$$W=\frac{wv+cN-CV-Cf-Rf}{V+f}.$$

For field-guns, the velocity of the recoil should not exceed 12 feet.

DIFFERENT KINDS OF CANNON.

126. **Gun.** In a technical sense, a gun is a heavy cannon, intended to throw solid shot with large charges of powder, for the purpose of attaining great range, accuracy, and penetration.

It may be distinguished from other cannon by its great weight and length, and by the absence of a chamber.

The gun is suited to fire hollow as well as solid projectiles; and the only limit to the charge is the strength of the projectile to resist rupture in the piece. The employment of shells in heavy cannon, after the manner of solid shot, constitutes the basis of what is known as General Paixhan's System of Artillery, and not the peculiar form of the gun, as is generally supposed.

The calibre of a gun is generally expressed in terms of the weight of a solid cast-iron ball of the size of the bore.

127. **Howitzer.** The howitzer is a cannon employed to throw large projectiles with comparatively small

charges of powder. It is shorter, lighter, and more cylindrical in shape than a gun of the same calibre; and it has a chamber for the reception of the powder, generally of a cylindrical form.

The chief advantage of a howitzer over a gun, is, that with less weight of piece, it can produce at short ranges a greater effect with hollow projectiles and case shot. It also affords the means of attaining an enemy sheltered from the direct fire of solid shot.

The calibre of a howitzer may be expressed in terms of the weight of solid shot, as in guns; or it may be expressed in terms of the diameter of the bore in inches.

128. **Mortars.** A mortar is a short and comparatively light cannon, employed to throw large, hollow projectiles, at great angles of elevation. It has a chamber, generally of a conical form, and of a size suited to a small charge of powder, compared to the weight of the projectile. The trunnions of all mortars are attached to the breach for convenience of elevation and depression. (See note at the bottom of page 162.)

Mortars are particularly intended to produce effect by the force with which the projectiles descend upon their objects, and by the force with which they explode. They are employed chiefly against inanimate objects, such as the roofs of buildings, magazines, and casemates, and the decks of ships of war.

RIFLE-CANNON.

129. **Definition.** A rifle is a fire-arm which has certain spiral grooves (or "rifles") cut into the surface of

its bore, for the purpose of communicating a rotary motion to a projectile around an axis coincident with its flight.

Without this rotation, or rifle-motion, as it is generally called, an oblong projectile will naturally turn over end for end, and thereby present a constantly varying surface to the resistance of the air. With it, and by virtue of the principle of inertia, the axis of rotation remains parallel to itself, and the point is constantly presented to the air. The object, therefore, of rifle-motion is to increase the range of the projectile by causing it to move through the air in direction of its length, or least resistance. If one side of a projectile be lighter than the other, the resistance of the air will be greater on this side, and deviation in a direction from it would follow, were it not that the sides are constantly changing places by rotation. Rifle-motion, therefore, also gives steadiness to a projectile by distributing the deviating forces uniformly around its line of flight.

130. **Rifles—how classified.** Military rifles may be divided into rifle-cannon and rifle-muskets, or small arms, which only differ in the practical application of the rifle principle. The yielding nature of lead renders the application of the rifle principle of easy accomplishment in the case of rifle-muskets, but such is not the case with rifle-cannon, where the projectiles are made of iron.

The principal question which now occupies the attention of persons engaged in improving this species of cannon, is to obtain the *safest* and *surest* means of causing the projectile to follow the spiral grooves as it passes along the bore of a rifled piece. Various plans have been tried to attain the proposed object, some of which are found to be successful. Nearly all may be ranged under the following classes, viz.:

1st. *By means of flanges, or giving a peculiar form to the projectile.* 2d. *By expansion.* 3d. *By compression.*

1*st Class.* Solid flanges projecting from the body of a projectile and so shaped as to fit the rifling of the bore, were the means first used to communicate the rifle-motion in cannon. These projections were in the form of ribs or rounded buttons of the same substance as the body of the projectile. This, however, being of a very unyielding nature, frequently led to the bursting of the piece, and buttons of zinc, copper, or bronze, firmly secured in mortices in the projectile, were adopted. These buttons were arranged in rows of two or more in such a way that each row entered freely into a corresponding groove of the rifling, in loading. Nearly all of the military powers of Europe have adopted the button system of rifling, in one shape or other.

If the bore of a gun be cut into a spiral form with a polygon, or curve for its base, and the projectile be shaped to fit it, it is evident that the projectile will receive the rifle-motion when fired. The Whitworth and Lancaster cannon are rifled in this manner, the cross section of the former being a hexagon with rounded corners, while the latter is an ellipse.

The principal advantages of the foregoing modes of rifling are that the projectiles are strong and that the rifle-motion is communicated to them with great certainty and regularity. They are, however, liable to produce a strain on the gun, and unless great care is taken to clean the bore, are sometimes difficult to load.

2*d Class.* In projectiles of the second class the body

is composed of a hard metal, as cast-iron, and there is attached to it, generally at the base, a cup, band, or other arrangement of soft metal, which is expanded by the action of the charge into the grooves of the gun when fired. Expanding projectiles are exclusively used in the U. S. service, principally for the reason that they are easy to load and do not clog the grooves in a manner to overstrain the piece. Expanding projectiles of different patterns can be generally used in the same gun, a point of great importance during the late war when the service depended on private manufacturers for its supply. Expanding projectiles cannot be fired with as heavy a charge of powder as others, for fear of breaking, nor are they so sure to receive the rifle-motion. Parrott, Hotchkiss, Schenkle, Dyer, and many other projectiles belong to this class.

3d Class. Projectiles of the 3d class are principally used in breach-loading cannon. They are generally covered with soft metal in such manner as to enter the chamber freely. The bore being smaller than the projectile, a portion of the soft metal covering is forced into the grooves, thus compelling the projectile to follow the direction of the rifling.

Very few rifle-cannon are now made on the breech-loading principle, as they are necessarily weaker than muzzle-loading guns.

Centering System. In consequence of windage, which is necessary in all muzzle-loading guns, the axis of the projectile does not always coincide with that of the bore in firing. This leads to inaccuracy of fire.

A projectile is said to be *centered* when the grooves of the rifling are so constructed as to bring the axis of the projectile in line with that of the bore when the piece is fired.

There are several ways of accomplishing this; the most noted perhaps is the Armstrong "shunt rifling," described in the appendix.

131. **Form of groove.** The form of a rifle groove is determined by the angle which the tangent at any point makes with the corresponding element of the bore. If the angles be equal at all points, the groove is said to be *uniform.* If they increase from the breech to the muzzle, the grooves are called *increasing*, if the reverse, *decreasing* grooves.

Twist is the term generally used by gun-makers, to express the inclination of a groove at any point, and is measured by the length of a cylinder corresponding to a single revolution of the spiral; this, however, does not convey a correct idea of the inclination of a groove. A correct measure of the inclination of a rifle groove at any point, is the tangent of the angle which it makes with the axis of the bore; and *this is always equal to the circumference of the bore divided by the length of a single revolution of the spiral, measured in the direction of the axis.*

132. **Uniform groove.** To construct the development of a uniform groove, let $A\ B$ be the base of a rectangle, equal to the circumference of the bore; $B\ C$, the height equal to the length of a single revolution of the spiral, measured on the axis of the bore. The diagonal $A\ C$ is the development of an entire revolution of the required groove. On the line $A\ D$ lay off the distance

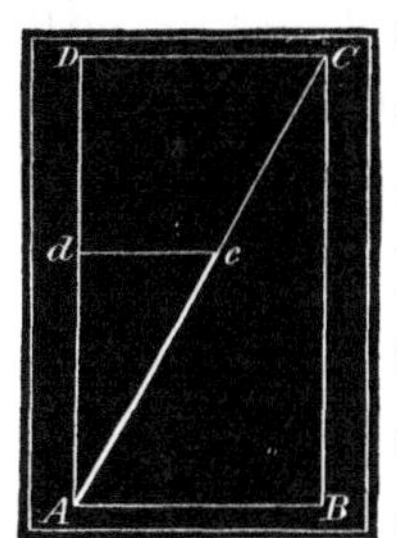

Fig. 40.

$A\ d$, equal to the length of the bore; at d erect a perpendicular, and the line $A\ c$ will be that portion of the development of the groove which lies on the surface of the bore.

133. **Variable groove.** Variable grooves are constructed by wrapping a curve around the surface of the bore. The curve generally selected for this purpose, is the arc of a circle.

To construct the development of an increasing groove which shall be the arc of a circle: The known conditions of construction are the length of the bore, and the inclination, or *twist*, at the breech and at the muzzle. The quantity to be determined is the radius of the generating circle.

Suppose the problem solved, and let $B\ P$ represent the element of the bore passing through the extremity of the groove at the muzzle; $P\ C$, the tangent to the groove at this point; A the starting point of the curve; and $A\ D$ the tangent to the groove at this point.

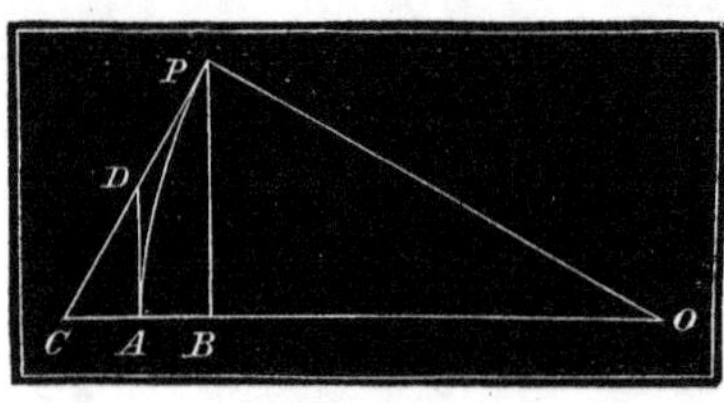

Fig. 41.

The angle of the tangent and element at the breech, is generally made zero, and is so considered in this particular case. The perpendiculars at A and P are radii of the required circle, and their intersection, O, will be its centre.

To determine the length of the radius $A\ O$, and the versed sine of the arc $A\ P$: From the nature of a circle, the angle $CPB = AOP$; $BO = \frac{BP}{\tan. BOP}$; and $OP = \sqrt{BP^2 + BO^2}$; or, by substituting for BO

its value, $OP=\sqrt{BP^2+\frac{BP^2}{\tan.^2CPB}}$, and $AB=\sqrt{BP^2+\frac{BP^2}{\tan.^2CPB}}-\frac{BP}{\tan.BOP}$.

134. **Method of cutting grooves.** The practical method of cutting grooves in cannon is essentially the same as in small arms. It consists in moving a rod, armed with a cutter, back and forth in the bore, and at the same time revolving it around its axis. If the velocities of translation and rotation be both uniform, the grooves will have a uniform twist; if one of the velocities be variable, the grooves will be either increasing or decreasing, depending on the relative velocities in the two directions.

135. **Comparative advantages.** The comparative advantages of uniform and variable grooves, depend on the means used to connect them with the projectiles. If the bearing of the projectile in the grooves be long, and the metal of which it is made be unyielding, it will be unsafe, if not impracticable, to employ variable grooves; and if the metal be partially yielding, a portion of the force of the charge will be expended in changing the form of that part of the projectile which projects into the grooves, as it moves along the bore.

When the portion in the grooves is so short that its form will undergo but slight alteration, the increasing groove may be used with advantage, as it diminishes the friction of the projectile when it is first set in motion, and thereby relieves the breech of the piece of a portion of the enormous strain which is thrown upon it. If the twist be too rapid toward the muzzle, there will be danger of bursting the piece in the chase.

It is claimed by some, that the variable groove is well adapted to expanding projectiles with short bearing surfaces; but the uniform groove, being more simple in its construction, and nearly as accurate in its results, is generally preferred for military fire-arms, both large and small.

136. **Number, width, and shape of grooves.** The width of a groove depends on the diameter of the bore, and the peculiar manner in which the groove receives and holds the projectile.

Wide and shallow grooves are more easily filled by the expanding portion of a projectile than those which are narrow and deep; and the same holds true of circular-shaped grooves, when compared to those of angular form. An increase in the number of grooves increases the firmness with which a projectile is held, by adding to the number of points which bear upon it.

It has been suggested that rifle-cannon, intended for flanged projectiles, should have *four* grooves; as a greater number increases the difficulties of loading, and a lesser number does not hold the projectile with sufficient steadiness.

For expanding projectiles, an odd number of grooves is generally employed, for, as this places a groove opposite to a *land*, less expansion will be required to fill them. The number of grooves used in the 3-inch field-gun is *seven*, and the number used in 4½-inch siege-guns is *nine*. The number of grooves in the 4-inch Armstrong gun is *fifty*.

137. **Initial velocity of rotation.** Let V be the initial velocity of the projectile, or space which it would pass over in one second, in the direction of flight,

moving with the velocity with which it leaves the piece, and l the distance passed over by the projectile in making one revolution; therefore, $\frac{V}{l}$ will be the number of revolutions in one second, and $2\pi\frac{V}{l}$ the *angular velocity* of the projectile at the muzzle. The velocity of rotation of a point on the surface is given by the expression,

$$rw = 2\pi r\frac{V}{l},$$

in which r is its distance from the axis of motion, and w is the angular velocity.

138. **Inclination of grooves.** The object of rifle-grooves being to communicate an effective rotary motion to a projectile throughout its flight, it remains to determine what velocity of rotation, or inclination of grooves, is necessary for different projectiles.

The velocity of rotation will depend on the form and initial velocity of the projectile, the causes which retard it, and the time of flight; therefore, *there is a particular inclination of grooves which is best suited to each calibre, form of projectile, charge of powder, and angle of fire.*

It is proposed to investigate the effect of the length and calibre of the projectile, on the inclination of the grooves.

139. **Effect of length.** It has been noticed that if very long projectiles be fired from the rifle-musket, they are less accurate than the ordinary projectile, the length of which is less than two calibres. Mr. Whitworth states that he has known a bullet twice this length, turn over

end for end, within six feet of the muzzle of the English rifle-musket, the calibre of which is nearly the same as that of the American rifle-musket.

This instability undoubtedly arises from the want of sufficient rotation around the long axis. What increase of angular velocity must, therefore, be given to compensate for a given increase of length of an oblong projectile?

The resistance which a projectile offers to angular deflection, when rotating around a *principal* axis, is proportional to the moment of its quantity of motion taken with reference to this axis, or

$$Mk^2w,$$

M being the mass, k radius of gyration, and w the angular velocity.

Let this expression represent the moment of the quantity of motion around the long axis, and $k_{,}$ and $w_{,}$ the radius of gyration, and angular velocity, around a short axis, and suppose the angular velocities w and $w_{,}$ to be such that the resistance to a deflection from the axes shall be equal, we have

$$Mk^2w = Mk_{,}^2w_{,},$$

and by reduction,

$$w_{,} = \frac{k^2}{k_{,}^2}w.$$

Hence, if we determine by experiment the value of w, the angular velocity necessary to give *practicable* stability of rotation, we can determine the value of $w_{,}$, and consequently the superior limit of the deflecting forces.

Substituting the value of $w_{,}$ in a similar expression for any other projectile, we can determine the angular velocity, and from this the inclination of grooves necessary to give the second projectile steadiness in flight.

The foregoing method of determining the relation between the lengths of two rifle-projectiles, and the inclination of grooves necessary to give them equal steadiness of flight, is true only under the supposition that they preserve throughout their range the relative angular velocities with which they started. It is necessary, therefore, to consider the causes which affect rotation.

140. **Effect of resistance of the air.** The cause which retards the rotary motion of a rifle-projectile, is the friction of the air on its surface; and its retarding effect will be equal to its moment divided by the moment of the projectile's quantity of motion.

Let f be the friction on a unit of surface; s, the surface of the projectile; p, the distance of the resultant moment of the friction from the axis of motion; k, the radius of gyration; and v, the mean velocity of the projectile during its flight. The pressure of the air on the projectile is nearly proportional to the square of its velocity: $f \cdot s\ v^2$ will therefore represent the friction on the projectile, and $f\ s\ v^2\ p$ will be its moment.

The moment of the quantity of motion of the projectile is Mk^2w, M being the mass, and w the angular velocity.

The expression for the angular retardation is, therefore,

$$\frac{fsv^2p}{Mwk^2}.$$

To find the angular velocity that it is necessary to give to another projectile, that it may experience the same retarding effect, place this expression equal to a similar one, answering to this projectile, and we have:

$$\frac{fsv^2p}{Mk^2w}=\frac{fs_{,}v_{,}^2p_{,}}{M_{,}k_{,}^2w_{,}}.$$

Reducing, and recollecting that the surface is proportional to the square, and the mass to the cube of the mean diameter, we have:

$$\frac{v^2p}{dk^2w}=\frac{v_{,}^2p_{,}}{d_{,}k_{,}^2w_{,}};$$

or,

$$w : w_{,} :: \frac{v^2p}{dk^2} : \frac{v_{,}^2p_{,}}{d_{,}k_{,}^2}.$$

Hence, if the angular velocity necessary for one rifle-projectile be known, the angular velocity necessary for another of similar form and material, but of different size, may be determined by calculation.

Suppose the two projectiles to be round shot, and moving with the same mean velocity, through the same extent of trajectory, the proportion reduces to

$$w : w_{,} :: \frac{1}{d^2} : \frac{1}{d_{,}^2}.$$

But the angular velocity is inversely proportional to the length of the twist; it follows, therefore, *that the "length of twist" of grooves, for round shot, moving through equal lengths of trajectory, and with equal mean velocities, should be directly as the squares of their diameters.*

141. **Position of centre of gravity.** The further the centre of gravity of a projectile is in rear of the centre of figure, or resistance of the air, the greater will be the lever arm of the deviating force, and, consequently, the greater must be the inclination of the grooves, to

resist deviation. A conical projectile, of the same length and diameter, requires a greater inclination of grooves than a cylindrical projectile; and the same will hold true of other forms, as they approach one or the other of these extreme cases.

142. **Limit of inclination.** The friction of the projectile as it passes along the grooves, increases with their inclination; its effect will be to diminish the range, and increase the strain on the piece. It is easily to be seen that the inclination may be carried so far as to break the projectile, or rupture the piece.

143. **Most suitable inclination.** The inclination of grooves for a rifle-cannon, best suited to a given projectile, has not yet been determined by experience; and the consequence is, that a wide diversity of "twists" is employed in different services, and by different experimenters. Colonel Cavalli, in his experiments in Sweden, obtained good results from twists of one turn in 12 feet, and one turn in 35 feet, in a 32-pdr. gun; and a still greater variety of twists have been employed in our own service.

For a projectile one and a half diameters long, and 6-pdr. calibre, excellent practice has been obtained with a twist of 25 feet; and in a certain form of the Armstrong gun, the twist is 12 feet for a bore 4 inches in diameter.

The twist of the new wrought-iron rifle-gun for field service, is 10 feet, and the twist of the new siege gun is 15 feet. The calibre of the former is 3 inches, and the latter 4½ inches.

USES TO WHICH CANNON ARE APPLIED.

Having discussed the general principles which govern the construction of all cannon, it is now proposed to consider the peculiarities which arise from the uses to which the several kinds are applied.

144. **Field-cannon.** Field-cannon are intended to be used in the operations of an army in the field; they should, therefore, have the essential quality of mobility. They are divided into light and heavy pieces. The former are constructed to follow the rapid movements of light troops and cavalry. The latter are employed to follow the movements of heavy troops, to commence an action at long distance, to defend field-works and important positions on the field of battle, &c.; hence they are said to constitute "batteries of position."

Formerly the light pieces of the field service were the 6-pdr. gun and 12-pdr. howitzer; and the heavy pieces were the 12-pdr. gun, and 24-pdr. and 32-pdr. howitzers.

At the commencement of the late war in this country, these pieces were set aside for arming field works, block houses, etc., and their places were supplied with the light 12-pdr. gun (smooth bore) and the 3-inch rifle-gun. The regulations prescribe that, as a general rule, one-third of the pieces of a field-battery be rifles and the remainder smooth bores. Of course this proportion is subject to be modified by the character of the operations and the nature of the country. The country in which most of our late military operations were conducted, was either broken in surface or heavily wooded, and the most effective fighting was done at moderate ranges, at which the light 12-pdr., with its heavy shell and case-shot, was found more destructive than the 3-inch rifle-gun.

Rifle-guns. The 3-inch rifle-gun is made of wrought-iron after a plan invented at the Phœnixville Iron Works, Penn. This plan consists simply in wrapping boiler plate around an iron bar so as to form a cylindrical mass of a certain diameter. The whole is placed in a furnace and brought to welding heat and then passed between rollers to unite it solidly together. The trunnions are afterwards welded, or "jumped" on, and the piece is bored and turned to the proper size and shape.

The form of this piece is shown in the accompanying figure, and is the same in its general character as all the guns of the Rodman pattern.

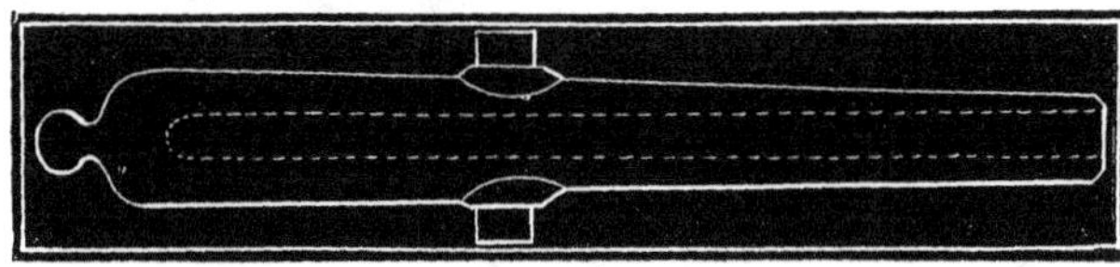

Fig. 41 (*a*).

The following are its principal weights, dimensions, etc.:

Weight, - - - -	820 lbs.
Diameter of bore, - - -	3 in.
Length of bore, - - -	21½ diam.
Number of grooves, - - - -	7
Depth of grooves, - -	0.075 in.
Twist, - - - - - - -	11 ft.
Projectile, weight of - - -	10 lbs.
Powder, - - - - - -	1 lb.

The projectiles used are shells with time and percussion fuzes, and case and canister shot. Hotchkiss,

Schenkle, and Dyer projectiles are all suited to the rifling of this gun.

A 3-inch Parrott rifle-gun, using Parrott projectiles, is also employed in the U. S. field service. This is a cast-iron piece reinforced with a band of wrought-iron, similarly to other guns of the Parrott system, (see appendix).

145. **Napoleon Gun.** In 1856 it was proposed to increase the power of the light and diminish the weight of the heavy field artillery, by the introduction of a single piece of medium weight and calibre, (see par. 76).

Fig. 42.

Form. The form of the new piece is shown in fig. 42. It has no chamber, and should therefore be classed as a gun. Its exterior is characterized by the entire absence of moulding and ornament; and in this respect it may be at once distinguished from the old field cannon. The first reinforce is cylindrical; and it has no second reinforce, as the exterior tapers uniformly with the chase from the extremity of the first reinforce. The size of the trunnions and the distance between the rimbases are the same as in the 24-pdr. howitzer, in order that both pieces may be transported on the same kind of carriage.

Dimensions, &c. The *diameter of the bore* is that of a 12-pdr. The *length of bore* is 13¾ calibres. The *weight* is 100 times the projectile, or 1,200 lbs. The *charge of powder* is the same as for the heavy 12-pdrs. (pattern of 1840), or $2\frac{1}{2}$ lbs. for solid and case shot, and 2 lbs. for canister shot. It has, therefore, nearly as great range and accuracy as the heaviest gun of the old system; and, at the same time, the recoil and strain on the carriage are not too severe. The new gun and carriage weigh about 500 lbs. more than the 6-pdr. and carriage; still, it has been found to possess sufficient mobility for the general purposes of light artillery.

The effect of this change is to simplify the *materiel* of field artillery, and to increase its ability to cope with the rifle-musket, principally by the use of larger and more powerful spherical case-shot. The principal objection to an increased calibre for light field-guns, is the increased weight of the ammunition, and the reduction of the number of rounds that can be carried in the ammunition chests.

MOUNTAIN AND PRAIRIE CANNON.

146. **Mountain howitzer.** Mountain artillery is designed to operate in a country destitute of carriage-roads, and inaccessible to field artillery. It must, therefore, be light enough to be carried on pack-animals.

Fig. 43.

The piece used for mountain service is a short, light 12-pdr. howitzer, weighing 220 lbs. See fig. 43. The form of the chamber is cylindrical, and suited to a charge of $\frac{1}{2}$ lb. of powder. The projectiles are shells and case-shot.

It is discharged from a low two-wheel carriage, which

serves for transportation whenever the ground will permit. When the piece is packed, the carriage is packed on a separate animal.

The mountain howitzer is also employed for prairie service, and in defending camps and frontier forts against Indians, in which case it is mounted on a light four-wheel carriage, called "the prairie carriage."

In the Mexican war, the mountain howitzer was found useful, from the facility with which it could be carried up steep ascents, and to the tops of flat-roofed houses, in street-fighting.

SIEGE AND GARRISON CANNON.

Siege cannon are intended for attacking, and *garrison* cannon for defending, inland fortifications and the land fronts of sea-coast fortifications.

They comprise *guns*, *howitzers*, and *mortars*.

147. **Siege guns.** A siege gun is constructed to throw a solid projectile with the highest practicable velocity, in order to penetrate the masonry of revetments, and to diminish the curvature of the projectile's flight, thereby increasing its chances of hitting objects but slightly raised from the ground.

Rifled guns having been found to fulfill these conditions more fully than smooth bored guns, they have entirely superseded the latter for siege purposes.

A siege gun, properly so called, is one that can be mounted on a siege carriage which serves both for the purposes of transportation and for firing. Very large guns, as the 10-inch columbiad and 8 and 10-inch rifle-guns were used in the siege operations of the late war, but they are not technically siege guns, for no suitable carriages have been yet provided for their transportation over common roads. Whenever railroad or water communications will permit them to be transported, the largest guns may be employed in siege operations with great effect. The two siege guns now in use are the 4½ inch and the 30-pdr. rifles.

Four and a half in.-rifle-gun. The 4½ inch rifle-gun is made of cast-iron, cooled from the exterior; the great length and small size of the bore rendering the water-cooling process impracticable. Its form is similar to that of the 3-inch field-gun (see fig. 41 *a*).

Weight, Dimensions, &c. The principal weight and dimensions of this gun are as follows, viz.:

Weight, - - - -	3,450 lbs.
Diameter of bore, - - -	4.5 in.
Length of bore, - - -	26½ diam.
Number of grooves, - - - -	9
Depth of grooves, - - -	.075 in.
Twist of grooves, - - - -	15 ft.
Weight of projectile, - - -	30 lbs.
Weight of powder, - - -	3.25 lb.

This piece is mounted on the 12-pdr. siege carriage, slightly modified.

Thirty-pdr. rifle-gun. This gun is made of cast-iron reinforced, after the plan of Captain Parrott, with a band of wrought-iron. Its bore is 4.2 inches diameter, and its length about 28 diameters. Its weight is 4,200 lbs., and it is carried on the 18-pdr. siege carriage. The weight of its projectile and charge of powder are the same as for the 4½-inch gun. It has no preponderance, and has been found to be a very accurate and reliable gun in service.

148. **Siege howitzer.** The *siege howitzer*, is principally employed for *ricochet* firing, and for the purpose of battering the earth and fragments of masonry which are left standing after the fire of the breaching guns has ceased.

Dimensions, &c. The form of this howitzer is shown in fig. 44. Its bore is 8 inches in diameter, and it has an elliptical chamber, and the maximum charge is 4 lbs. of powder. As this piece is sometimes fired over the heads of men in the advanced trenches, no sabot or cartridge-block should be used in it except for canister.

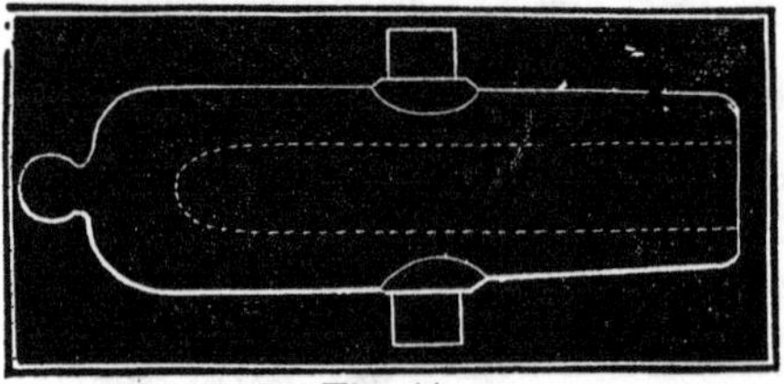

Fig. 44.

The size of the trunnions and the distance between the rimbraces are made the same as in the 24.pdr. gun, that it may fit the 24-pdr. siege carriage, in which case a quoin is used for elevating instead of the ordinary screw. It would be an improvement in the working of this piece were it without preponderance. Its weight is 2,550 lbs. It has the advantage over the old siege howitzer that the projectile always rests in contact with the powder, however small the charge, and the consequence is that the ranges are greater for the smaller charges.

This piece fires mortar shells, spherical-case, grape, and canister shot. By taking off the wheels of the carriage, and reversing the piece in its trunnions, and supporting the whole with strong timbers, this piece may be used as a mortar, in which capacity it gives very long ranges.

149. **Siege mortars.** Siege mortars comprise the *common mortars* and *Coehorn mortar*. The stone mortar was formerly used for siege purposes, but is now laid aside.

Dimensions. The form of the common siege mortars is shown in fig. 45. There are two sizes, the 8 and 10 inch, so called from the diameters of their bores. They have no preponderance, and elevation and depression are effected by a lever the point of which acts in a ratchet (*a a*) cast on the breech. The fulcrum of this lever is attached to the rear transom of the mortar-bed. The chamber is elliptical as in the siege howitzer, and for the reason given in the description of that piece, the new mortars give longer ranges than the old ones. The bores of these mortars are about two diameters long, and the weights about 22 times that of the projectiles, respectively. To prevent the primers from pulling out of the vent, there should be a grooved pulley attached to the *clevis-lug* (*b*) for the lanyard to pass over.

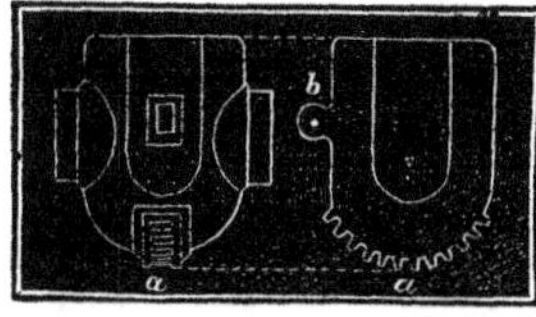

Fig. 45.

Vertical field of fire. The vertical field of fire lies between 30° and 60°. The angle at which mortars are usually fired is 45°. This gives very nearly the maximum range for a given charge of powder.

Natural line of sight. The exterior form of the siege mortars is cylindrical; consequently the natural line of sight is parallel to the axis of the bore,—a position of great convenience in aiming. The line of sight should be permanently marked on the piece while it is in the boring mill.

Object. Siege mortars are used to attain those portions of a work, by vertical fire, which are defended against the direct and ricochet fires of guns and howitzers, such as the covered way, the ditch with its communications, and the roofs of magazines, casemates, etc.

Projectiles. The projectile principally used in mortars is the shell, which for the 8-inch weighs 44 lbs., and for the 10-inch, 88 lbs.

General Bormann, of the Belgian service, some years ago proposed to convert the 10-inch mortar shell into a spherical case shot by filling it with balls about the size of 12-pdr. canister shot, and a bursting charge sufficient to rupture the shell; the fuse to be timed so that the projectile would burst about 50 feet from the ground. The effect of such projectiles at the siege of Petersburgh is thus described by General Abbott, the commander of the siege batteries:

"This battle was probably the first in which spherical case shot from heavy mortars was used. The expedient of putting thirty 12-pdr. canister shot under the bursting charge of the 10-inch shells was of great utility, their steady fire keeping quiet the most dreaded flanking batteries of the enemy's line."

From its lightness, and consequent mobility, the 8-inch mortar may be usefully employed in reaching an enemy sheltered by temporary field works.

151. **Coehorn mortar.** The Coehorn mortar, so called from after its inventor, General Coehorn, is a very small bronze mortar, designed to throw a 24-pdr. shell to distances not exceeding 1,200 yds. Its weight is 164 lbs., its maximum charge ½ lb. of powder, and it is mounted on a wooden block furnished with handles, so that two

men can easily carry it from one part of a work to another.

In the late war this piece was much used in the field against troops covered by rifle-pits. At Fort Wagner, General Gillmore says that it followed close on the heels of the sappers and did good service against the enemy who were not sheltered against vertical fire.

If fired with friction primers, this mortar should be provided with a shield of sheet-iron so placed as to prevent the fragments of the primers from flying among the gunners.

SEA-COAST CANNON.

152. **Object.** Sea-coast cannon are mounted in sea-coast batteries for the defence of harbors, roadsteads, etc., against vessels of war.

The efficiency of all cannon depends on their calibre, combined with facility of manœuvre, or rapidity of fire. As sea-coast cannon generally occupy permanent positions, the weight of the piece and projectile is not a serious objection to an increase of calibre, provided the proper mechanical facilities are supplied for moving them with celerity. Projectiles of 10-inch calibre can be handled by two men. Those of 15-inch calibre require the aid of machinery to lift them quickly to the muzzle of the piece.

153. **Large cannon.** It is proposed to mount special cannon of very large calibre at certain points commanding the entrance to the most important harbors. The intention in such cases is that the projectile shall contain powder enough to constitute a *mine*, and destroy an

enemy's vessel at a single shot. As the fire of such pieces is necessarily slow, and the speed of steam vessels very great, it is evident that success depends on the certainty with which a single shot strikes the object. A system of obstructions, such as rafts or torpedoes, should therefore be combined with them.

A 20-inch gun has lately been made and mounted in accordance with this idea, for the defence of the New York Harbor. It weighs 117,000 lbs., carries a solid shot weighing 1,080 lbs., with a charge of 100 lbs. of powder.

154. **Projectiles.** The most effective projectiles that can be brought to bear against wooden ships are shells and hot shot. The destructive superiority of the former was well attested in the Crimean war, and particularly in the naval engagement of Sinope, when the entire Turkish fleet was destroyed by Russian shells in about one hour's time. In the sea-fight between the Kearsarge and Alabama, the victory of the former was won by the great destructive power of its 11-inch shells.

Modern mechanical skill has succeeded in covering vessels of war with plates of wrought-iron which are proof against shells and solid projectiles of less calibre than 8 inches. Rifle projectiles of less calibre have the power to penetrate such plates, but they do not make the irregular holes and produce the shattering effects of larger round shot,—holes that cannot be plugged, and injuries that cannot be repaired in action.

155. **Kinds of sea-coast cannon.** Sea-coast cannon comprise *guns*, *columbiads*, *howitzers* and *mortars*.

The solid shot pieces of the sea-coast service are the 32-pdr. and 42-pdr. guns, and the 8-inch, 10-inch, 13-inch, and 15-inch columbiads and Rodman guns.

The form of the sea-coast guns in service is shown in fig. 26. Those to be made hereafter will have no *base-ring*, nor *swell of muzzle.*

Dimensions, &c., of the guns. The *mean length of bore* is about 16 calibres. The *mean weight* is about 200 times the solid shot. The *natural line of sight* being intercepted by the reinforce, an artificial one is formed by affixing a sight to the swell of the muzzle. The *charge of powder* is ordinarily $\frac{1}{4}$ the weight of the solid shot. The *projectiles* employed are solid shot, shells, and case-shot. The *maximum angle* of elevation when mounted in barbette is 11°, and in casemate it is 9°.

156. **Columbiads.** The columbiads are a species of sea-coast cannon which combine certain qualities of the gun, howitzer, and mortar; in other words, they are long, chambered pieces, capable of projecting solid shot and shells, with heavy charges of powder, at high angles of elevation, and are, therefore, equally suited to the defence of narrow channels and distant roadsteads.

The columbiad was invented by the late Colonel Bomford, and used in the war of 1812 for firing solid shot. In 1844 the model was changed, by lengthening the bore and increasing the weight of metal, to enable it to endure an increased charge of powder, or $\frac{1}{6}$ of the weight of the solid shot.

Six years after this, it was discovered that the pieces thus altered did not always possess the requisite strength. In 1858, they were degraded to the rank of shell-guns,

NOTE.—All of our serviceable sea-coast guns have been lately rifled on the Parrott system. They are served with a charge of powder which weighs one-fifth, and a projectile which weighs twice that of the corresponding round ball. After extended trials it was only found necessary to band the 42-pdrs. to enable them to endure this charge.

to be fired with diminished charges of powder, and their places supplied with pieces of improved model. The changes made in forming the new model, consisted in giving greater thickness of metal in the prolongation of the axis of the bore, which was done by diminishing the length of the bore itself; in substituting a hemispherical bottom to the bore, and removing the cylindrical chamber; in removing the swell of the muzzle and base-ring; and in rounding off the corner of the breech.

From the fact that all the trial pieces have successfully endured very severe tests, it is to be inferred that the defects of the previous model arose from the presence of a cylindrical chamber, and a deficiency of metal in the prolongation of the bore.*

In 1860, the model proposed by Captain Rodman was adopted for all sea-coast cannon. This model is shown in fig. 46; it does not differ, however, in its essential particulars from the model of 1858.

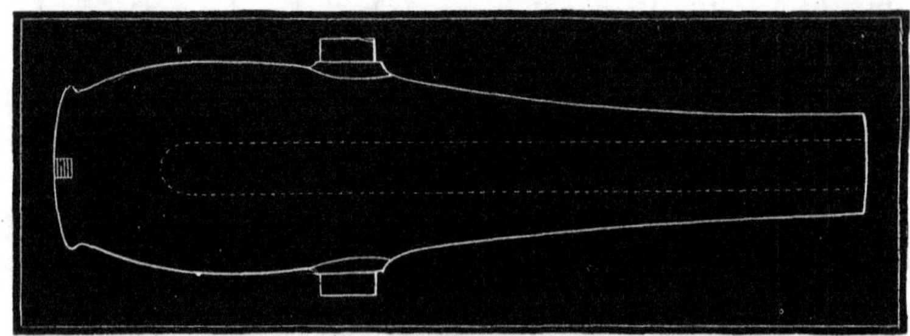

Fig. 46.

* Six of the trial pieces (three 8-in. and three 10-in.), made at different foundries, endured successfully upward of 1,000 service-rounds. Two 10-inch pieces (one cast hollow, and the other cast solid) were fired 2,500 rounds with solid shot and 14 lbs. of powder, and 1,632 rounds with 18 lbs. of powder and solid shot. The only injury sustained was the enlargement of the bore by the cutting action of the gas as it passed over the shot. The enlargement of the bore of the solid-cast piece was much greater than in the hollow one.

Dimensions. The following are the principal dimensions, etc., of the new Rodman guns:

Length of bore,	8-inch,	about	14 diameters.
	10-inch,	"	12 diameters.
	15-inch,	"	11 diameters.
Weight,	8-inch,		8,500 lbs.
	10-inch,		15,000 lbs.
	15-inch,		50,000 lbs.
Charge, shot and shells,	8-inch,		10 lbs.
	10-inch,		15 lbs.
	15-inch,		40 lbs.

In firing against iron-clad vessels, the foregoing charges may be increased fully fifty per cent.

All of the new sea-coast cannon are made without preponderance, and elevation and depression are effected by a lever, the point of which works in a ratchet cut in the breech of the piece at right angles to the trunnions. The fulcrum upon which the lever rests is attached to the rear transom of the gun-carriage. These cannon can be depressed 5° and elevated 39°.

Besides the old 32 and 42-pdr. sea-coast guns which have been rifled for temporary purposes, there are now mounted on our forts a large number of 100-pdr., 200-pdr. and 300-pdr. Parrott guns, firing charges of 10, 16, and 25 lbs. of powder, respectively. Trials are now being made with 10 and 12-in. rifle-guns of the Rodman pattern.

157. **Howitzer.** An iron piece, similar in form to the 24-pdr. field howitzer, is employed to flank the ditches of permanent works, principally with canister shot. This piece is sometimes called a *garrison* howitzer.

158. **Mortar.** The sea-coast mortars are similar in shape and construction to the siege mortars (fig. 45), but

being intended for greater range are heavier in proportion to the weight of the projectiles used in them.

The 13-inch weighs 17,000 lbs.; the 10-inch, 7,300 lbs. The maximum charge is one-tenth of the weight of the projectiles, respectively. The bore of the 13-inch mortar is 2.7 diameters in length; that of the 10-inch mortar is 3.25 diameters. Sea-coast mortars are sometimes used for siege purposes, but they are properly intended for striking the decks of vessels in a vertical direction, hence penetrating to the bottom, causing them to sink.

MANUFACTURE OF CANNON.

160. **Where made.** Cannon for the United States' service are made by private founders. The material and product of the casting are under the supervision of an ordnance officer, who receives the pieces only after they have satisfied all the conditions imposed by the regulations of the service.

The foundries for making cast-iron cannon are at Cold Spring, N. Y., South Boston, Mass., Pittsburgh and Reading, Pa., and Providence, R. I. Bronze cannon are made at Chicopee and South Boston, Mass., and Cincinnati, Ohio. Wrought-iron field cannon are made at Phœnixville, Pa. A foundry for casting 15-inch and smaller cannon is being erected at the Washington Navy Yard. There is at this Navy Yard a foundry for casting bronze howitzers which has been in successful operation for many years. Besides these, there are several private establishments where special cannon are made.

The manufacture of cannon is a valuable branch of the work of these foundries, but the orders of the government are not sufficiently regular and extensive to support them, and much other work is attended to; but the experience thereby obtained in the properties and manner of treating metals, is useful in the improvement of ordnance.

The several operations of manufacturing cannon are, *moulding*, *casting*, *cooling*, and *finishing*.

161. **Moulding.** Moulding, in general terms, is the process by which a cavity of the form of the gun is obtained, by imbedding a wooden model in sand, and then withdrawing it.

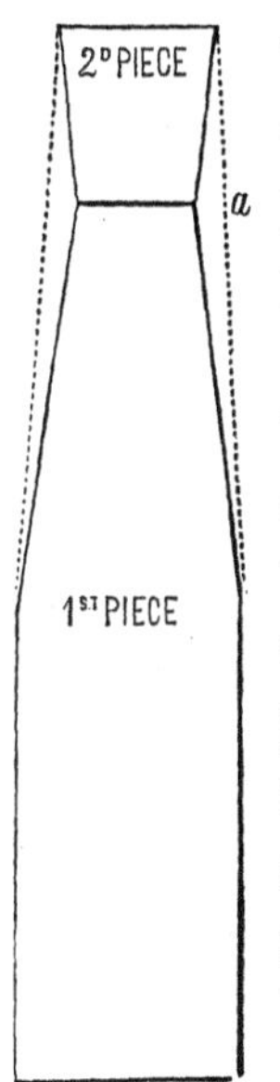

Fig. 49.

The wooden model is technically called the "*pattern;*" and the sand is confined in a box, which is divided into two or more parts, for convenience in withdrawing the pattern.

The pattern of the piece to be cast, somewhat enlarged in its different dimensions, is composed of several pieces of hard wood, well seasoned, or, for greater durability, of cast iron. The *first piece* of the model comprises the body of the piece from the base-ring to the chase-ring; the swell of the muzzle, and the sprue, or dead-head, are formed of the *second piece* (fig. 49); the breech, of the *third;* and the trunnions, of *fourth* and *fifth* pieces. See figs. 50 and 51.

Fig. 50.

The sprue, usually called "*the head*," is an additional length given to the piece, for the purpose of receiving the scoria of the melted metal as it rises to the surface, and furnishing the extra metal needed to feed the shrinkage. Its weight also increases the density of the lower portions of the piece.

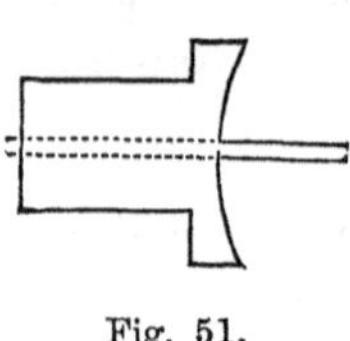
Fig. 51.

The breech is slightly lengthened in the direction of the knob of the cascable, to form a square projection by which the piece can be held, when being turned and bored.

The best material for the mould is dry, hard, angular, and refractory *sand*, which must be moistened with water in which strong clay has been stirred, to make it sufficiently adhesive. When not sufficiently refractory, the sand is vitrified by the high temperature of the melted metal, and protuberances—not easily removed—are formed on the casting. When not sufficiently coarse and angular, the materials cannot be so united as to preserve the form of the moulds.

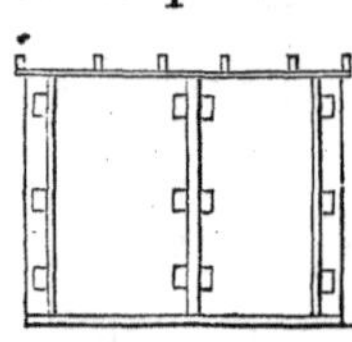
Fig. 52.

The mould is formed in a case of cast-iron, and termed the "*box*," or the "*flask*," consisting of several pieces, each of which has flanges perforated with holes for screw-bolts and nuts, to unite the parts firmly. Fig. 52.

To form the mould, the pattern for the sprue and muzzle, previously coated with pulverized charcoal, or coke moistened with clay-water, to prevent adhesion, is placed vertically on the ground, muzzle-part up, and carefully surrounded by the corresponding parts of the jacket. When properly adjusted, the sand, prepared as above, is rammed around it. The model for the

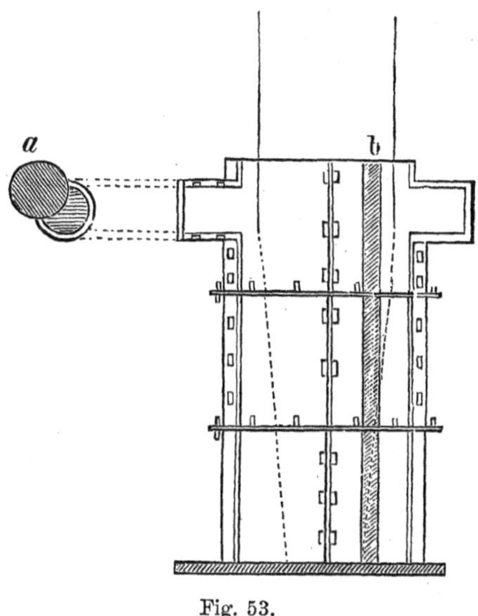

Fig. 53.

body of the piece is then placed on the top of this, and the corresponding parts of the jacket correctly secured on, and filled in succession with the moulding composition. The patterns for the trunnions and rimbases are bolted to the model of the piece, and when the sand is rammed firmly around these, the bolts are withdrawn, this part of the mould completed, and the end-plates screwed on. Fig. 53.

After completing the mould for the body of the piece, the model for the cascable is properly adjusted and the mould completed.

Care is taken to cover each portion of the model with the coke-wash mentioned above, and to sprinkle dry sand upon the top of the mould in each piece of the jacket, to prevent adhesion, so that the portions of the mould may be separated.

In the body of the sand, a channel (*b*) for the introduction of the metal, is formed in the same manner as the mould cavity. It enters at the bottom of the mould, to prevent the bottom from being injured by the falling metal, and in an oblique direction, to give a circular motion to the metal as it rises in the mould, and thereby prevent the scoria from adhering to the sides.

When the mould is completed, the parts of the flask are carefully taken apart, and the pieces of the model withdrawn from the mould contained in them. If any portions of the mould be injured in withdrawing the model, they are repaired, and the interior of the mould is covered with coke-wash; after which the several parts are placed in an oven to be gradually and perfectly dried. When this is accomplished, the parts are carried to a pit, where they are united and secured in a vertical position, with the breech below. Any portion of the sand broken off during the movements and adjustments, should be replaced, and the whole of the interior covered with coke-wash.

The object of coke-wash is to prevent the sand from adhering to the melted metal, which, when prepared, is made to flow in at the entrance of the side-channel. As the metal rises in the mould, a workman agitates it with a long pine stick, to cause the scoria and other impurities to rise to the surface, and bring them toward the centre of the mould to prevent their entering the cavities for the trunnions.

162. **Cooling.** After the mould is placed properly in the pit, it is usual to surround the box with sand, at

least as high as the trunnions of the gun. This is done to prevent rapid cooling. With guns as heavy as 24-pdrs., this sand is not removed for three days; and as the gun is heavier the time is prolonged, and is from 7 to 8 days for the 10-in. columbiad. At the proper time this sand is removed, and the gun, still imbedded in the box and sand of the mould proper, is hoisted out, and the box taken off, and, when nearly cold, the gun cleaned of the sand. For the method of cooling from the interior,* see section 102.

163. **Boring and turning.** A cannon is bored by giving it a rotary motion around its axis, and causing a rod armed with a cutter to press against the metal in the proper direction.

The piece, supported in a rack, is carefully adjusted, with its axis horizontal, and made to revolve on this axis, by machinery attached to the square knob on the cascable. After adjustment, the sprue-head is first to be cut off. This is effected by placing a cutter opposite the point at which the section is to be made, and pressing it against the metal whilst the piece is turning. The head being cut off, and the cutter removed, the boring is commenced by placing the boring-rod, armed with the first cutter, called the *piercer*, in the prolongation of the axis of the piece, and pressing it against the metal. The piercer is used till it penetrates to the bottom of the chamber, after which, a second cutter, or *reamer*, is attached to the boring-rod; and with this the boring is made complete to the round part of the chamber. The reamer is then removed, and its place sup-

* All cast-iron cannon for the land service are now required to be cast hollow, and cooled from the interior, after the plan of Captain Rodman.

plied by the *chamber-cutter*, which gives the necessary form and finish to that part of the bore. In hollow-cast cannon the piercer is dispensed with.

Whilst the boring is taking place, the workman contrives to finish the turning of all the exterior of the piece except the portion between the trunnions, which is afterward planed off in another machine.

These operations having been completed, the piece is placed in the trunnion-machine, and the trunnions are turned down to the proper size.

Care is taken to make the trunnions of the same diameter, and perfectly cylindrical. Their axes should be in the same right line, perpendicular to the axis of the piece, and intersecting it.

Boring the vent. Whilst in the trunnion-lathe, the axis of the piece is inclined to the horizon at the angle the vent is to make with it. A drill is placed vertically over the point where the vent is to be bored, and pressed against the metal whilst a rotary motion is given to it by hand or machinery.

The time required to finish a cannon, ready for inspection, depends upon its size, or from *three* to *four* weeks for a 24-pdr. gun, and *six* weeks for an 10-inch gun.

INSPECTION OF CANNON.

164. **Inspecting instruments.** These are used to verify the dimensions of cannon, and to detect the presence and measure the size of cavities in the metal.

The star-gauge is an instrument for measuring the diameter of the bore at any point.

The cylinder-staff is used to measure the length of

the bore. It is supported by a rest of a T form at the muzzle, and the extremity inserted in the gun is armed with a *measuring-point* and *guide-plate.*

The cylinder gauge is a cylinder of cast iron, turned to the exact or true diameter of the bore. When used, it is attached to the end of the cylinder-staff.

The searcher consists of four flat springs turned up at the end, and attached to a socket which is screwed on to the end of the cylinder-staff. It is used to feel for cavities in the surface of the bore.

The *trunnion-gauge* verifies the diameters of the trunnions and rimbases.

The *trunnion-square* is used to verify the position of the trunnions with regard to the bore.

The *trunnion-rule* measures the distance of the trunnions from the rear of the base-ring.

Calipers, for measuring exterior diameters.

A *standard-rule*, for verifying other instruments.

The *vent-gauges* are two pointed pieces of steel wire, 0.005 inch greater and less than the true diameter of the vent, to verify its size.

The *vent-searcher*, is a hooked wire, used to detect cavities in the vent.

A *rammer-head*, shaped to the form of the bottom of the bore, and furnished with a staff, is used to ascertain the interior position of the vent.

A *wooden rule* to measure exterior lengths.

A *mirror;* a *wax taper;* *bees-wax.*

Rammer, sponge, and priming-wire.

Figure and letter stamps, to affix the required marks.

165. **Inspection.** The objects of inspecting cannon are to verify their dimensions, particularly those which

affect the accuracy of fire, and the relation of the piece to its carriage, and to detect any defects of metal and workmanship, that would be likely to impair their strength and endurance.

Cannon presented for inspection and proof, are placed on skids, for the convenience of turning and moving them easily. They are first examined carefully on the exterior, to ascertain whether there are any flaws or cracks in the metal, whether they are finished as prescribed, and to judge, as well as practicable, of the quality of the metal. They must not be covered with paint, lacquer, or any other composition. If it be ascertained that an attempt has been made to conceal flaws or cavities, by plugging them, the gun is rejected without further examination.

After this preliminary examination, the inspector proceeds to verify the dimensions of the piece.

The *interior of the bore* is first examined by reflecting the sun's rays into it from a mirror; or by a lighted wax-taper placed at the end of a long rod, and inserted into the bore. The *searcher* is then introduced, and pushed slowly to the bottom of the bore, and withdrawn, turning it at the same time; if one of the points hangs, the position of the hole is marked on the outside of the gun, by noticing its distance from the muzzle, and its position in the bore; the size and figure of the cavity, are found by taking an impression of it in *wax*, placed on the end of a hook.

The *cylinder gauge*, screwed on the staff, is then pushed gently to the bottom of the cylindrical part of the bore, and withdrawn; it must go to the bottom, or the bore is too small.

The *bore* is then measured with the *star-gauge.* The measurements should be made at intervals of ¼ inch in the part of the bore occupied by the shot; at intervals of one inch in rear of the trunnions, and of about one calibre from the trunnions to the muzzle.

The position of the *trunnions*, with regard to the axis of the bore, and to each other, is next ascertained.

To *verify the position of the axis of the trunnions*, set the trunnion-square on the trunnions, and see that the lower edges of its branches touch them throughout their whole length; push the slide down till it touches the surface of the piece, and secure it in that position by the thumb-screw; turn the piece over, and apply the trunnion-square to the opposite side, and if, when the point of the slide touches the surface of the piece, the lower edges of the branches rest on the trunnions, the axis of the trunnions is in the same plane with the axis of the bore; if they do not touch the trunnions, their axis is above the axis of the bore by half the space between; and if the edges touch the trunnions, and the point of the slide does not touch the surface of the piece, their axis is below the axis of the bore. If the *alignment of the trunnions* be accurate, the edges of the trunnion-square will fit on them, when applied to different parts of their surface; their diameter and cylindrical form, and the diameter of the rimbases, are verified with the trunnion-gauge.

To *ascertain the length of the bore*, screw the *guide-plate* and *measuring-point* on the cylinder-staff, and push them to the bottom of the bore; place a *half-tompion* in the muzzle, and rest the staff in its groove;

apply a *straight-edge* to the face of the muzzle, and read the length of the bore on the staff.

The *exterior lengths* are measured by the *rule*, or by a *profile*, the accuracy of which is first verified.

The *exterior diameters* are measured with the *calipers* and *graduated rule.*

The *position of the interior orifice of the vent* is found from the mark on the *rammer-head* by the *vent-gauge*, inserted in the vent while the rammer-head is held against the bottom of the bore; two impressions are taken. The position of the exterior orifice of the vent is also verified. The vent is examined with the *gauges*, and with the *vent-searcher*, to ascertain if there be any cavities in it.

In *mortars*, the dimensions of the *conical chambers* and the *form of the breech*, may be verified with *patterns* made of plate-iron. After the *powder-proof* the bore is washed and wiped clean, and the bore and vent again examined, and the diameter of the bore remeasured.

The results of each of the measurements and examinations are noted on the inspection report against the number of the gun.

166. **Bronze cannon.** That the finished bore of a bronze piece may not be injured by the proof-charge, it is bored out under size, from .04 to .05 inch, and, after proof, reamed out to the true size. When the powder-proof is finished, the bore should be cleaned and examined; the vent should be stopped up with a greased wooden plug, the muzzle raised, and the gun filled with water, to which pressure should be applied to force it into any cavities that exist; or the water should be allowed

to remain in the bore twenty-four hours. The *bore* must then be sponged dry and clean, and viewed with a *mirror* or *candle*, to discover if any water oozes from cracks or cavities, and also, if any enlargement has taken place. The quantity that runs out of a crack or honey-comb will indicate the extent of the defect; and if it exceed a few drops, the piece should be rejected, although the measured depth of the cavity may not exceed the allowance.

After the bore has been reamed out to its proper size, its dimensions are again verified, and an examination of the bore and vent is made, to detect any defects which may have been caused or developed by the proof.

Whitish spots show a separation of the tin from the copper, and, if extensive, should condemn the piece.

A *great variation from the true weight*, which the dimensions do not account for, shows a defect in the alloy.

Bronze cannon should be rejected for the following sized cavities or honey-combs:

Exterior. Any hole or cavity 0.25 in. deep in front of the trunnions, and 0.2 in. deep at or behind the trunnions.

Interior. From the muzzle to the reinforce, any cavity 0.15 deep. Any cavity from the reinforce to the bottom of the bore.

In all other respects, the inspection of cast-iron and bronze cannon are alike.

PROOF OF CANNON.

167. **Powder-proof.** *Gunpowder* for proving ordnance

should be of the best quality, ranging not less than 250 yds. by the eprouvette. The *cartridge-bags* are made of woollen stuff or paper, the full diameter of the bore or chamber. They are filled by weight, and if not filled at the place where the guns are proved, each bag should be enveloped in a paper cylinder and cap, marked with the weight of powder and its proof-range.

The *shot* must be smooth, free from seams and other inequalities that might injure the bore of the piece, and they must be of the true diameter given in the tables.

168. **Iron cannon.** *Guns and howitzers* are laid with the muzzle resting on a block of wood, and the breech on the ground or on a plank, giving the bore a small elevation.

Mortars are mounted on strong wooden frames or beds, at an elevation of 45°, supported by the trunnions.

Guns, &c. Guns, howitzers and columbiads, are fired three times, with a solid shot and a charge of powder somewhat greater than the service charge.

In proving new guns, compound shot, or a cylinder with hemispherical ends, of the true diameter of the shot, and equal in weight to the two shot, should be used instead of them.

Should any of the guns proved at one time, fail to sustain the above proof, the remainder shall be rejected, if made of the same metal and treated in the same manner.

Mortars are proved in the same manner as the above, with the exception that shells filled with sand are used in place of shot.

169. **Bronze cannon** are fired three times with solid shot and a charge of powder *one-third* the weight of the

shot. If the piece has been in service, or if it be new, and its bore be of the true size, the shot should be wrapped in cloth or strong paper, to save the bore as much as possible from injury.

170. **Inspection marks.*** All cannon are required to be weighed, and to be marked as follows, viz.: the *number of the gun*, *the initials of the inspector's name*, on the face of the muzzle—the numbers, in a separate series, for each kind and calibre at each foundry; the initial letters of the name of the *founder* and the *foundry*, on the end of the right trunnion; the *year of fabrication*, on the end of the left trunnion; the *foundry number*, on the end of the right rimbase, above the trunnion; the *weight of the piece in pounds*, on the base of the breech; the letters U. S., on the upper surface of the piece, near the end of the reinforce.

The *natural line of sight*, when the axis of the trunnions is horizontal, should be marked on the base-ring and on the swell of the muzzle, whilst the piece is in the trunnion-lathe.

Cannon *rejected* on inspection, are marked XC, on the face of the muzzle; if condemned for erroneous dimensions which cannot be remedied, add XD; if by powder-proof, XP; if by water-proof, XW.

INJURIES CAUSED BY SERVICE.

171. **External.** The only external injury of importance, is the bending of the trunnions of bronze cannon by long firing.

* In cannon modelled in 1861, all the marks are placed on the face of the muzzle.

172. **Internal.** Internal injuries arise from the separate actions of the powder and the projectile. They increase in extent with the calibre, whatever may be the nature of the piece, but are modified by the material of which it is made.

Injuries from the powder. The injuries from the powder generally occur in rear of the projectile. They are,

1st. The *enlargement* of that portion of the bore which contains the powder, arising from the compression of the metal. This injury is more marked when a sabot or wad is placed between the powder and projectile, and is greatest in a vertical direction.

2d. *Cavities*, produced by the melting away of a portion of the metal by the heat of combustion of the charge.

3d. *Cracks*, arising from the tearing asunder of the particles of the metal at the surface of the bore. At first a crack of this kind is scarcely perceptible, but it is increased by continued firing until it extends completely through the side of the piece. It generally commences at the junction of the chamber with the bore, as this portion is less supported than others.

4th. *Furrows*, produced by the erosive action of the inflamed gases. This injury is most apparent where the current of the gas is most rapid, or at the inner orifice of the vent, and on the surface of the bore, immediately over the seat of the projectile.

The wear of the vents of bronze cannon is obviated by inserting a copper vent-piece (par. 84). The effect of continuous firing on the vents of iron cannon is to produce a uniform enlargement of the inner orifice, and to seriously weaken the piece. The appearance of a vent thus enlarged, is irregular and angular, with its greatest diameter in the direction of the axis of the bore.

To obviate the serious consequences that result from this injury, Captain Dahlgren has placed in his naval guns two vents, each a short distance from, and on opposite sides of, the vertical plane passing through the axis of the piece. One of them is filled with melted zinc; the other is used until it becomes so much enlarged as to endanger the safety of the piece; it is then filled with zinc, and the first one is opened.

Injuries from the projectile. The injuries arising from the action of the projectile occur around the projectile, and in front of it. They are,

1st. *The lodgement.* This is an indentation in the lower side of the bore, produced by the pressure upon the ball by the escape of the gas through the windage, before the ball has moved from its seat. The elasticity of the metal, and the *burr*, or *crowding up*, of the metal in front of the projectile, cause it to rebound, and being carried forward by the force of the charge, to strike against the upper side of the bore, a short distance in front of the trunnions. From this it is reflected against the bottom, and re-reflected against the top of the bore, and so on until it leaves the piece.

The first indentation is called the *lodgement;* the others, *enlargements.* In pieces of ordinary length, there are generally three enlargements, when this injury first makes its appearance, but their number is increased as the *lodgement* is deepened and the angle of incidence increased. Bronze pieces are considered unserviceable when the depth of the lodgment is .18 in., and the depth of an enlargement is .16 in.

The effect of this bounding motion, is to alternately raise and depress the piece in its trunnion-beds, and to diminish the accuracy of fire, until finally, the piece becomes unfit for service.

It is principally from this injury that bronze guns become unserviceable. Mortars and howitzers are not much affected by it.

The principal means used to obviate this injury, are to wrap the projectile with cloth or paper (as the cylinder cap of the cartridge used with field-guns), and to shift the seat of the projectile. The latter may be done by a wad, or lengthened sabot, or by reducing the diameter and increasing the length of the cartridge. The last of these methods is considered the most practical as well as the most effective; and it has the additional advantage of diminishing the strain on the bore, by increasing the space in which the charge expands before the ball is moved.

The French bronze siege-guns, which formerly were rendered unserviceable in 600 service-rounds, now endure, by this method, 2,500 service-rounds.

2d. *Scratches*, or furrows made upon the surface of the bore by rough projectiles, or by case-shot. This is not a serious injury.

3d. *Cuts*, made by the fragments of projectiles which break in the bore.

4th. *Wearing away of the lands of rifle-cannon*, especially at the driving edges.

5th. *Enlargement of the muzzle*, arising from the forcing outward of the metal by the striking of the projectile against the side of the bore, as it leaves the piece. By this action, the shape of the muzzle is elongated in a vertical direction.

6th. *Cracks on the exterior.* These are formed by the compression of the metal within, generally at the chase, where the metal is thinnest. This portion of a bronze gun is the first to give way by long firing, whereas, cast-iron cannon are burst in rear of the trunnion, and the fracture passes through the vent, if it be much enlarged.

Cast-iron cannon. The principal injuries to which cast-iron cannon are liable are the wearing away of the metal of the bore above and below the projectile, and at the interior corners of the vent.

In guns which have seen much service the enlargements thus occasioned have been known to exceed one inch in both cases. It has been seen that the strength of cast-iron cannon is diminished by repeated firing, and that there is a limit beyond which they should not be used. For American cannon this limit has been fixed at one thousand service-rounds. The number of times which an iron piece has been fired may be approximately determined by the size of the bore, and vent if it be not bushed. The first is taken with the "star guage," and the second by an impression in wax.

Slight cracks in the surface of the bore, particularly about the seat of the charge, indicate the approaching fracture of a cast-iron gun.

Wrought-iron cannon. The injuries to which wrought-iron cannon are most subject, are the enlargement of the bore by the extension or compression of the metal around it, and the rapid enlargement of slight cracks and cavities by the flame of the powder.

CHAPTER IV.

ARTILLERY CARRIAGES.

173. **Classification.** Artillery carriages may be divided into two classes, viz., those employed for the immediate service and transportation of cannon, as *gun-carriages* and *mortar-beds*, and those employed for the transportation of ammunition, implements, and materials for repairs, as *caissons*, *mortar-wagons*, *forges*, and *battery-wagons.*

The points to be considered in the construction of all carriages are those which relate to the *draught*, those which refer to the *load* to be transported, and (in the particular case of artillery carriages) those which relate to the service of the piece.

Under the first head will be considered the *horse* as a motive power, the *harness* and its mode of *attachment* to the carriage, and the *wheel*, in its relation to friction, &c.

The horse transports his load in two distinct ways: 1st, as a *pack*-horse; 2d, as a *draught*-horse.

174. **Pack-horse.** The load, gait, journey, forage, intervals of rest, &c., of a work-horse should be so proportioned that he will be no more fatigued one day than another.

It has been determined by experience, that a pack-horse, travelling at a walk, over a good road, can carry from 220 to 300 lbs., 30 miles in 10 hours; or if he moves at a trot, 175 lbs. over the same distance.

The daily work of a pack-horse is equal to that of five men, under the same circumstances. If the road be hilly the advantage will be in favor of the men.

The above data suppose that the animal is regularly fed on the service-ration. If he be fed on grass alone, an allowance must be made for its quality and abundance.

In some respects the mule is a superior pack-animal to the horse. His peculiar build gives him, in proportion to his weight, a greater power to transport a load on his back; besides this, the mule eats less than the horse, and is more sure-footed.

175. **Draught-horse.** The force exerted by a draught-horse may be divided into two parts, viz., that which overcomes the inertia and friction of the carriage and sets it in motion, and that which is necessary to overcome the resistances which recur along its path. The first, being of momentary duration, may approximate the utmost strength of the animal; its intensity should be known in order to give the necessary strength to the harness.

If Q represent the mean force (in lbs.) exerted by a horse, in a unit of time, in drawing a load over a road, the length of which is l, Ql represents the quantity of work performed. The direction of the force is taken parallel to the plane along which the load moves. If it make an angle, a, with this plane, the work will be decomposed into two components, Ql cos. a, which is parallel to the plane, and Ql sin. a, which is perpendicular to it: the latter transfers a portion of the load from the ground to the horse's shoulders, thereby increasing his friction, and to a certain extent the power of traction.

Momentary effort. Careful experiments have been made in France to determine the proportion of those two components most favorable to the exercise of the horse's power. It was found that the most suitable angle for the traces of an unloaded horse, with the ground, was from 10° to 12°; and for a horse that carried his driver, from 6° to 7°; or, in other words, a draught-horse should *carry* $\frac{1}{5}$ of his load on his back.

Continuous effort. The relation between the weight of a loaded carriage and the force to be expended by the horse to keep it in motion, depends upon so many circumstances that it is impossible to give a general expression for its determination. It can only be determined by direct experiment in each particular case.

NATURE OF CARRIAGE.	NATURE OF ROAD.	GAIT.	VALUE OF m
Spring carriages.	Pavement in good condition.	Slow walk.	$\frac{1}{46}$
		Fast walk.	$\frac{1}{36}$
		Slow trot.	$\frac{1}{24}$
		Fast trot.	$\frac{1}{15}$
		All gaits.	$\frac{1}{25}$
	Slightly sandy.	All gaits.	$\frac{1}{22}$
	Very sandy.	"	$\frac{1}{10}$
Field artillery carriages.	Turf.	Walk.	$\frac{1}{22}$
	Newly ploughed up and dug over.	"	$\frac{1}{10}$

The foregoing table embraces the results of some experiments on this point, in which m is the ratio of the weight of the loaded carriage and the force of traction; whence it is seen that a carriage moving over a rough or paved road, meets with a resistance which increases rapidly with its velocity; but over a smooth or sandy road, the resistance to draught is independent of the velocity.

The load allotted to an artillery horse is less than that usually drawn by a horse of commerce, for the reason that allowance must be made for bad roads, bad forage, rapid movements, and forced marches. They are as follows:

Light artillery horse,	700	lbs., including carriage.
Heavy field artillery,	850	" " "
Siege artillery,	1,000	" " "

The above is based on the rapidity of movement required in the different services.

An ordinary draught-horse can draw 1,600 lbs. 23 miles in a day.

Usually, a horse can draw 7 times as much as he can carry; hence, all material of war should be transported on carriages, if practicable.

HARNESS.

176. **Requirements.** The best method of attaching horses to a carriage is that which enables each one to perform a given amount of work with the least fatigue; or, in other words, no horse should be restrained by the

efforts of another, and the direction of the traces should be most favorable for draught.

Besides these conditions, artillery-harness should be so constructed that it can be put on and taken off promptly, by night as well as by day, in all states of the weather, and in cases of danger, when the drivers would be liable to lose their presence of mind. The fall of one horse should not interfere with another; and a dead or wounded horse should be easily replaced, whatever may be his position in the team. The absence of some of the horses, the unhitching or cutting of some of the traces should not arrest the movement of the carriage. Finally, the drivers, who are mounted for the better command of their horses, should not be incommoded by the pole of the carriage.

177. **Modes of attachment.** There are three general modes of attaching horses to artillery carriages, and upon the employment of any one of which depends the construction of the harness.

In the first method the wheel-horse is placed between two shafts, by which he guides and regulates the motion of the carriage.

The horses may be arranged in single or double file. The former arrangement was much in vogue in artillery before the days of Gribeauval, but at present is only employed in the mountain service.

This method has the merit of being well suited for drawing heavy loads over smooth roads, but is not adapted to rapid movements over ordinary roads, as much of the tractile force is lost by the continued change in the line of traction incident to long columns. The force thus lost is expended in a great measure on the shaft-horse, which, by constant fatigue, is soon rendered unserviceable.

In the English light artillery the horses are arranged in double file, the *off* wheel-horse being placed in shafts.

178. **Gribeauval's method.** In the second method, the horses are arranged in double file—a wheel-horse being placed on each side of the pole, which is attached to the front axle-tree.

The pole is supported and kept steady by the pressure of the body of the carriage on the *sweep-bar*, which projects in rear of the front axle-tree. The leading horses are attached to the *swing-tree*, which is fastened to the pole, and the wheel-horses are attached to a *movable splinter-bar*, the centre of which is in the axis of the pole. The object of making a splinter-bar movable is to equalize the draught between two horses, one of which works more *freely* than the other.

This system of attachment is used in most carriages of commerce, and so far as the draught alone is concerned, is superior to all others. It is also used in all siege-carriages and baggage-wagons of the military service, except that in the former the splinter-bar is fixed.

179. **Present method.** In field-carriages of late pattern the *sweep-bar* is omitted, to facilitate attaching and detaching the rear carriage in time of action; and the pole is supported by two yokes attached to the collars of the horses. The wheel-horses are attached to a fixed splinter-bar, which is strong and simple in its construction; and the traces of the leading horses are attached directly to those in the rear, giving a continuous line of traction, communicating directly with the carriage. This

method of attaching artillery-horses in line is extremely simple, and at the same time it fulfils nearly all the conditions requisite for artillery harness. Its principal defect, however, is that, from the want of a sweep-bar the weight of the carriage-pole is borne on the necks of the wheel-horses, which is a serious inconvenience in long marches.

How composed. Artillery harness is composed of the *head-gear*, to guide and hold the horse; the *saddle*, for the transportation of the driver and his valise; the *draught-harness*, which enables the horse to move the carriage forward; and the *breeching*, which enables him to hold it back, stop it, or move it to the rear.

Head-gear. The head-gear is composed of the *bridle*, by which the horse is guided, and the *halter*, by which he is held when detached from the carriage.

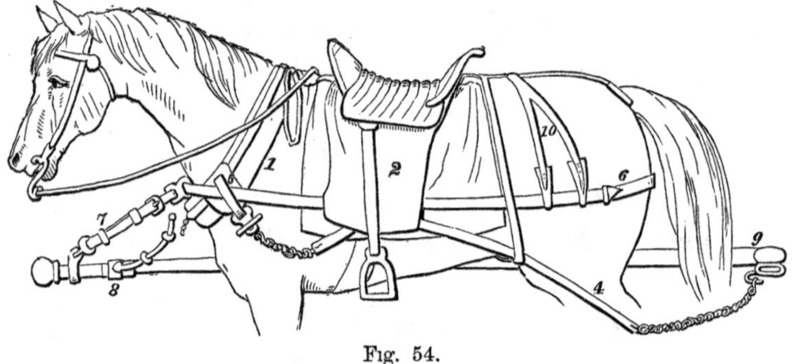

Fig. 54.

Saddle. A riding-saddle (2, fig. 54) is placed on each *near* horse, for the driver, and a *valise*-saddle on each *off* horse.

Draught-harness. This is composed of a collar (1), which serves as a cushion for the hames to rest upon, without injuring the horse's shoulders. The *hames* are two curved pieces of iron, which embrace the collar, and are fastened together at the top and bottom. To each hame is attached a stout leather *tug* (5), which terminates in an iron ring, to which the trace is attached. The *traces* are stout leather straps, terminated at each end with chains, and are used in pulling the carriage. The chain at the rear extremity is used to shorten or lengthen the trace at will. The forward chain plays back and forth in the ring of the tug, which makes the wheel-horse independent of the leading horses. The *pole-yoke* (8) is supported by a chain attached to the hame-clasp, and a ring which slides along the yoke. The branches of the yoke are jointed to a collar near the extremity of the pole, in such manner that they can only play in a plane passing through the axis of the pole. This arrangement enables the horse to keep the pole steady without constraining his motion.

Breeching. The breeching is composed of the *breech-strap* (6), *breast-strap*, and *hip-strap* (10). The breech-strap and breast-strap united, completely encircle the horse. They are attached to the *pole-strap* (7) by an iron loop. The hip-strap sustains the breeching as it passes around the horse's flanks.

The harness of the leading horses has no breeching; in all other respects, it is similar to that of the wheel-horses.

180. **Preservation, &c.** A storehouse for harness should be well ventilated—not too dry, but free from dampness. The different articles should be arranged in bundles, according to kind and class, without touching the wall or each other. Harness should be examined four times a year, at least. The leather parts are brushed and greased with neatsfoot oil as often as con-

dition requires; if they have a reddish hue, add a little lampblack to the oil. The hair side of the leather should be wet with a sponge dipped in warm water, and the oil applied before the surface is dry. The iron parts which are not japanned should be covered with tallow.

WHEEL.

181. **Nomenclature.** All artillery carriage wheels are similarly constructed; they differ, however, in the size and strength of certain parts, depending on the size of the carriage to which they are attached.

The principal parts are (fig. 55), the *nave* (2), the *nave-bands* (3), the *nave-box* (1), the *spokes* (4), the *felloes* (5), and the *tire* (6).

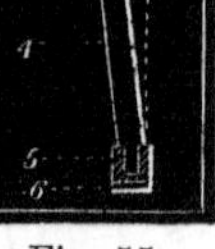

Fig. 55.

The nave constitutes the central portion of the wheel, and distributes the pressure of the axle-arm to the spokes. It is generally made of a single piece of wood, and strengthened by four iron bands called the nave-bands. It is also pierced with a conical hole for the axle-arm; and to diminish wear and friction, it is lined with a box of brass or cast-iron, called the nave-box. The spokes serve to transmit the pressure of the load to the rim of the wheel. In all artillery carriages there are *seven* felloes and *fourteen* spokes. The felloes are the wooden segments which form the rim, and are joined together at their ends by wooden pins, or *dowels.* The tire is a strong band of iron, shrunk tightly around the felloes, to hold them together, and protect the rim from wearing away by contact with the ground.

182. **Dish.** The spokes are fastened to the nave and felloes by means of mortices and tenons, and in a direction oblique to the axis of the nave. Thus situated, they constitute the elements of a conical surface, which is called the *dish*—the principal object of which is to give stiffness to the wheel, and enable it to offer greater resistance to the lateral vibrations of the load, in passing over uneven ground.

The height of the dish will therefore depend on the nature of the ground; and in artillery carriages, which are required to pass over a great variety of ground, it is about two inches.

The dish gives elasticity to the wheel, and increases its durability; it permits the axle-tree to be made shorter, and therefore stronger; it relieves the linch-pin of a certain amount of pressure, which it transfers to the shoulder-washer—the wheel is, therefore, less liable to come off in travelling; for a given length of axle-tree, it allows a greater width of carriage-body; and finally, it throws the mud clear of the carriage.

The stiffness of a carriage-wheel may be increased by placing every alternate mortice in the nave nearer the shoulder of the axle-tree: this gives one half of the spokes a greater dish than the other half. This plan, however, does not answer for artillery carriages.

183. **Friction.** The object of a carriage-wheel is to diminish the resistance opposed to draught, by transferring the friction from the ground to the axle-arm.

When a carriage is at rest, the lowest element of the axle-arm is supported on the bottom of the nave-box. To set the carriage in motion, the friction along the elements of contact, arising from the weight of the carriage, must be overcome, and the axle-arm must rise in the nave-box as though it were moving up an inclined plane tangent to the surface of the box. When this is done, the weight of the loaded axle-arm causes the wheel to revolve around the point of contact with the ground, and the constant repetition of these conditions produces motion.

Let P be the weight resting on the element of contact; p the weight of the wheel; X the force necessary to produce motion; r the radius of the box; and f the co-efficient of friction between the arm and box. When the wheel is well greased, this co-efficient is about 0.180.

To determine the force acting parallel to the ground which will move the wheel, we have the resultant of P and X, equal to $\sqrt{P^2+X^2}$, and the friction arising from it, equal to $f\sqrt{P^2+X^2}$. The pressure on the ground is $P+p$; and if the wheel slips, the friction on the ground will be $F(P+p)$, F being the co-efficient of friction. The points of these resistances, being at the distances r and R from the centre of rotation, respectively, they will counteract each other when

$$rf\sqrt{P^2+X^2}=FR(P+p).$$

If the wheel turns, there is no slipping on the ground, and

$$fr\sqrt{P^2+X^2}<FR(P+p),$$

from which it results that

$$X=fr\frac{\sqrt{P^2+X^2}}{R}=\frac{frP}{\sqrt{R^2-f^2r^2}}.$$

So long as the wheel turns, the draught is not affected by the friction on the ground, since the value of X is independent of F; but if F becomes so small that $fr\sqrt{P^2+X^2}$ becomes equal to or greater than $FR(P+p)$, the wheel will no longer turn, but slide as the runner of a sled. This occurs on ice, or when the wheels are locked; in which case the draught is proportional to the friction on the ground.

From the expression for the value of X we see *that the resistance which a wheel offers to motion, increases with the radius of the axle-arm, and decreases with the radius of the wheel.*

When the radii are nearly equal, the wheel becomes a roller—a machine much used in modifying the friction of fortress carriages.

184. **Rolling-friction.** In the theoretical expression of the force necessary to move a wheel, rolling-friction has been omitted, as it is very small when the wheel is inelastic, and the ground is very hard. The experiments of Coulomb show that this kind of friction does not increase the draught of an artillery carriage more than $2\frac{1}{4}$ lbs.

When the wheel penetrates the ground, it will experience the same resistance as though it were moving upon an inclined plane whose inclination increases with the depth of penetration; and Edgeworth found, in experiments with two-horse carriages, that the force necessary to move a wheel is six times greater than the theoretical force. This difference arises from the compressibility of the soil, and the flexibility of the wheel. On railroads, where the wheels and track are made of iron, the actual and the theoretical draught are very nearly the same; and on the best roads it is about *five* times more than on a railroad.

The depth of the *rut*, or track, made by a wheel, may

be reduced by making the felloes broader; this increase will also cause a wheel to pass more easily over rough ground. Rumford found by experiment that a 7-inch felloe required *one-tenth* more tractile force than one of 12 inches breadth, on a pavement, *one-twelfth* on a hard road, and *one-seventh* on a sandy road.

185. **Size of wheel.** The saving of tractile force arising from increasing the diameter of a carriage-wheel, is limited by the height of the horse, for if the centre of the nave be higher than his shoulders—the point at which the traces are attached—the line of traction will be inclined downward, and if he be moving up hill, or on level ground, the vertical component of the tractile force will increase the friction of the wheel, and diminish the hold of the horse upon the ground. If he be moving down hill, the same cause diminishes the friction of the wheels, and consequently increases the difficulty of holding back.

Large wheels surmount ordinary obstacles more easily than small ones, and penetrate less into yielding ground.

Weight of wheel. The wheels of gun-carriages should be as light as possible, to prevent the axle-tree from being bent in the first instant of the recoil, before their inertia is overcome.

186. **Kinds.** To make it practicable to replace broken wheels in the field, there should be as few kinds as possible for each service. In the field service there are two sizes, called Nos. 1 and 2; and in the siege service but one. The No. 2 wheel is stronger than No. 1, and is used on the heaviest carriages. Both wheels, however, have the same height (58 inches) and the same size of

nave-box, that they may be interchanged if necessary. The siege wheel is 60 inches in diameter.

GUN-CARRIAGES.

187. **General conditions.** Gun-carriages are designed to transport cannon from one point to another, and to support them when fired. A suitable gun-carriage, therefore, should allow the piece to be easily and promptly pointed in the direction of its object; it should be capable of being served by the smallest number of men, and transported with the greatest ease; its recoil, under fire, should be restrained within suitable limits; and it should have sufficient strength and stability to resist overturn or injury from the greatest service-charge.

The injury to the carriage arising from the recoil of the piece, increases with the square of the velocity of the recoil, which is dependent on the relation between the weight of the carriage and the weight of the piece. Generally speaking, the piece should be heavier than the carriage.

188. **Principal parts.** Artillery carriages, like the cannon which they support, are classified into *field*, *mountain*, *prairie*, *siege*, and *sea-coast* carriages. The sea-coast carriages not being required for the transportation of their pieces, differ materially from the others in their construction.

The principal parts of all other artillery carriages (fig. 56) are, the *stock* (1), the *cheeks* (2), the *axle-tree* (3), the *wheels* (4), and the *elevating screw* (5).

The stock. The stock is a long rectangular piece of

Fig. 56.

wood, the front end of which is attached to the axle-tree, while the rear end rests upon the ground when the piece is fired, to form, with the wheels, the three necessary points of support.

When the carriage is in travelling condition, the stock connects the front and rear wheels, and constitutes the basis of the carriage-body. It is employed as a point of support for the elevating screw, and to give, with the assistance of a handspike inserted in its rear end, the proper direction to the piece in aiming.

The cheeks. The cheeks are two thin but strong pieces of wood, attached, one to each side of the head of the stock, to sustain the trunnions of the piece. The notches into which the trunnions fit are lined with iron plates, called the trunnion-bed plates.

The axle-tree. The axle-tree is composed of two parts—the axle-tree proper, which is made of wrought iron, and the wooden body, which encases all the iron portion between the wheels, in such a manner as to distribute the pressure of the head of the stock and cheeks uniformly over it, and prevent it from being bent by the shock of the discharge.

The extremities of the axle-tree, or *arms*, are accurately turned to a conical shape to fit the nave-boxes. The conical shape gives lightness and stiffness to the arm, and facilitates putting on the wheel. The lower elements of the arms being horizontal, the pressure is normal to the surface of the arm, and there is no undue tendency of the wheel to slip off.

The wheel is secured by a linch-pin; and the extremities of the nave are protected from wear by two rings of iron, called the *linch-washer* and the *shoulder-washer*.

Nave-box. The inner surface of the nave-box is enlarged about its middle portion, forming a recess, or receptacle, for the lubricating material.

So long as the wheel is kept well greased, there is but little difference between the friction and wear of brass and cast-iron nave-boxes, but when the grease is exhausted, there is a superiority in favor of brass. Cast iron, however, is most generally employed, on account of its cheapness.

The relation between the draught of greased and ungreased wheels, is determined by experience to be,

		WITH GREASE.	WITHOUT GREASE.
On horizontal ground,	Wooden axle-trees,	65	108
	Iron "	$56\frac{1}{2}$	$61\frac{2}{3}$
Inclined ground, 1-24,	Wooden "	$96\frac{2}{3}$	$162\frac{1}{2}$
	Iron "	95	100

Irons. The remaining iron parts of a gun-carriage may be divided into three classes, viz.:—1st. Those which serve to connect and strengthen the principal parts, before enumerated, as the *assembling-bolts*, *straps*, and *bands*. 2d. Those which protect the wood-work from wearing away at certain points, as the *trunnion-plates* (6), the *trail-plate and shoe* (7), and the *wheel-*

guard plate (8). 3d. Those employed to fasten the implements to the carriage. The number of the pieces of the last class depends upon the character of the service to which the piece belongs. In the field, mountain, and prairie carriages, all the implements necessary for the use of the piece are carried upon the carriage. The implements of the siege-carriage are carried in store-wagons.

189. **Forces acting on a gun-carriage.** As the axis of the bore intersects the axis of the trunnions, the entire force of the charge, acting on the bottom of the bore, is communicated to the carriage at the trunnion-beds. The carriage being constructed symmetrically with regard to the axis of the piece, we are at liberty, in the following discussion, to suppose that the wheels, trunnion-beds, and trail, are all situated in the same plane, and that the resultant of the force of the charge is applied at the point where the axis of the trunnions pierces this plane.

The action of the force of the charge is to move the carriage along the surface of the ground (supposed to be horizontal), to press the wheels and trail upon the ground, and to rotate the carriage around the point of contact of the trail with the ground.

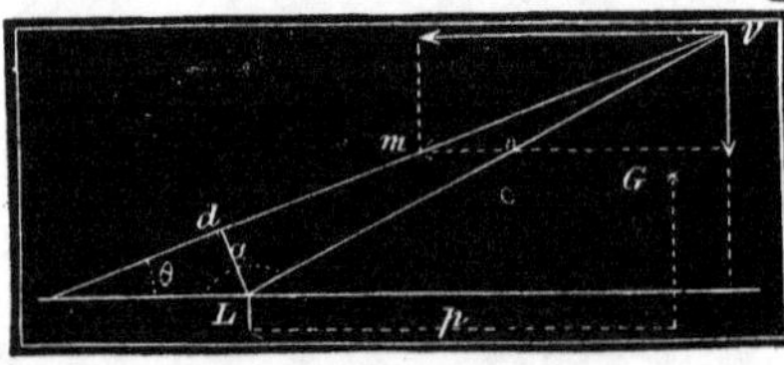

Fig. 57.

Let v be the position of the axis of the trunnions, and mv represent the amount and direction of the force of the recoil, and θ the angle of fire. Let L be the point of contact of the trail and ground, a the distance of this point from the trunnions, α the angle which the line joining these two points makes with the horizontal, G the position of the centre of gravity, and p its horizontal distance from the point L.

If mv be the force of the recoil, R and C the pressures exerted by it upon the wheel and trail, respectively, we have the relation

$$mv \sin. \theta = R + C.$$

The horizontal component acts to overcome the friction of the wheel and trail, and to set the carriage in motion. By making f the unit of friction, and MV the quantity of motion impressed on the carriage, we have

$$mv \cos. \theta = f(C+R) + MV;$$

or, by substituting the value of $R+C$ from the above equation, and solving with reference to V, we have

$$V = \frac{mv(\cos.\theta - f\sin.\theta)}{M},$$

which is the velocity of recoil.

As the unit of friction of the wheel and trail are not exactly the same, the foregoing equation will not give a strictly correct value for V for field and siege carriages, but it will be correct for fortress-carriages and mortar-beds, which do not move on wheels, in recoil.

The force mv also acts to rotate the carriage around the point L with an effect proportional to its lever arm Ld, which is equal to a sin. dvL; but sin. dvL=sin. $(180^\circ - (\alpha + \theta),)$ and the moment of the force of the charge, with reference to the trail, is mva sin. $(180^\circ - (\alpha + \theta))$.

This moment being equal to the moment of the weight of the piece, and the moment of the quantity motion impressed upon the carriage, or P, we have

$$mva \sin. (180^\circ - \alpha - \theta) = Wp + P.$$

But $P = \frac{Wk^2w}{g}$; k being the radius of gyration of the gun and carriage taken with reference to the trail, g the force of gravity, and w the angular velocity of the gun and carriage.

Substituting this value of P in the above equation, and reducing, we have

$$w = g\frac{mva \sin(180^\circ - a - \theta) - Wp}{Wk^2}.$$

With this relation we can discuss, by giving different values to θ, a, a, and p, the effect of the angle of fire, length of trail, position of trunnions, and centre of gravity, on the stability of the carriage, or the resistance which it offers to overturning by the force of the charge acting at the centre of the trunnions.

LIMBER.

190. **Object.** Thus far a gun-carriage has been considered only in relation to the fire of the piece, or as a *two*-wheel carriage. To suit it to the easy and rapid transportation of its load, it must be converted into a *four*-wheel carriage, which is done by attaching it to another two-wheel carriage called a *limber*.

191. **Construction.** The field-limber is composed (see fig. 58), of an *axle-tree* (1), a *fork* (2), two *hounds* (3 3), a *splinter-bar* (4), two *foot-boards* (5 5), a *pole* (6), a *pintle-hook* and *key* (7), two *pole-yokes* (8), and a *pole-pad* (9).

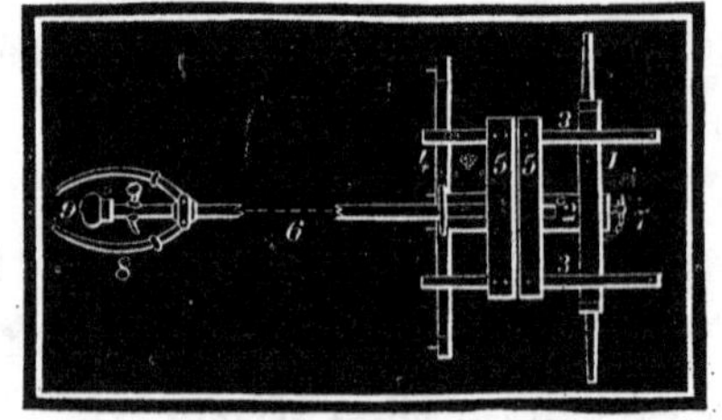

Fig. 58.

A side view of this limber is also shown in fig. 59, together with the manner of attaching the rear carriage to the pintle-hook.

The axle-tree. The limber axle-tree is made of iron, imbedded in a body of wood, as in the case of the gun-carriage.

The fork. The fork constitutes the middle portion of the limber, and is the portion to which the pole is attached. It is formed of a single piece of wood, one end of which is mortised into the axle-body, and secured by the pintle-hook bolts, and the other is cut into the shape of a fork, to receive the tenon of the pole.

The hounds. The hounds are two wooden rails which are bolted to the axle-body and splinter-bar. They serve to support the ends of the limber-chest and foot-boards, and also to transmit the draught of the horses to the axle-tree. The chest is secured by a stay-plate, situated at the bottom of the cut in the fork, and two stay-pins, which pass through holes near the rear ends of the hounds.

The splinter-bar. The splinter-bar is a piece of wood placed cross-wise with the pole, and is firmly secured to the fork and hounds. It has four hooks, to which the traces of the wheel horses are attached.

The pole. The pole, or tongue, is employed to regulate the motion, and give direction to the carriage. The point of attachment of the rear carriage being near the axle-tree, and there being no sweep-bar, the weight of the pole is mostly supported by the collars of the rear horses; it should therefore be made of strong, light wood—ash is generally used for this purpose.

As the pole is liable to be broken in service, the

method of attaching it to the fork should be such that the fragments can be promptly removed, and a new pole inserted.

The foot-boards. The foot-boards are secured to the fork and hounds in a proper position for the feet of the cannoniers to rest upon, while riding upon the limber-chest.

The pintle-hook. The pintle-hook is a stout iron hook firmly fastened to the rear of the axle-tree, for the purpose of attaching the rear carriage. This mode of attachment is *simple*, *strong*, and *flexible*—qualities which are essential to rapid movements and great endurance. The point of the hook is perforated with a hole for the *pintle-key*, which prevents the carriages from separating while in motion.

In the old system of field-carriages, the operation of limbering and unlimbering was so difficult, that a rope, called a "prolonge," was used to connect the gun-carriage and limber, in action. This implement is still retained, but the same necessity does not exist for using it.

192. **Turning.** All field-carriages should admit of being turned in the shortest possible space. This depends upon the size of the front wheels, the distance between the front and rear axle-trees, the position of the pintle, and the thickness of the stock at the point where the front wheel strikes it. Notwithstanding that the front wheels are made higher in the present system of field-carriages than the Gribeauval system, which preceded it, the carriages of the former have greater facility of turning, in consequence of the diminished thickness of the stock.

193. **Track.** By *track* is understood the distance between the furrows formed by the wheels in the ground.

It is important that the track should be the same for all carriages likely to travel the same road, in order that the wheels of one carriage may follow in the furrows formed by those of its predecessor, and thereby prevent a loss of tractile force. The track of artillery carriages is 5 feet, and the extreme length of the axle-tree is $6\frac{1}{2}$ feet for field, and $6\frac{3}{4}$ feet for siege-carriages.

194. **Load.** As the forward wheels of a carriage form the ruts, they should support a smaller portion of the load than the rear wheels: in field-carriages, the proportion is as *two* to *three*.

195. **Length of Stock.** The length of the stock determines the distance between the front and rear wheels. The longer this distance is, the greater will be the space required to turn the carriage in, and the greater will be the effort necessary to pull the carriage over a sharp elevation of the ground.

196. **Wheels.** All wheels of an artillery carriage should be of the same height, to permit of interchange, and to make the line of traction parallel to the ground.

197. **Locking Wheels.** The work of holding back a carriage, on descending ground, devolves on the pole-horses. When the descent is very steep, and the load large, they are relieved of a portion of this work by attaching a chain to one of the rear wheels, in such a manner as to prevent it from turning, and thereby changing the friction on the axle-arm to friction on the ground. In field-carriages, one end of the locking-chain is secured to the stock by an assembling-bolt, and the other is passed around the felloe, and secured to itself

by a key. In siege-carriages, where the load is much heavier, a shoe is attached to the chain, upon which the wheel rides. This prevents the tire from being worn and the wheel from being strained; at the same time, the operation of locking and unlocking can be performed without stopping the carriages.

FIELD–CARRIAGES.

198. **Kinds.** The carriages pertaining to the field service, are the *gun-carriage*, the *caisson*, the *travelling-forge*, and the *battery-wagon*. The same limber is used for all the field-carriages, with the exception of the interior arrangement of the *chest*, which is adapted to the kind of the carriage to which the limber is attached.

199. **Gun-carriages.** Field-carriages are characterized by great lightness, strength, and mobility. They are,

The 6-*pdr. gun* and 12-*pdr. howitzer carriage.*

The 12-*pdr. gun* (*light*) and the 24-*pdr. howitzer carriage.*

The 12-*pdr. gun* (*heavy*) and the 32-*pdr. howitzer carriage.**

These carriages are of similar construction, the only difference being in the size and strength of the several parts. The first is mounted on light, or No. 1 wheels, and the second and third on No. 2, or heavy wheels. Attached to each carriage are the following named implements, viz., two rammers and sponges, two trail-handspikes, one worm, one sponge-bucket, one tar-bucket, one watering-bucket.

* The 10-pdr. Parrott and the 3-in. rifle guns are mounted on the 6-pdr. carriage, and the 20-pdr. Parrott rifle-gun is mounted on the 12-pdr. (heavy) carriage.

200. **Caisson.** The caisson is used to transport ammunition; and in light field-batteries, there is one caisson to each piece, in heavy batteries there are two. The ammunition is contained in three chests—two mounted on the body, and one on the limber. The number of rounds for each chest varies with the calibre of the piece, as follows, viz.:

6-pdr. gun, and 3-inch rifle-gun, .	50
12-pdr. gun, ...	32
12-pdr. howitzer, ...	39
24-pdr. howitzer, ...	23
32-pdr. howitzer, ...	15

The whole number of rounds for each piece may be ascertained by multiplying the above numbers by four.

The caisson is composed of a *body*, and a limber. See fig. 59. The body is composed of one *middle* and two *side rails* (1), one *stock* (2), and one *axle-tree* (3). It

Fig. 59.

carries two ammunition-chests (4, 5), a *spare wheel* (6), which fits upon an iron axle-arm attached to the rear end of the middle rail, one *spare pole* (7), fastened to the under side of the stock, and a *spare handspike*. The spare articles are needed to replace broken parts.

The caisson also carries a *felling-axe*, *shovel*, and *pick-axe*, to remove obstructions, repair roads, &c., a *tarpau-*

lin strapped on to the limber-chest, a *tar-bucket*, and a *watering-bucket.*

201. **Travelling-forge.** The travelling-forge is a complete blacksmith's establishment, which accompanies the battery for the purposes of making repairs and shoeing horses. It consists of a body, upon which is constructed the bellows-house, &c., and the limber, which supports the stock, in transportation. The body (see fig. 60) is composed of two *rails* (1), a *stock* (2), and an *axle-tree* (3). The *bellows-house* is divided into the *bellows-room* (4), and the *iron-room* (5). Attached to the back of the house is the *coal-box* (6), and in front of it is the *fire-place* (7). From the upper and front part of the bellows, an *air-pipe* (8) proceeds in a downward direction to the *air-box*, which is placed behind the fire-place. The *vise* (9) is permanently attached to the stock, and the *anvil*, when in use, is supported on a stone or log of wood, and when transported is carried on the hearth of the fire-place. The remaining tools are carried in the limber-chest. When in working order, the point of the stock is supported by a prop (10).

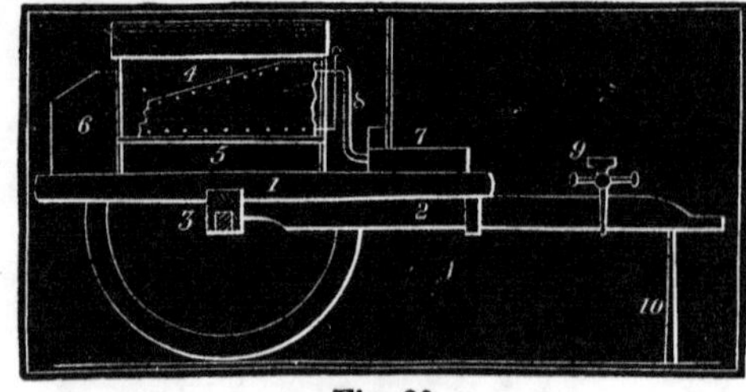

Fig. 60.

202. **Battery-wagon.** The battery-wagon is employed to transport the tools and materials for repairs. Among the tools are those for carriage-makers, saddlers, armorers, and laboratorians' use, scythes and sickles for cutting forage and spare implements for the service of the piece.

The *body* (1) of the battery-wagon (see fig. 61) is a large rectangular box, covered with a roof of painted

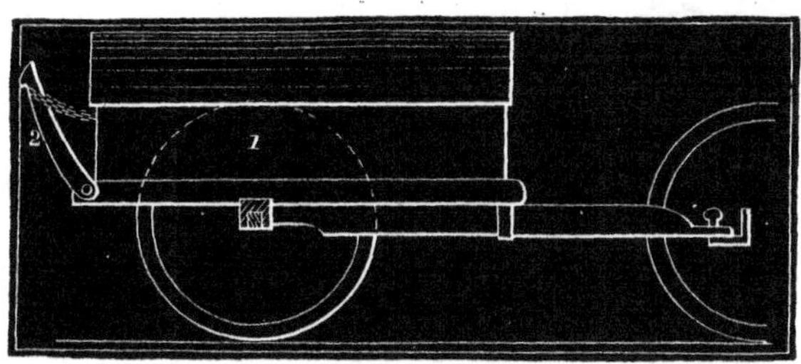

Fig. 61.

canvas; and to the back part is attached a *rack* (2) for carrying forage. The bottom of the body is formed of one *middle* and two *side rails*, resting on a stock and axle-tree, as in the travelling-forge.

The tools and materials of the battery-wagon are carefully packed in the manner prescribed by the Ordnance Manual, in order that no difficulty may be experienced in finding a particular article when wanted. The smaller articles are carried in boxes properly lettered and numbered.

The travelling-forge and battery-wagon are not confined to the service of field-batteries, but are used with siege and sea-coast carriages, as occasion may require.

MOUNTAIN-CARRIAGE.

203. **Requirements, &c.** The mountain-howitzer carriage should be light enough to be carried on the back of a pack animal, and the axle-tree should be short enough to permit it to pass through very narrow defiles.

It differs in construction from the field-carriage,

Fig. 62.

inasmuch as the stock and cheeks (1) (fig. 62) are formed of the same piece, by hollowing out the head of the stock. The wheels are thirty-eight inches in diameter, and the axle-tree is made of wood, the arms being protected from wear by *skeans*, or strips of iron.

The distance between the wheels is about equal to their diameter. It is arranged for draught by attaching a pair of shafts to the trail. The pack-saddle and its harness are constructed to carry severally, the howitzer and shafts, the carriage, or two ammunition chests, or it enables an animal to draw the carriage, with the howitzer mounted upon it.

A portable forge accompanies each mountain battery, and is so constructed that it can be enclosed in two chests, and carried, with a bag of coal, upon the pack-saddle.

PRAIRIE–CARRIAGE.

204. **Description, &c.** The prairie-carriage is designed to carry the mountain-howitzer, and is similar to the mountain-carriage in the form and combination of its parts; but being exclusively intended for draught, the axle-tree is made of iron, the wheels are made higher, and the distance between them greater than in the mountain-carriage. It has a limber, and is drawn by two horses abreast, as in field-carriages. The ammunition is packed in mountain ammunition chests, two of which are carried on the limber, and the remainder in a covered cart, of peculiar construction, or packed on animals, as in the mountain service.

SIEGE–CARRIAGES.

205. **Kinds of.** The siege-carriages are

The 24-pdr. gun and 8-inch howitzer carriage.

The 18-pdr. smooth, 30-pdr. rifle do.

The 12-pdr. smooth, 4½-in. rifle do.

The mortar-wagon.

The limber.

The mortar-bed.

206. **Gun-carriage.** The construction of the siege-gun carriage is similar, in most of its details, to the field-gun carriage. It differs, however, in the greater strength of the parts, and in the mode of attaching to the limber, and by the absence of the parts used for carrying the implements.

The position of the trunnion-beds is such that when the carriage is limbered up, the weight of the piece is thrown too much on the rear wheels for convenience of transportation; another set of trunnions is therefore

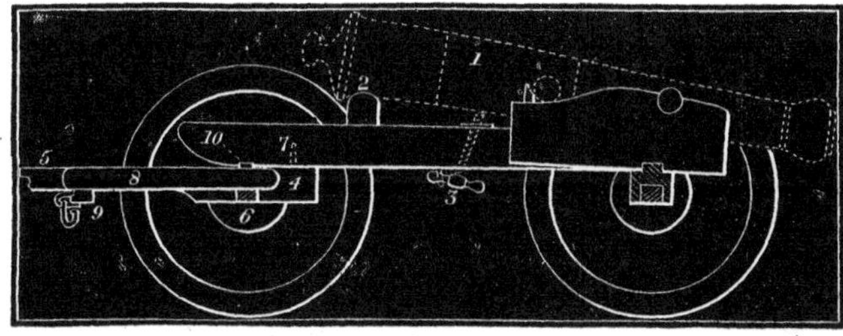

Fig. 63.

formed at the rear end of the cheeks, by enlarging the heads of the cheek-bolts, and the piece is shifted to them in transportation. They are called the "*travelling trun-*

nions." See (1) fig. 63. The breech of the piece rests in a groove formed in a block of wood, called the "*bolster*" (2); and the elevating screw is disposed of by reversing it in its nut. To prevent it from unscrewing by the motion of the carriage, one of the handles is slipped through a leather loop attached to the under side of the stock.

207. **Limber.** The same kind of limber is used for all siege-carriages. It is composed of a *fork* (4), the *pole* (5), the *axle-tree* (6), the *pintle* (7), the *hounds* (8), the *splinter-bar* (9), and the *friction-circle* (10), one end of which is only represented in the figure.

The fork constitutes the main part of this carriage, and to it are attached the pintle, the pole, the splinter-bar, the axle-tree, and the friction-circle.

As this carriage is not subjected to the shock of firing, the axle-tree is not imbedded in wood to give it stiffness, as in the gun-carriage.

The pintle is placed far enough in rear of the centre of the axle-tree to enable the weight of the stock of the gun-carriage to act as a counterpoise to the pole, and give it steadiness when the carriage is in motion. The friction-circle acts as a sweep-bar for the *shoe* of the trail to rest upon when the limber turns around its pintle. The attachment of the two carriages is secured by a lashing *chain and hook.*

208. **Mortar-wagon.** The mortar-wagon is employed to transport siege projectiles, mortars and their beds, and spare guns.

It is composed of a limber and body. The body consists of two middle-rails, united so as to form the stock, and two side-rails. These pieces rest upon the axle-tree, and are strongly connected together by cross pieces of wood and straps of iron. At the rear of the body is placed a windlass, which aids in mounting guns and mortars. Stakes are placed around the sides of the body, to sustain the side and end boards which are used in transporting projectiles.

209. **Mortar-bed.** The lightness of the mortar, and the high angle under which it is fired, render it unsafe to be fired from a carriage; it is, therefore, mounted on a bed, which rests directly on a platform.

The siege mortar-bed is made of wrought-iron, and put together after the manner described in the article on sea-coast carriages. The form is shown in fig. 64. The different parts are the cheeks and the two transoms which connect the cheeks together. There are two projections at the end of each cheek, underneath which the hand-spikes are placed for moving the bed in aiming; also one manœuvring bolt, for the same purpose. The head of this bolt is shown at (*a*); (*b*) represents the elevating bar in position. The sea-coast mortar-beds have an ex-centric truck wheel for facility of manœuvre.

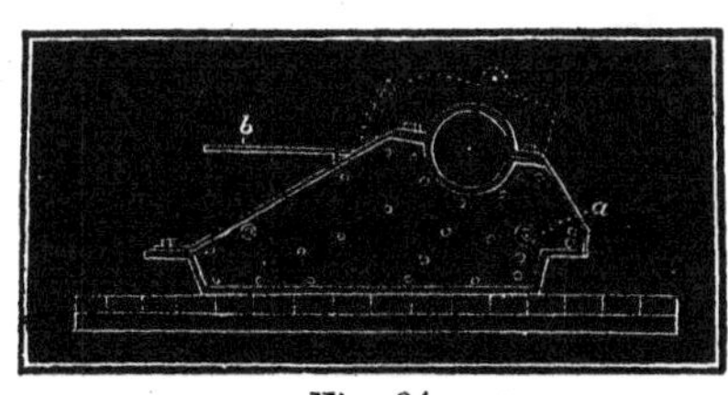

Fig. 64.

210. **Platform.** To insure accuracy of fire with heavy guns and mortars, it is absolutely necessary that their carriages and beds should rest upon solid and substantial platforms.

The platforms for siege-pieces, being transported with an army, should have the greatest lightness, compatible with strength to endure the shocks of long-continued firing. They are composed of a certain number of pieces of wood; and in order that these pieces may be carried on the backs of soldiers from the depot to the battery, the weight of the heaviest piece should not exceed fifty pounds. Siege-platforms consist of *sleepers* (1), (fig. 65), and *deck-plank* (2). The general direction of the sleepers is parallel to the axis of the piece, and the deck-plank at right angles to it; this disposition of the parts offers the greatest resistance to the recoil of the carriage. The deck-planks are fastened together at their edges by dowels; the outer planks are secured by iron eye-pins, one at each end of a sleeper. The platform is secured in its place by driving stakes around the edges.

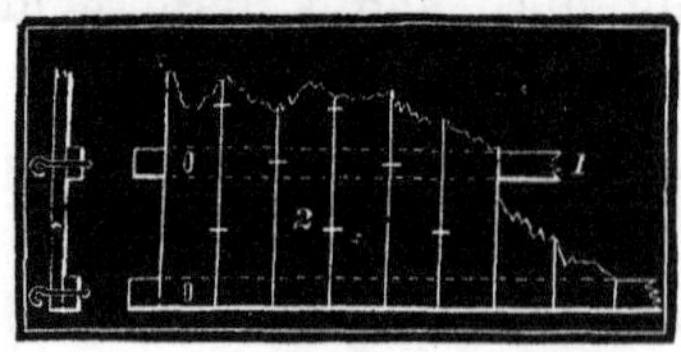

Fig. 65.

There are two principal platforms for the siege-service, viz., the *gun*-platform, and the *mortar*-platform. The former is composed of twelve sleepers and thirty-six deck-planks; the mortar-platform of six sleepers and eighteen deck-planks.

A simple and strong mortar-platform, called the *rail-platform* may be used where trees or timber can be easily procured. This is composed of three sleepers and two rails, secured by driving stakes at the angles and at the rear ends of the rails. The rails are placed at the proper distance apart to support the cheeks of the bed.

SEA-COAST CARRIAGES.

211. **Classification.** Sea-coast carriages are divided into *barbette*, *casemate*, and *flank-defence* carriages, depending upon the part of a work in which they are mounted.

212. **Material.** Heretofore, nearly all sea-coast carriages were made of wood; but in consequence of the great difficulty of preserving this material from decay, especially when exposed to the dampness of casemates, it has been determined to replace it by wrought-iron; and strong, cheap, and manageable carriages have been devised and tested for this service.

The principal feature in the construction of the new carriages, is a peculiar combination of boiler-plate and rolled beams, which gives, with requisite lightness, great strength and stiffness to the important parts.

213. **Carriage.** All sea-coast carriages are composed of the *gun-carriage* (1), and the *chassis* (2), (fig. 66, which shows a 10-in. front pintle barbette carriage).

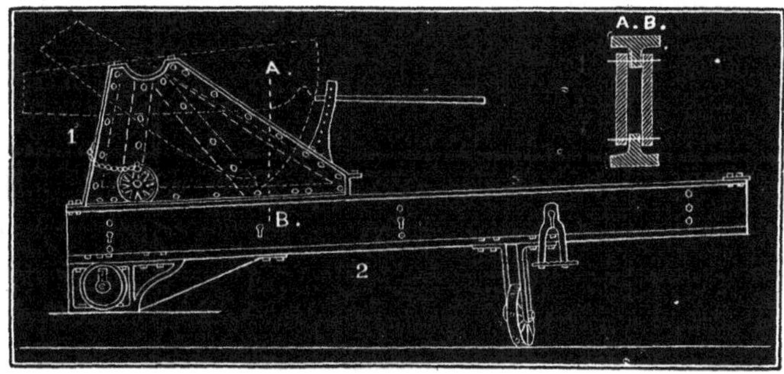

Fig. 66.

Gun-carriage. The purpose of the gun-carriage being to support the piece, it should be so constructed that the piece can be elevated or depressed, in aiming; and run into and out of battery, in firing. The term "in battery," as applied to sea-coast guns, refers to the position which the piece occupies when it is ready to be fired—in casemate pieces the muzzle must be in the throat of the embrasure, and in barbette pieces, directly over the superior slope of the parapet.

The gun-carriage is composed of two *cheeks*, held together by two plates of boiler iron called the *front* and *rear transoms*.

Each cheek is formed of two pieces of boiler iron cut to a triangular shape, separated at the edges by interposing the vertical portion or web of a T shaped bar. The horizontal branches project over each side to form a double rim, which gives stiffness to the cheeks. Flat bars of iron are also placed between the plates at suitable intervals to stiffen the cheeks in the direction in which the weight and recoil of the piece bear upon them. All these parts are held together by screw bolts.

The motion of the carriage to and from battery is regulated by a pair of *manœuvring wheels* which work on an eccentric axletree, placed underneath and a little in front of the centre of the trunnions. The 13 and 15-in. carriages have two pairs of these wheels.

When it becomes necessary to check the recoil of the gun-carriage, the wheels are thrown out of gear by means of a handspike, and the forward part of the carriage moves on sliding friction. When it is necessary to move to battery, the wheels are thrown into gear and the carriage moves on rolling friction.

Elevating apparatus. All sea-coast cannon having no preponderance are elevated and depressed by a lever, the point of which works in a ratchet cut in the breech of the piece. The fulcrum is made of cast-iron and rests on the rear transom of the gun-carriage. It has several notches for adjusting the position of the elevating lever. This apparatus is exceedingly simple, and affords the means of changing the elevation with ease and rapidity.

The piece is pointed, 1st, by two fixed sights, one attached to the breech and the other to a lug between the trunnions; 2d, by a brass graduated arc attached to the breech, parallel to the ratchet, and a pointer attached to the fulcrum. By the fixed sights the piece is brought in line with the object, and the arc and pointer mark the elevation necessary to strike it.

Chassis. The chassis is the movable railway along which the gun-carriage moves to and from battery. It is composed of two wrought-iron rails inclined 3° to the horizon and united by transoms, as in the gun-carriage. In addition to the transoms, there are certain diagonal braces to give stiffness to the chassis.

For the 10-inch and smaller carriages, the chassis rails are single beams of rolled iron, 15 inches deep; for all calibres above, the rails are made of long rectangular pieces of boiler plate and T iron, in a similar manner to that of the cheeks of the gun-carriage. That the carriage may be moved horizontally in the operation of aiming the piece, the chassis is supported on traverse wheels, which roll on circular plates of iron, fastened to a bed of solid masonry, called the *traverse circles.*

The motion of the gun-carriage is checked front and rear, by pieces of iron bolted to the top of the rails, called *hurters* and *counter-hurters;* and it is prevented from slipping off sideways by *friction rollers* and *guides*, which are bolted to the cheeks and transoms.

Pintle. The pintle is the central piece around which the chassis is traversed. It is formed of a stout cylinder of wrought-iron inserted in a block of stone, if the battery be a fixed one, or it is secured to cross pieces of timber bolted to a platform firmly imbedded in the ground, if it be of a temporary nature. In casemate batteries the pintle is placed immediately under the throat of the embrasure, and the chassis is connected with it by a stout strap of iron, called the *tongue.* The muzzle of the piece when in battery, is situated in the throat of the embrasure—a position which, taken in connection with that of the pintle, gives the greatest horizontal field of fire.

The centre pintle carriage, or the one in which the pintle is in the centre of the chassis, is preferable for barbette batteries to the one in which the pintle is in the centre of the front transom, as with an equal space it affords a greater horizontal field of fire, or affords a better opportunity to use heavy traverses of earth to protect the pieces from an enfilading fire.

Manœuvring. Sea-coast carriages are at present manœuvred by handspikes applied to holes in the circumference of the traverse and manœuvring wheels. This mode is defective, inasmuch as power is wasted by continually overcoming the inertia of the gun and carriage when the handspikes are readjusted. It is proposed to supersede this by the continuous motion produced by cranks, and the experiments that have been made indicate that the heaviest pieces may be moved with much greater ease than heretofore.

Much attention is now being paid towards obtaining a more effective cover for barbette guns and the men that work them. It is understood that several plans to this end are under consideration by the Engineer Department, and among them are fixed and revolving turrets, the latter similar to those used on the monitors, except that they are to be moved by man rather than steam power.

214. **Kinds of carriages.** The sea-coast carriages of the present system are:

The	15-inch Rodman gun, - -	Barbette.
"	10-inch " " - - -	"
"	8-inch " " - - -	"
"	13-inch " " - - -	Casemate.
"	10-inch " " - - -	"
"	8-inch " " - - -	"
"	24-pdr. Flank Defence Howitzer,	"

There are many 8-inch howitzers, and 42 and 32-pdr. guns of the old system mounted in sea-coast batteries, but they are being replaced by those of larger calibre as fast as they can be provided by the private founders.

CHAPTER V.

MACHINES AND IMPLEMENTS.

215. **Object.** Artillery machines are employed to mount and dismount cannon from their carriages, and to transport artillery material from one part of a work to another. They comprise the *gin*, the *sling-cart*, the *casemate-truck*, the *hand-cart*, the *lifting-jack*, and the *lever-jack*.

216. **Gin.** A gin is a tripod formed of ash or spruce poles. Two of the poles are joined together by two *cross-bars* of wood or iron (1, 2), see fig. 67, and are called the *legs*. The third is called the *pry-pole*, and is used in elevating the gin to its proper position. The hoisting apparatus is supported by a *clevis* which is secured by the bolt which unites the legs and pry-poles; it consists of two sets of iron blocks, through which is rove a rope called the *fall*, and which is wound around the *windlass* (3). The windlass is worked by two handspikes which fit into brass sockets, one at each extremity of the windlass; the operation of the handspikes is made continuous by the action of the pawl of the socket on the ratchet of the windlass. The piece to be raised is attached to the hook of the lower block, by a stout rope called a

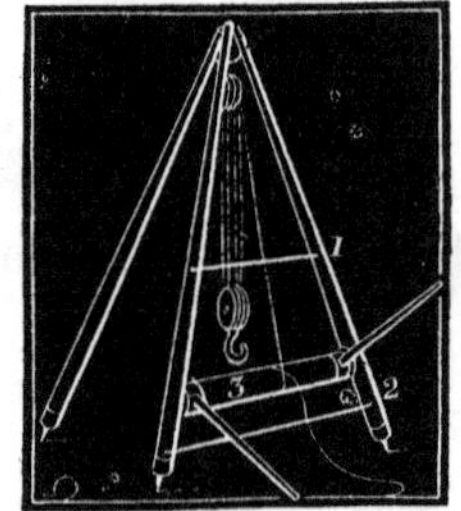

Fig. 67.

sling, which passes around the knob of the cascable and a piece of wood projecting from the muzzle.

Kinds. There are three kinds of gins, viz.: the *garrison gin*, the *casemate gin*, and the *field and siege gin.* The last mainly differs from the others in the smaller size and lesser strength of its parts; the casemate gin is not so tall as the garrison gin, on account of the lowness of the casemate roofs under which it is used; in all other respects the two are alike.

217. **Sling-cart.** A sling-cart is composed of two wheels of large diameter, an *axle-tree*, a *tongue*, and the *hoisting apparatus;* and is intended to transport cannon and their carriages. There are two kinds, viz.: the *wooden* sling-cart and the *iron* sling-cart.

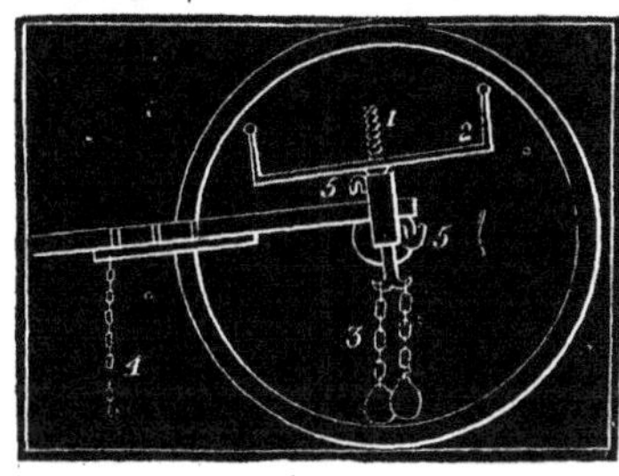

Fig. 68.

The wheels of the wooden sling-cart are eight feet in diameter, (figure 68.) The hoisting-apparatus is a *screw* (1), which passes vertically through the wooden axle-tree, and is worked by a *nut* with long handles (2). The lower part of this screw is terminated with two hooks, to which are fastened the *chains and trunnion-rings* (3). The breech of the piece is sustained by the *cascable chain* (4). A piece may be also raised by surrounding it with a chain, fastening the chain to the *hooks* (5, 5), and depressing the tongue, which acts as the long arm of the lever.

Iron sling-cart. The iron sling-cart is smaller than the wooden one, and is employed to transport cannon in siege-trenches, &c. (figure 69.) The weight is at-

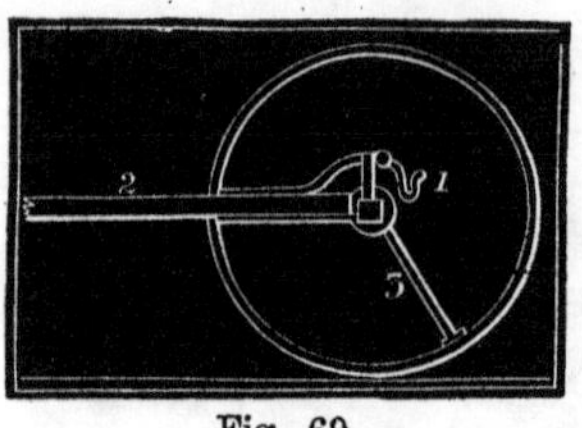

Fig. 69.

tached by a chain or rope to the *hook* (1), and is raised by pressing down the *tongue* (2), which is made of wood. When sling-carts are to be moved, they are attached to the pintle of a field or siege limber.

218. **Casemate truck.** The casemate truck (see fig. 70) is composed of a stout frame of wood mounted on three barbette traverse-wheels, and is employed to move cannon and carriages through posterns and along casemate galleries.

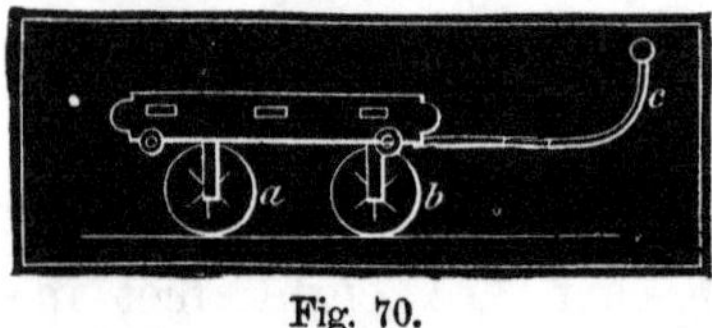

Fig. 70.

Two of the wheels are placed at *a*, and one at *b;* the latter turns around a vertical axis, when the direction of the truck is changed by the *handle* (*c*). The rings shown in the figure are for the purpose of attaching *drag-ropes.*

219. **Lifting-jack.** The lifting-jack is a small but powerful screw, worked by a geared nut. It is useful when the space for manœuvring is small, and the number of men is limited. If the weight to be raised is sufficiently high, the lifting power is applied at the top (*a*), fig. 71; if it be low, it is applied at the foot (*b*).

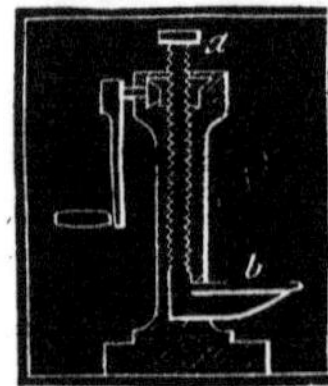

Fig. 71.

220. **Lever-jack.** The lever-jack is another but less powerful apparatus for lifting. It consists of a *lever* of wood (*a*), fig. 72, resting on a *bolt* (*c*), which passes through holes in two *uprights* (*b*). The height of the bolt is varied by passing it through different holes in

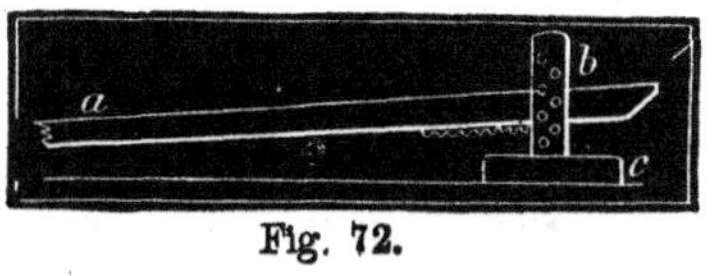

Fig. 72.

the uprights (eight in number), and the power of the lever is regulated by a notched piece of cast iron screwed to the under side of the lever.

221. **Trench cart.** The trench cart is a common hand-cart, employed principally in the transportation of ammunition in siege-trenches. In the Crimea, trained pack-mules were employed in transporting ammunition and other supplies, from the depots to all parts of the trenches.

IMPLEMENTS AND EQUIPMENTS.

Artillery implements and equipments are employed in loading, pointing, and firing cannon, and in the manœuvre of artillery carriages.

222. **Loading.** The implements for loading cannon are,

1st. The *rammer-head* (1) (fig. 73), is a short cylindrical piece of beech or other tough wood, fixed to the end of a long stick of ash (3), called a *staff*, and is employed to push the charge to its place in the bore or chamber of a cannon.

Fig. 73.

2d. The *sponge* is a *woollen* brush (2), attached to the end of a staff, for the purpose of cleaning the interior of cannon, and extinguishing any burning fragments of the cartridge that may remain after firing.

In the field and mountain services, the rammer-head and sponge are attached to the opposite ends of the same staff; in the siege and sea-coast services they are attached to separate staves.

To protect the sponge from the weather, it should, when not in use, be enclosed in a *cover* made of canvas and painted.

3d. The *ladle* is a copper scoop (1), fig. 74, attached to the end of a staff for the purpose of withdrawing the projectile of a loaded piece. Ladles are only used for siege and sea-coast cannon, as field and mountain cannon can be unloaded by raising the trail of the carriage, which permits the projectile to slip out by its own weight.

Fig. 74.

4th. The *worm* (fig. 74), is a species of double cork-screw, (2), attached to a staff, and is used in field and siege cannon to withdraw a cartridge.

5th. The *gunner's haversack* is made of leather, and suspended to the side of a cannonier by a shoulder-strap. It is used to carry cartridges from the ammunition-chest to the piece, in loading.

6th. The *pass-box* is a wooden box closed with a lid, and carried by a handle attached to one end. It takes the place of the haversack in siege and sea-coast service, where the cartridge is large.

7th. The *tube-pouch* is a small leather pouch attached to the person of a cannonier by a waist-belt. It contains the friction-tubes, lanyard, priming-wire, thumb-stall, &c.

8th. The *budge-barrel* is an oak barrel bound with copper hoops. To the top is attached a leather cover, which is gathered with a string, after the manner of the mouth of a bag. It is employed to carry cartridges from the magazine to the battery, in siege and sea-coast services.

9th. The *priming-wire* is used to prick a hole in a cartridge for the passage of the flame from the vent. It is a piece of wire, pointed at one end, and the other is formed into a ring which serves as a handle.

10th. The *thumbstall* is a buckskin cushion, attached to the finger to close the vent in sponging.

11th. The *fuze-setter* is a brass drift for driving a wooden fuze into a shell.

12th. The *fuze-mallet* is made of hard wood, and is used in connection with the setter.

13th. The *fuze-saw* is a 10-inch tenon saw for cutting wooden or paper fuzes to a proper length.

14th. The *fuze-gimlet* is a common gimlet, which may be employed in place of the saw to open a communication with the fuze composition.

15th. The *fuze-auger* is an instrument for regulating the time of burning of a fuze, by removing a certain portion of the composition from the exterior. For this purpose it has a movable graduated scale, which regulates the depth to which the augur should penetrate.

16th. The *fuze-rasp* is a coarse file employed in fitting a fuze-plug to a shell.

17th. The *fuze-plug reamer* is used to enlarge the cavity of a fuze-plug, after it has been driven into a projectile, to enable it to receive a paper fuze.

18th. The *shell-plug screw* is a wood screw with a handle; it is used to extract a plug from a fuze-hole.

19th. The *fuze-extractor* is worked by a screw, and

is a more powerful instrument than the preceding; it is used for extracting wooden fuzes from loaded shells.

20th. The *mortar-scraper* is a slender piece of iron with a spoon at one end, and a scraper at the other, for cleaning the chamber of a mortar.

21st. The *gunner's sleeves* are made of flannel or serge, and are intended to be drawn over the coat-sleeves of the gunner, and prevent them from being soiled while loading a mortar.

22d. The *funnel* is made of copper, and is used in pouring the bursting charge into a shell.

23d. The *powder-measures* are made of copper, of cylindrical form, and of various sizes, for the purpose of determining the charges of shells and cannon, by measurement.

24th. The *lanyard* is a cord, one end of which has a small iron hook, and the other a wooden handle. It is used to explode the friction-tubes with which cannon for the land service are now fired.

25th. The *gunner's pincers*, *gimlet*, and *vent-punch* are instruments carried in the tube-pouch for removing ordinary obstructions from the vent.

26th. The *shell-hooks* is an instrument constructed to fasten into the ears of a shell, for the purpose of lifting it to the muzzle of the piece.

223. **Pointing.** The implements for pointing are:

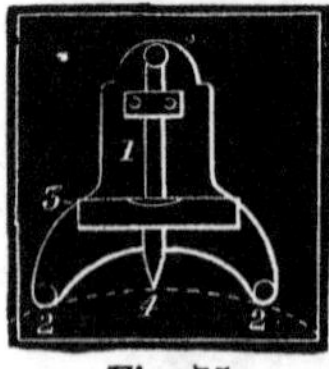

Fig. 75.

1st. The *gunner's level* (fig. 75), is an instrument for determining the highest points of the breech and muzzle of a cannon when the carriage-wheels stand on uneven ground. It is made of a *brass plate* (1), the lower edge of which is terminated by two *steel points* (2 2) which rest upon the surface of the piece. A *spirit-level* (3) is attached to the plate with its axis parallel to the line joining the points of contact. When the level is in position, the vertical *slide* (4) is pressed down with the finger to mark the required point.

2d. The *tangent-scale* (fig. 76) is a brass plate, the lower edge of which is cut to the curve of the base-ring of the piece, and the upper edge is formed into offsets which correspond to differences of elevation of a quarter of a degree. It is used in pointing, by placing the curved edge on the base-ring, with the radius of the offset corresponding with the highest point of the ring, and sighting over the centre of the offset and the highest point of the swell of the muzzle.

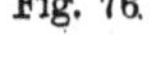

Fig. 76.

3d. The *breech sight* (fig. 77), is a more accurate form of the tangent scale. It consists of a *vertical scale* (1), graduated to degrees and eighths of degrees, and a curved *base* (2), which rests upon the breech of the gun. A slide is attached to the vertical piece, which has a small hole or notch cut on its upper edge, through which the aim is taken. The slide is fixed at any point by a thumbscrew.

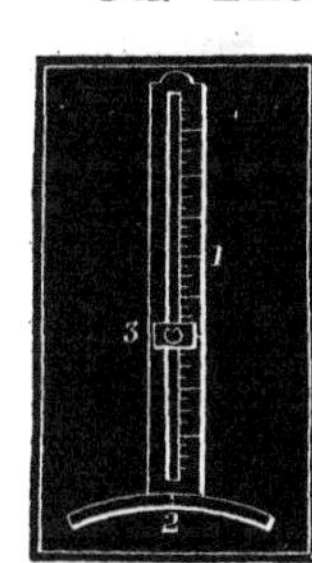

Fig. 77.

4th. The *pendulum hausse* (fig. 78), is used to point field-pieces, and at the same time to obviate the error which arises when the wheels of the carriage stand on uneven ground. It is composed of a *scale* (1), arranged as a pendulum, a *suspension piece* (2), and a *seat* which is screwed to the breech of the gun. A slot is cut in

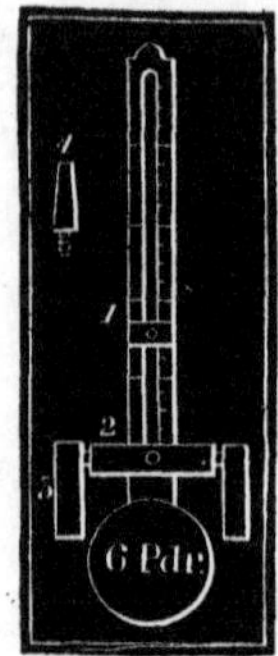
Fig. 78.

the suspension piece into which the scale is inserted, and fastened by a pivot, which allows it to vibrate in a lateral direction. The scale also vibrates in a longitudinal direction, as the journals of the suspension piece are free to turn in the grooves cut in the seat to receive them, thus assuming a vertical position independently of the surface of the ground on which the carriage stands. In order that the line of metal which passes through the centre of motion of the pendulum may be parallel to the axis of the piece, a *front sight* (4) is screwed into the swell of the muzzle, the height of which is equal to the dispart of the piece.

The length *of each graduation is equal to the distance between the front and rear sights, multiplied by the tangent of the corresponding angle.* This rule applies in graduating all breech-sights.

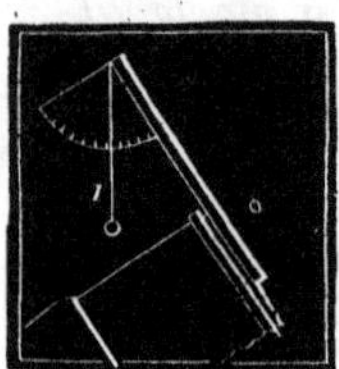
Fig. 79.

5th. The *gunner's quadrant* (fig. 79) is a wooden instrument for measuring the angles of elevation and depression of cannon, and particularly of mortars. The figure explains the nature of the instrument and its mode of application. The plumb-line and bob (1), when not in use, are carried in a hole formed in the end of the long branch, and covered with a brass plate.

224. **Manœuvre.** The principal manœuvring implements are—

1st. The *trail handspike* (fig. 80, 1), which is made of wood, and attached to the trail of a field-carriage for the purpose of giving direction to the piece in aiming.

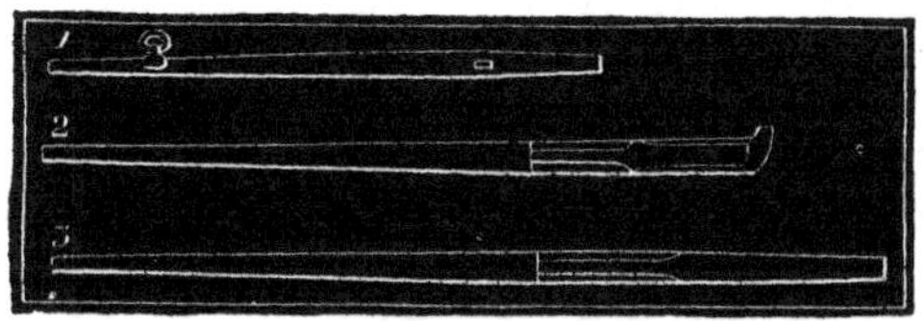

Fig. 80.

When the carriage is limbered, the handspike is attached to the cheek by means of a ring and hook.

2d. The *manœuvring handspike* (3) (fig. 80) is also made of wood, but it is longer and stouter than the preceding; it is used for siege and sea-coast carriages and gins.

3d. The *shod handspike* (2) (fig. 80) is made of wood, armed with an iron point, which is turned up in a way to prevent slipping on the platform. It is particularly useful in the service of mortars and sea-coast carriages.

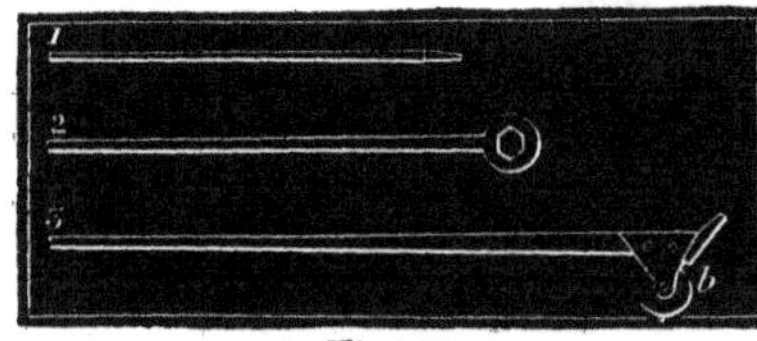

Fig. 81.

4th. The *truck handspike* (1) (fig. 81) is made of iron, and is employed to work the manœuvring wheels of sea-coast carriages, by inserting it in the holes formed in the circumference of the wheels.

5th. The *eccentric handspike* (2) (fig. 81) is used to throw the eccentric axis of the manœuvring wheels of sea-coast carriages into and out of gear, and for this purpose it has a head with a hexagonal hole which fits upon the extremities of the eccentric axle-tree.

6th. The *roller handspike* (3) (fig. 81) supplies the place of rear manœuvring wheels in certain of the new sea-coast gun-carriages. It is operated by inserting the point of the handspike under the heel of the carriage-shoe, and pressing down the long arm of the lever; in this way the weight of the rear portion of the carriage is thrown upon the roller (b), which moves upon the rail of the chassis.

7th. The *prolonge* is a stout hemp rope, occasionally employed in field service to connect the lunette of the carriage and pintle-hook of the limber when the piece is fired. It is terminated at one end with a hook, at the other with a toggle, and has two intermediate rings, into which the hook and toggle are fastened when it is necessary to shorten the distance between the carriages.

8th. The *sponge-bucket* is made of sheet-iron, and is attached to field-carriages; it is used for washing the bore of the piece.

9th. The *tar-bucket* is also made of sheet-iron, and is used to carry the grease for the wheels.

10th. The *watering-bucket* is made of sole-leather, riveted at the seams, and is used to water horses. Gutta Percha watering-buckets are sometimes used.

11th. The *water-buckets* are made of wood, and bound with iron hoops. There are two kinds, one for the travelling-forge, and the other for the service of garrison-batteries.

12th. The *drag-rope* has a hook at one end, a loop at the other, and six wooden handles placed about four feet apart. It is used whenever it may be necessary to employ a number of men in hauling loads, or extricating a carriage from a difficult part of a road.

13th. The *men's-harness* is similar to the drag-rope, except that the rope is stouter, and the handles are replaced by leather loops which pass over the shoulders of the men, to enable them to exert their strength to advantage.

14th. The *bill-hook*, or hand-bill (fig. 82), is used for cutting twigs.

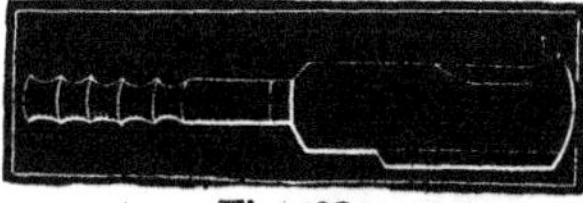

Fig. 82.

15th. The *screw-jack* (fig. 83) is a lifting-machine, composed of a screw worked by a *movable nut* (2), supported on a *cast-iron stand* (3). It is useful in greasing carriage-wheels.

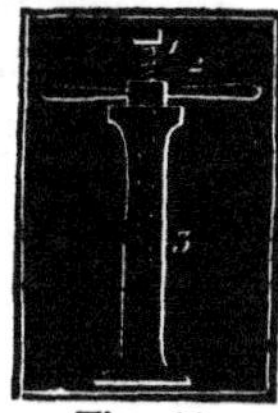

Fig. 83.

PRESERVATION AND REPAIRS.

225. **Preservation.** Carriages, implements, &c., are preserved in dry and airy store-houses. Each kind should be piled so as to occupy the least space, and each pile should be marked with the nature and number of the contents, and should be convenient of access. Care should be taken to place strips of board under the wheels and trails of carriages, if they are stowed away on a ground floor.

226. **Repairs.** Carriages are repaired in the field from the spare parts and materials carried in the battery-wagons.

DESTRUCTION, ETC., OF MATERIAL.

227. **Carriages, &c.** When it is necessary to abandon artillery material, it must be destroyed, to prevent it from falling into the hands of the enemy. Caissons should be blown up, or water poured over the contents. Carriages should be formed into a pile, and burned; or, if it be only necessary to prevent them from being moved, the spokes of the wheels and poles may be cut off with

an axe or saw. The implements should be taken away or destroyed.

228. **Cannon.** Cannon may be permanently or temporarily disabled. The first is done by bursting, bending the chase, breaking off the trunnions, and scoring the surface of the bore.

To burst an iron piece, load it with a heavy charge of powder, and fill the bore with sand, or with shot, and fire it at a high elevation. To bend the chase of a bronze cannon, fire one piece against another, muzzle to muzzle, or the muzzle of one to the chase of another, or light a fire under the chase and strike on it with a sledge-hammer. To break off a trunnion of an iron cannon, strike on it with a heavy hammer, or fire a shotted gun against it. To score the surface of the bore, cause shells to burst in it, or fire broken shot from it with high charges.

Cannon are temporarily disabled to prevent them from being immediately used by the enemy, and when they are expected to be retaken. This operation is accomplished by means of a *spike.*

To spike. A spike is made of hardened steel, with a soft point that it may be clinched on the inside: a nail without a head, or a bit of ramrod, may be used in place of a regular spike. To spike a piece, drive in the spike flush with the outer surface of the vent, and clinch it on the inside with a rammer. To prevent the spike from being blown out, wedge a shot in the bottom of the bore by wrapping it with cloth or felt, or by means of iron wedges, driven in with a bar of iron; wooden wedges might be easily burned out by means of a charcoal fire lighted with a pair of bellows.

When a piece is likely to be retaken, a *spring-spike* is used, having a shoulder to prevent its being too easily extracted.

To unspike. To unspike a cannon, attempt to drive the spike into the bore with a punch; if this succeeds, and the bore be obstructed, introduce powder into the vent to force the obstacle out. If it do not succeed, and the spike be not screwed or riveted in, and the bore be not obstructed, put in a charge of powder equal to one-third the weight of the shot, and ram junk-wads over it with a handspike, laying on the bottom of the bore a strip of wood, with a groove on the under side for a strand of quick-match, by which fire is communicated to the charge; in a bronze piece, take out some of the metal at the upper orifice of the vent, and pour sulphuric acid into the groove before firing. If this method, several times repeated, do not succeed, unscrew the vent-piece, if it be a bronze cannon; or, if an iron one, drill out the spike, or drill a new vent.

To drive out a shot. To drive out a shot wedged in the bore, unscrew the vent-piece, if there be one, and drive in wedges so as to start the shot forward, then ram it back again in order to seize the wedges with a hook; or pour powder in, and fire it, after replacing the vent-piece. In the last resort bore a hole in the bottom of the breech, drive out the shot, and stop the hole with a screw.

MATERIALS FOR CARRIAGES, &c.

229. **Timber.** Timber and wrought iron are the

principal materials used in the construction of artillery carriages and machines. Of the former are:

White oak. The bark of white oak is white, the leaf long, narrow, and deeply indented; the wood is of a straw-color, with a somewhat reddish tinge, tough, and pliable. It is the principal timber used for ordnance purposes, being employed for all kinds of artillery-carriages.

Beech. The white and red beeches are used for fuzes, mallets, plane-stocks, and other tools.

Ash. White ash is straight-grained, tough, and elastic, and is therefore suitable for light carriage-shafts; in artillery, it is chiefly used for sponge and rammer staves, sometimes for handspikes, and for sabots and tool-handles.

Elm. Elm is used for felloes and for small naves.

Hickory. Hickory is very tough and flexible; the most suitable wood for handspikes, tool-handles, and wooden axle-trees.

Black walnut. Black walnut is hard and fine-grained; it is sometimes used for naves, and the sides and ends of ammunition-chests; it is exclusively used for stocks of small arms.

Poplar. White poplar, or tulip-wood, is a soft, light, fine-grained wood, which grows to a great size; it is used for sabots, cartridge-blocks, &c., and for the lining of ammunition-chests.

Pine. White pine is used for arm-chests and packing-boxes generally, and for building purposes.

Cypress. Cypress is a soft, light, straight-grained wood, which grows to a very large size. On account of the difficulty of procuring oak of a suitable kind in the Southern States, cypress has been used for sea-coast and garrison carriages. It resists better than oak the alternate action of the heat and moisture to which sea-coast carriages are particularly exposed in casemates; but being of inferior strength, a larger scantling of cypress than oak is required for the same purpose; and on account of its softness, it does not resist sufficiently the friction and shocks to which such carriages are liable.

Basswood. Basswood is very light, not easily split, and is an excellent material for sabots and cartridge-blocks.

Dogwood. Dogwood is hard and fine-grained, suitable for mallets, drifts, &c.

230. **Selection of trees.** The principal circumstances which affect the quality of growing trees are *soil*, *climate*, and *aspect*.

Soil. In a moist soil, timber grows to a larger size, but is less firm, and decays sooner, than in a dry, sandy soil; the best is that which grows in a dark soil, mixed with stones and gravel; this remark does not apply to the poplar, willow, cypress, and other light woods which grow best in wet situations.

Climate. In the United States, the climate of the Northern and Middle States is most favorable to the growth of timber used for ordnance purposes, except the cypress.

Aspect. Trees growing in the centre of a forest, or on a plain, are generally straighter and freer from limbs than those growing on the edge of the forest, in open ground, or on the sides of hills, but the former are, at the same time, less hard. The aspect most sheltered from the prevalent winds is generally most favorable to the growth of timber. The vicinity of salt water is favorable to the strength and hardness of white oak.

Selection. The selection of timber trees should be made before the fall of the leaf. A healthy tree is indicated by the top branches being vigorous and well covered with leaves; the bark is clear and smooth, and of uniform color. If the top has a regular, rounded form; if the bark is dull, scabby, and covered with white and red spots, caused by running water or sap, the tree is unsound. The decay of the topmost branches, and the separation of the bark from the wood, are infallible signs of the decline of the tree.

231. **Felling.** The most suitable season for felling timber is that in which vegetation is at rest, which is the case in midwinter and midsummer. Recent experiments incline to give preference to the latter season, say the month of July; but the usual practice is to fell trees for timber between the first of December and the middle of March.

The tree should be allowed to attain full maturity before being felled; this period, in oak timber, is generally at the age of seventy-five to one hundred years, or upward, according to circumstances. The age of hard wood is determined by the number of rings which may be counted in a section of a tree.

The tree should be cut as near the ground as possible, the lower part being the best timber; the quality of the wood is, in some degree indicated by the color, which should be nearly uniform in the heart-wood, a little deeper toward the centre, and without sudden transitions. Felled timber should be immediately stripped of its bark, and raised from the ground.

232. **Defects of trees.** *Sap.* The white wood next to the bark, which very soon rots, should never be used, except that of hickory. There are sometimes found rings of light-colored wood surrounded by good hard wood; this may be called the *second sap;* it should cause the rejection of the tree in which it occurs.

Brashwood. This is a defect generally consequent on the decline of the tree from age; the pores of the wood are open, the wood is reddish-colored, it breaks short, without splinters, and the chips crumble to pieces. This wood is entirely unfit for artillery carriages.

Dead-wood, &c. Wood which died before felling should, generally, be rejected; so should *knotty trees*, and those which are covered with tubercles or excrescences.

Cross-grained wood. Wood in which the grain ascends in a spiral form, is unfit for use in large scantlings; but if the defect is not very decided, the wood may be used for naves and for some light pieces.

Splits, &c. Splits, checks, and cracks, extending toward the centre, if deep and strongly marked, make wood unfit for use, unless it is intended to be split. *Wind-shakes* are cracks separating the concentric layers of wood from each other; if the shake extends through the entire circle, it is a serious defect. The *centre-heart* is also to be rejected, except in timber of very large size, which cannot, generally, be procured free from it.

233. **Seasoning and preserving timber.** As soon as practicable, after the tree is felled, the sapwood should be taken off, and the timber reduced, either by sawing

or splitting, nearly to the dimensions required for use. Pieces of thickness, or of peculiar form, such as those for the bodies of gun-carriages and for chassis, are got out with a saw; smaller pieces, as spokes, are split with wedges. Naves should be cut to the right length, and bored out, to facilitate seasoning and to prevent cracking. Timber of large dimensions is improved by *immersion in water* for some weeks, according to size, after which it is less subject to warp and crack in seasoning.

Seasoning. To season or dry timber, it should be piled under shelter, in such manner as to allow a free circulation, but not a strong current of air, around each piece. The piles should be taken down and put up again at intervals, varying with the length of time the timber has been cut.

The seasoning of timber requires from two to eight years, according to size. Oak timber loses about *one-fifth of its weight* in seasoning, and about one-third of its weight in becoming perfectly dry.

234. **Bill of timber.** Timber for the ordnance department is purchased in pieces of the size required to make each part. A list of the pieces for a certain kind of carriage, including the contents of each piece, in board-measure, is called a *bill of timber.* The unit of board-measure is a board one foot square, and one inch thick.

235. **Wrought iron.** None but the best wrought iron should be employed in ordnance constructions. Large and peculiar-shaped pieces, as *axle-trees*, *trunnion-plates*, &c., as well as those requiring great strength, are made from *hammered shapes*, furnished by the iron manufacturer, according to prescribed patterns; other parts are made of rolled iron. Axle-trees are proved by sup-

porting them at the arms, and dropping a heavy weight upon the middle of the body.

236. **Construction.** Timber for gun-carriages is now, almost entirely, worked into shape by machinery; the operations are *sawing*, *planing*, *turning*, *mortising and tenoning*, *dove-tailing*, &c.

In joining together the different pieces of a carriage, regard should be had to the character of the fibre of the wood, and the effect of drying in changing the form of the piece. If a piece be supported at both ends, as in the cases of carriage-stocks, chassis-rails, &c., the greatest convexity of the fibre should be placed uppermost; if in the middle, as in cases of hounds of limbers, side-rails of caissons, &c., it should be placed downward.

When the pieces are to be united in pairs, as cheeks, side-rails, &c., use such pieces as have nearly the same curvature of fibre.

In drying a piece of timber, the sap-wood shrinks more than the heart, and the effect will be to warp in the direction of the sap; therefore, to prevent the joint, formed by the two pieces which constitute a carriage-stock, from opening, the heart-wood should be placed on the outside. To prevent the cheeks from warping inward, place the heart-wood on the inside. In hounds and side-rails, the heart side should be placed on the outside, as this will have a tendency to tighten the joints.

When pieces are to be joined, the surfaces of contact and the dowels should be covered with a good coat of white-lead. Bolts and bolt-holes should be well covered with tallow moistened with neat's-foot oil.

The surface of holes for elevating screws and pintles

should be painted. If woodwork is to be painted immediately, it should have a priming coat of lead before the irons are put on; if not, it should receive a good coat of linseed oil.

237. **Painting.** For service, the wood-work of carriages and machines is painted, in addition to the priming of lead-color, with two coats of olive paint; the iron-work, with one coat of lead, and one coat black paint. Great care should be observed to protect iron fortress-carriages against the corroding influence of the sea-coast atmosphere; the best means remains to be determined by experience; at present they are covered with one coat of hot linseed oil and three coats of a reddish brown paint.

238. **Models, &c.** The models, &c., of all ordnance "*matériel*," are determined by the ordnance board, subject to the revision of the chief of ordnance, and the final approval of the secretary of war. When a model has been duly approved, copies, or drawings of it, are sent to the different arsenals of construction, and from these, patterns and gauges are made for the guidance of the workmen. Patterns are generally made of well-seasoned mahogany, and bound with strips of brass; gauges are made of sheet iron or steel. To secure uniformity of work at the different arsenals, it is made a part of the duty of the inspector of arsenals to see that the patterns correspond with the originals; and it is always the duty of the officers stationed at an arsenal, to see that the work, as it progresses, corresponds with the patterns, and that none but suitable materials are used.

All arsenals may be divided into two classes, viz., arsenals of deposit and construction, and arsenals of deposit and repairs. The arsenals of construction are at West Troy, N. Y.; Watertown, Mass.; Washington, D. C.; Pittsburgh, Pa.; and Fort Monroe, Va.

Ordnance Department. The Ordnance Department, U. S. Army, is composed of *one* Chief of Ordnance, with the rank of Brig.-General, *three* Colonels, *four* Lieut. Colonels, *ten* Majors, *twenty* Captains, *sixteen* First Lieutenants, *ten* Second-Lieutenants, *thirteen* Ordnance Store-keepers, and as many enlisted men of ordnance, consisting of sergeants, corporals, and first and second class privates, as the War Department may deem necessary for guard and other duty at the arsenals. It is the duty of this Department to furnish all ordnance and ordnance stores for the military service, either by manufacture at the arsenals and armories, or by purchase of private manufacturers.

CHAPTER VI.

SMALL-ARMS.

239. **History.** Ancient arms consisted of three kinds:

1st. *Hand-arms*, for close conflict.

2d. *Projectile arms*, to attain an enemy at a distance.

3d. *Defensive arms*, to protect the body.

240. **Hand-arms.** Hand-arms comprised the *war-club*, *battle-axe*, *pike*, *sword*, and *sabre*.

1st. The *war-club* was a stout stick, the large end of which was armed with blades, or points of metal; that used by foot-soldiers was from 7 to 8 feet long, and weighed from 20 to 30 pounds. It was extensively used in the middle ages, and is still employed by certain oriental cavalry.

2d. The *battle-axe* was at first made of stone or bone, and afterward of metal. This weapon was much used by the Celts and Gauls, but principally by the Franks, who hurled it with great skill and effect against an enemy.

3d. The *pike* was generally employed both by infantry and cavalry. That for the infantry was very long, as in the case of the Macedonian lance, or *sarissa*, the length of which was about 20 feet. In some countries this weapon continued to be used as late as the seventeenth century. The cavalry lance was shorter and lighter than the preceding; it is still used by certain kinds of cavalry.

The Roman *javelin* was a short pike, about $6\frac{1}{2}$ feet long, which was thrown against the enemy. The *spontoon*, or *half-pike*, was carried by French infantry officers as late as the time of Louis XV. The *halberd* and the musket with its bayonet fixed, are pikes.

4th. *Swords and sabres* have usually varied in character with the manner of fighting of different nations; for instance, the Gauls and Germans, who defended themselves with shields made of willow, or other light wood, made use of long and flexible swords, while the Greeks and Romans, who wore breastplates and helmets of metal, used short and stout swords.

The knights of the middle ages carried long and heavy swords, which they wielded with both hands.

241. **Projectile arms.** The principal projectile arms, before the invention of gunpowder, were the *sling*, *bow*, and *crossbow*.

1st. The *sling* was formed of a leather cap suspended by two cords; a stone was placed in the cap, a rapid rotary motion was communicated by the hand, one of the cords was set free, and the stone escaped in a tangential direction, and was thrown to a distance varying from 200 to 300 steps.

2d. The *bow* was formed of a piece of highly elastic wood, confined in a bent position by a strong cord attached to its extremities; it possessed the power of projecting arrows to long distances. This weapon played a very important part in ancient warfare, and continued to be used by civilized nations to a comparatively late period.

In the middle ages, it was said, that a skilful archer could fire twelve arrows in a minute, and strike a man at a distance of 100 yards. If certain English authors are to be believed, an archer who could not perform this feat, was disgraced. According to their statements, an arrow had sufficient force to penetrate an oak plank, two inches thick, at the distance of 200 yards.

3d. The *crossbow* was a bow attached to a stock having a channel to guide the arrow. It is said to have been introduced into Europe from Asia, during the Crusades, by Richard Cœur de Lion, who armed the English troops with it.

At present, the use of the bow is confined to barbarous tribes alone; the skill and dexterity with which it is managed by the prairie Indians of this country, make it an exceedingly formidable weapon at short distances.*

242. **Defensive weapons.** Armor, which was employed to protect the most exposed part of the body, naturally followed the introduction of offensive arms. At first it was made of plates of wood, the skins of certain animals, the scales of serpents, shells of turtles, &c., and subsequently of metallic plates, or of cloth folded in layers. The body was also protected in rear by movable obstacles, as shields, bucklers, &c.

Among the Romans and Greeks, the infantry of the line wore the helmet, the breastplate, a species of half-boot protected with iron, and the buckler; the cavalry, ordinarily, wore a cuirass formed of bands of leather, covered with plates of bronze.

In the time of Charlemagne, coats of mail, formed of small chains, were much worn. These were followed

* Certain Circassian officers of the Russian army carry bows and arrows to enable them in a surprise to kill the enemy's sentinels without alarming the camp.

by complete suits of metallic armor, which were worn until the introduction of fire-arms.

243. **Portable fire-arms.** Portable fire-arms were invented about the middle of the fourteenth century. At first they consisted simply of a tube of iron or copper, fired from a stand or support. They were loaded with leaden balls, and were touched off by a lighted match held in the hand. They weighed from 25 to 75 pounds, and consequently two men were required to serve them. The difficulty of loading these weapons, and the uncertainty of their effects, as regards range and accuracy, prevented them from coming rapidly into use, and the crossbow was for a long time retained as the principal projectile weapon for infantry.

Breech-loading small-arms, similar in principle to the cannon described in section 66, were introduced about the same time, but they were soon thrown aside for want of strength and solidity.

244. **Arquebuse.** The difficulty of aiming hand-cannon, arising from their great weight, was in a measure overcome by making them shorter, and supporting them on a tripod, by means of trunnions which rested in forks. The breech was terminated with a handle which was held in the right hand, while the match was applied with the left. Thus improved, this fire-arm was called the *arquebuse;* it was employed in sieges, and to defend important positions on the field of battle.

The next improvement in the arquebuse, was to make it lighter, and enclose it in a piece of wood, called the *stock*, the butt of which was pressed against the left shoulder, while the right hand applied the match to the vent. It was still very heavy, and in aiming, the muzzle rested in the crotch of a fork placed in the ground.

245. **Matchlock.** To give steadiness to the aim while applying the match to the priming, a species of lock was next devised, which consisted of a lever holding at its extremity a lighted match. In firing, the lever was pressed down with the finger until the lighted end of the match touched the priming. This apparatus, known as the *serpentine*, continued in use until it was replaced by the *wheel-lock*, which was invented in Nuremburg, in 1517.

The wheel-lock consisted of a grooved wheel of steel, which acted through a half-revolution on a piece of alloy of iron and antimony, placed near a priming-charge of powder. The sparks thus evolved fell upon and ignited the priming-charge.

246. **Pistol.** The first pistol was a wheel-lock arquebuse, so small that it could be held and fired from the extended hand. It was invented in 1545, in Pistoia, a city of Tuscany; hence its name.

247. **Petronel.** The petronel was a wheel-lock arquebuse of larger calibre and lighter weight than its predecessors. To diminish the effect of the recoil, the butt of the stock was much curved, and had a broad base, which was pressed against the breastplate of the cuirass when the piece was fired. Two sizes of the petronel were used, one for infantry and one for cavalry.

248. **Musket.** The musket was first introduced by the Spaniards, under Charles V. The original calibre of the musket was such that eight round bullets weighed a pound; the piece was, consequently, so heavy that it was necessary to fire it from a forked

stick inserted in the ground. The size of the bore was finally reduced to eighteen bullets to the pound; and from this arm was derived the late smooth-bored musket.

249. **Rifles.** It is generally stated that the rifle was invented by Gaspard Zoller, of Vienna, and that it first made its appearance at a target practice at Leipsic, in 1498. The first rifle grooves were made parallel to the axis of the bore, for the purpose of diminishing the friction of loading forced or tightly-fitting bullets. It was accidentally discovered, however, that spiral grooves gave greater accuracy to the flight of the projectile, but the science of the day was unable to assign a reason for this superiority, and the form, number, and twist of the grooves, depended on the caprice of individual gun-makers.

About 1600, the rifle began to be used as a military weapon for firing spherical bullets. In 1729, it was found that good results could be attained by using oblong projectiles of elliptical form. The great difficulty, however, of loading the rifle, which was ordinarily accomplished by the blows of a mallet on a stout iron ramrod, prevented it from being generally used in regular warfare. The improvements which have been made in the last thirty years, principally by officers of the French army, have entirely overcome this difficulty, and rifles are now almost universally used in place of smooth-bored arms.

The rifle has ever been a favorite weapon in this country, arising, doubtless, from the peculiar circumstances which surrounded its early settlers and pioneers, and on more than one occasion has it proved, in the hands of irregular troops, a formidable weapon to its enemies. Until 1855, the mass of the American infantry was armed with smooth-bored muskets, but since that time it has been wholly armed with rifles.

250. **Bayonet.** In spite of the advantages which fire-arms possessed, they, like the arms which preceded them, were not suited to resist the charge of cavalry. The bayonet, and firing in closed ranks, were unknown; the most skilful captains of the age, however, sought to combine fire-arms with pikes, in such a manner that one might afford protection to the other. The French army was thus arranged in six ranks, four with muskets and two with pikes; on the introduction of the bayonet, it was reduced to four, and finally to three ranks.

The bayonet derived its name from Bayonne, where it was first made, in 1640. At first it was formed of a steel blade attached to a handle of wood, which was inserted into the bore of the barrel, except in the operations of loading and firing. Thirty years afterward the wooden handle was replaced by a hollow socket, which fitted over the muzzle of the barrel; this change rendered the musket at all times a pike as well as a fire-arm, and led to the formation of modern infantry.

251. **Flint-lock.** The flint-lock was derived from the wheel-lock, by substituting flint for the alloy of iron and antimony, and a steel battery for the wheel. It was generally introduced into the French army in 1680, and continued to be used in all military services, until about 1842, when it gave way to the percussion-lock.

252. **Loading.** In proportion as fire-arms were improved, rapidity of fire increased. In 1703, the loading of the musket was performed in twenty-six times, and

the fire of infantry was necessarily slow. In 1744, the employment of fine powder for priming was dispensed with, and the cartridge (said to have been the invention of Gustavus Adolphus) was adopted in its place.

HAND-ARMS.

253. **Classification.** Hand-arms are divided into three classes, depending on their mode of operation.

1st. *Thrusting*-arms, which act by the point.

2d. *Cutting*-arms, which act by the edge.

3d. *Thrusting and Cutting* arms, which act either way.

254. **General principles.** The object of all hand-weapons is to penetrate, directly, the person of an enemy. They may be divided into three distinct parts, viz.: 1st. The point, or edge, which attains the object; 2d. The body, or blade, which constitutes the mass of the weapon, and transmits the force of the hand to the object; and, 3d. The handle, or point of application of the motive force.

The mechanical principles to which they may be referred, are the lever and wedge.

255. **Thrusting-arms.** With a given force of the hand, acting against a given object, the penetration of a thrusting-weapon depends upon the power of the wedge formed at its point. The effect will be modified, however, by the position of the axis of the wedge, for if it do not coincide with the direction of the impelling force, there will be a component force which acts to turn the point to one side.

The blade of a thrusting-weapon should, therefore, be straight, and should taper to a point. To guide it easily, the centre of gravity should be found in or near the handle; this may be accomplished by grooving the blade, by making the handle heavy, or by adding a counterpoise to it.

Kinds. The principal thrusting-weapons are the *straight sword*, *lance*, and *bayonet*.

The straight sword, as well as other swords, are composed of the *blade*, the *hilt*, and the *guard*.

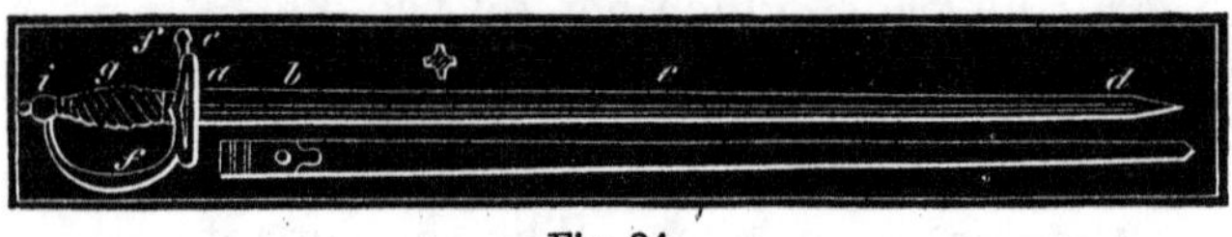

Fig. 84.

The *blade* (see fig. 84) is divided into the *point* (*d*), the *middle* (*c*), the *reinforce* (*b*), the *shoulder* (*a*), the *tang*, or portion which is inserted into the handle, and the *grooves*, the number of which is equal to the number of faces, or, from two to four.

The length of the blade varies from 30 to 33 inches, the width is from $\frac{1}{2}$ to $\frac{3}{4}$ of an inch, and the weight 1 to $1\frac{1}{2}$ pound.

The *hilt* is divided into the *knob* (*i*), and the *gripe* (*g*); the gripe is generally made of wood, covered with leather or sheet brass, and wrapped with wire to give it roughness, and prevent it from slipping in the hand.

The *guard* is composed of the *curved branch* and *cross-piece* (*f*), and the *plate* (*e*), all joined in one piece. The object of the guard is to protect the hand, the plate to ward off the point, and the branch, the edge of the enemy's sword.

The *wounds* made by thrusting-swords, particularly those with three or four concave sides, are very dangerous, as they close up externally and suppurate internally.

In experienced hands the straight sword is well adapted to encounter one of its kind, but it is too weak to parry the blows of a sabre. It is now but little used in this country, except for ornamental purposes; the sabre being preferred as a service weapon, even for infantry officers.

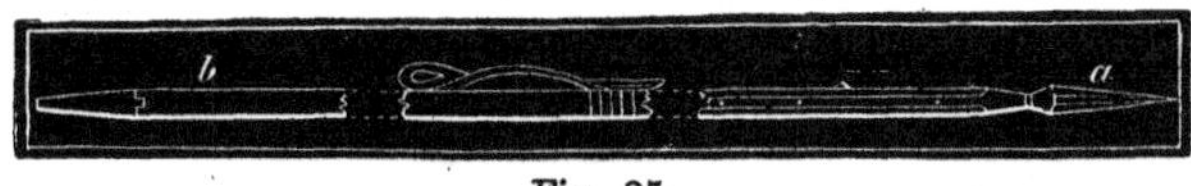

Fig. 85.

The lance. The lance, or pike, is composed of a sharp steel blade, fixed to the end of a long and slender handle of wood. (Fig. 85).

The *blade* is generally from 8 to 10 inches long, and, in order that it may combine stiffness with lightness, is grooved after the manner of the common bayonet, leaving three or four ridges. The base of the blade has a socket, and two iron straps, for securing it to the handle. Three small staples are sometimes fastened to the handle, below the blade, for the purpose of attaching a *pennon,* which serves as an ornament, and to frighten the enemy's horses.

The *handle* is made of strong, light, well-seasoned wood. The lower end is protected with a tip of iron, and a leather *loop* (*c*) is attached opposite the centre of gravity, to enable the arm to carry and guide the lance. The total length of a lance varies from $8\frac{1}{2}$ to 11 feet, and the weight is about $4\frac{1}{4}$ lbs.

Mode of carrying. On horseback, and when not in use, the lance may be carried in two ways: 1st. By placing the lower end in a leather boot attached to the stirrup, and passing the right arm through the leather loop; 2d. By placing the lower end in the boot, and strapping the handle to the pommel of the saddle. The first mode enables the horseman to take his lance with him when he dismounts, and is well suited to light lances. The second mode is necessary for heavy lances.

Advantages, &c. In the first shock of a cavalry charge, and in the pursuit of a flying enemy, the lance is a superior weapon to the sabre, as it has a greater penetration, and attains its object at a greater distance; but in the hand-to-hand conflict which follows a charge, the latter is superior to the former. Hence, it has been customary in certain services to arm a portion of both light and heavy cavalry with the lance. In the Russian service, the front rank of the cuirassiers, a species of heavy cavalry, is armed with the lance, and the rear rank with the long, straight, two-edged sabre; and in nearly every European service, the lancers constitute an important part of the cavalry organization. It is also a favorite weapon with the mounted Indians of this country.

Bayonet. The bayonet is a pointed blade attached to the end of a fire-arm, to convert it into a pike. The mode of attachment should be such that the bayonet will not interfere with the loading, aiming, and firing of the piece; and it should be so secure as not to be disengaged in conflict.

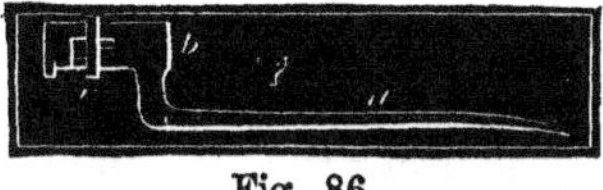
Fig. 86.

The *musket-bayonet* is composed of a *blade* (*a*), (fig. 86), a *socket* (*b*), and a *clasp* (*c*).

The *blade* of this bayonet is made of steel, 18 inches long, and, to give it lightness and stiffness, its three faces are grooved in the direction of the length. The grooves are technically called *flutes.* (See cross-section of blade.) The blade is joined to the socket by the *neck*, which should be strong, and free from defects of workmanship.

The *socket* is made of wrought iron, carefully bored out to fit the barrel of the piece easily, and at the same time closely. It is secured by a stud (brazed on the barrel), which fits into a crooked channel, or *groove*, cut in the socket, and by a movable ring called the *clasp.*

The Sword-bayonet. Short arms, such as carbines and musketoons, are sometimes furnished with bayonets of sufficient length to enable these arms to resist a charge of infantry or cavalry.

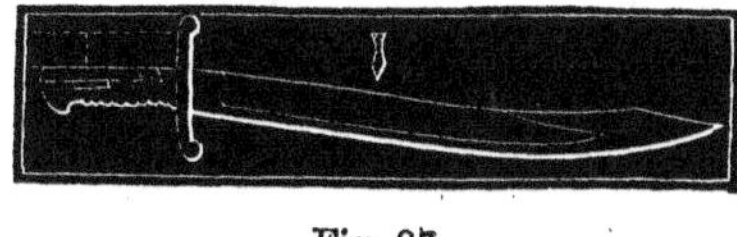
Fig. 87.

Such bayonets are generally made in the form of a sword. (Fig. 87.) The back of the handle has a groove which fits upon a stud on the barrel, and the cross-piece of the handle is perforated so as to encircle the muzzle end of the barrel. The bayonet is prevented from slipping off by a spring catch.

The handle is made of a solid piece of brass, with a hole running through it for the tang of the blade, which is secured by riveting down the point. The back of the blade is turned toward the barrel, and the body is bent outward, that neither may interfere with the hand in loading. Its length is about 23 inches, and its breadth $1\frac{1}{4}$ inches. The sword bayonet is too heavy to be carried habitually fixed to the barrel; ordinarily it is carried as a side-arm, for which purpose it is well adapted, as it has a curved cutting edge, as well as a sharp point. The ordinary bayonet, when not fixed, may be used as a poignard, for the personal defence of the soldier.

The bayonet contributes very much to the efficiency of a military fire-arm, particularly as it enables infantry to resist cavalry. Too much attention cannot be paid in teaching troops the use of this arm, and inspiring them with confidence in it, for it often decides the fate of a battle.

256. **Cutting-arms.** That edge of a cutting-arm will have the greatest penetration which opposes the fewest points to its object; a blade with a convex edge, will, therefore, have greater penetration than a straight one.

The effect of a cutting-blade will be modified by the manner it is applied to the surface of the object; an oblique stroke, for instance, will make a deeper cut than a direct one. If the edge of the sharpest blade be submitted to a microscope, it will present to the eye numerous asperities, which give it the appearance of the cutting edge of a saw; it is evident, therefore, that the motive force should act obliquely to the cutting edge of the blade, as that enables it to rupture the layers of flesh upon which it acts, in detail, and without expending its force upon the elasticity of several layers at once, which would be the case were it to act directly upon the object.

Form of the blade. When the curvature of a blade is convex on the cutting side, the part near the point makes a deeper cut when it is pushed from the hand that moves it, as will be the case with the blows delivered in a charge of cavalry. On the contrary, a *concave* cutting-edge, like that of a sickle, acts most favorably when it is drawn toward the person using it; such is the *yataghan* of the Arabs, the shape of which is that of an elongated letter S.

Handling. The facility of handling a sabre, and the effect of its blow, depend upon the relative positions of the *handle*, the *centre of gravity*, the *point of contact*, and the *centre of percussion.*

The nearer the centre of gravity is to the point of contact, the more powerful will be the blow; but the difficulty of handling increases with the distance of the centre of gravity from the handle. As the force of the blow is the important consideration in a sabre, and the facility of handling in a thrusting-sword, it is customary to make the point of the blade heavier, and the handle lighter, in the former than in the latter.

In certain light cavalry sabres, the centre of gravity is placed about three or four inches from the handle.

In order that no part of the force be lost, the point of contact should coincide with the centre of percussion; the position of the latter point, however, depends upon the weight of the soldier's arm, if motion takes place around the shoulder, and it therefore varies in particular cases.

Description. The principal cutting weapon is the sabre. The cutting edge is generally convex; and the degree of its curvature is the characteristic feature of the weapon. The nomenclature of the sabre is nearly the same as for the sword, the principal difference being in the structure of the guard, which is made lighter or heavier, as the sabre approximates the character of a cutting or thrusting weapon.

There are two kinds of sabres used in the United States' service, viz.: the *cavalry sabre*, and the *light-artillery sabre.*

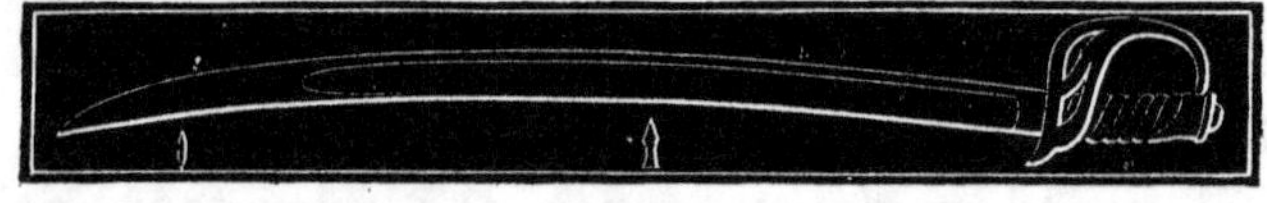

Fig. 88.

The *cavalry-sabre* (fig. 88), being used, to a certain extent, for pointing as well as cutting, has only a moderate degree of curvature, a long blade (36 inches), and a "basket-hilt" to protect the hand from the point of the enemy's sword, and to carry the centre of gravity toward the handle. The guard is composed of the *front*, *middle*, and *back* branches. The gripe is covered with calfskin, and bound with wire.

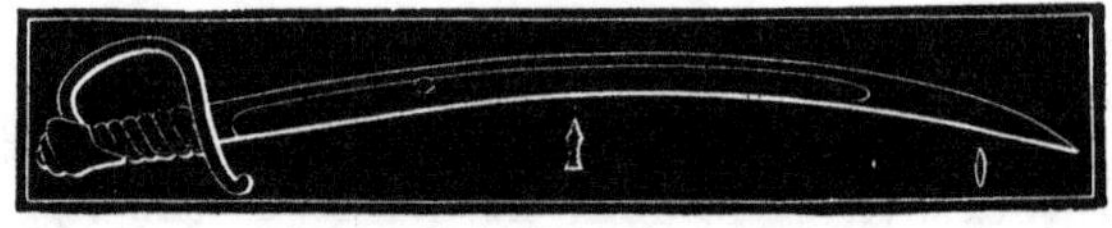

Fig. 89.

The *light-artillery sabre* (fig. 89), being used more particularly for hand-to-hand conflicts, has a shorter (32 inches) and more curved blade, and a lighter handle than the cavalry sabre. The guard is composed of a single piece of brass, terminating in a scroll.

The blades of all sabres are grooved, to give them lightness. See cross-sections of figs. 88 and 89.

257. **Thrusting and cutting arms.** In certain services

it is customary to arm the heaviest cavalry, or cuirassiers, with swords which are capable of coping with the bayonet or lance. The blades are long (from 36 to 40 inches), light, and straight, and they have a sharp point, and a single cutting edge. The hilt is heavy, and of the basket form.

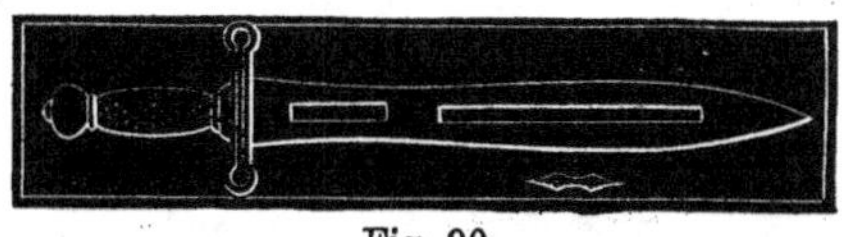

Fig. 90.

The only weapon of the thrusting and cutting class used in the United States' service is the foot artillery sword (fig. 90), which resembles the short Roman sword in its character. The blade has two cutting edges, is lightened toward the handle, and is 19 inches long. The guard is a simple cross-piece, formed of the same piece as the handle, which is made of brass.

258. **Scabbards.** The objects of the scabbard are to carry the sword, and protect the blade from injury. Scabbards are generally secured to a belt, which passes over one of the shoulders, or around the waist of the wearer.

For foot-troops, scabbards are made of leather, mounted with metal to protect them from wear. For scabbards for mounted troops, leather has not sufficient stiffness and strength, and steel is used in place of it. The principal objections to metal scabbards are, that they are heavy, dull the edge of the blade, and make considerable noise in striking against the equipments of the horse and rider.

In certain parts of Asia, these objections are overcome by making scabbards of wood, covered with leather or raw hide.

DEFENSIVE ARMOR.

259. **Cuirass and helmet.** The defensive armor of the present day consists of the *cuirass* and *helmet*—the former to protect the breast and back; and the latter, the head of the wearer.

The French cuirass (fig. 91) is composed of a *breastplate* (*a*), and a *back-piece* (*b*), joined together by straps. The thickness and form of the breastplate are such as to ward off small-arm projectiles beyond a distance of forty yards; this distance is assumed under the supposition that within it, the infantry soldier will be too busily engaged in preparing to defend himself against the cavalry soldier, with his bayonet, to fire his piece. The back-piece is only made of sufficient thickness to resist the stroke of a sword; it is presumed this will induce the wearer to present his front rather than his back, when he arrives within a short distance of his enemy. The middle of the breastplate is formed into a ridge, and the sides slope off to reflect projectiles coming from the front. The thickness at the ridge is .23 in.; from this, it tapers to the edges, where it is .078 in. The back-piece is .047 in. thick, throughout, and the weight of the entire cuirass is about 16.75 lbs. The edges are turned up to prevent the point of a sword from slipping off against the body.

Fig. 91.

The cuirass and helmet worn by the leading sapper in digging a siege-trench, are thick enough in all their parts to resist a bullet at the distance of 40 yards.

MANUFACTURE OF SWORD–BLADES, &c.

260. **Material.** Sword-blades are made of double shear steel, or what is better, of cast-steel. A material of great toughness and elasticity, as well as hardness, is made by forging together steel and iron, forming the celebrated *Damascus steel*, which is used for sword-blades, springs, &c. The damasked appearance is produced by the action of nitric acid and vinegar, which gives a black tint to the steel parts, whilst the iron remains white.

261. **Fabrication.** The operations of making sword-blades are: 1st. *The preparation of the skelp;* 2d. *Forging the blade;* 3d. *Tempering;* 4th. *Grinding;* 5th. *Retempering;* 6th. *Etching;* 7th. *Polishing.*

The skelp. The skelp is a tapering piece of steel about once and a half as thick, two-thirds as long, as the proposed blade, and it is formed out of a rectangular bar by a trip-hammer.

Forging. The first operation is to weld a piece on to the large end of the skelp to form the tang. The second, is to draw it out to the proper length and thickness. The number of heats required depends upon the length of the blade, and is generally from seven to eight. The third is to stamp the grooves. The fourth is to give the bevel to the cutting edge. This operation necessarily elongates this edge, and gives a curved shape to the blade, which should be regulated to suit the pattern. The blade is then cut off to the right length, and the tang finished.

Tempering. Forged blades are soft and flexible; it

is necessary, therefore, to give them a certain degree of hardness and elasticity. This operation is called hardening and tempering, and requires much tact on the part of the workman, in order to detect, by the eye, the temperature most suitable for the quality of steel employed.

To harden the blade, the workman holds it in the heat of a charcoal furnace, moving it back and forth to heat the several parts uniformly. When its color is cherry red, it is withdrawn and passed rapidly through moist iron filings, to render the heat still more uniform, and then plunged into cold water. It is immediately withdrawn and examined, to see if it be free from the coating of oxide which covers it when taken out of the fire.

The blade is now very hard and brittle, and oftentimes warped; it is necessary, therefore, that it should be partially annealed or tempered; and, for this purpose, it is again placed in the furnace until the surface is coated with blue oxide. In this condition it is soft, and in a condition to be straightened. This is quickly done, and the blade is plunged into cold water, which gives it the proper degree of hardness and elasticity.

Grinding. Grinding is done on *grindstones* which revolve with great rapidity—the object being to reduce all parts of the blade to the proper size.

Retempering. The blade is partially bent and annealed by grinding. To correct these defects it is again heated to the proper color, straightened, and plunged into cold water. If it have lost too much of its hardness, and be too much bent, it should be hardened and tempered as described before grinding.

Etching. Etching is the process of marking the blade with ornamental devices, name of maker, &c. It is done by heating the blade slightly, and covering the portion to be marked with a thin layer of beeswax; the design is marked on the wax with a steel point, leaving the metal bare; after this, the design is covered with powdered sea-salt, and then moistened with nitric acid. The acid acts only on the bare parts of the metal, leaving them in a rough state, while the unaffected part of the blade remains bright.

Polishing. The object of polishing the blade is to remove the marks of the grindstones, and give it a smooth finish. The polishing apparatus is a rapidly revolving hard-wood wheel, and the polishing material is oil and emery. Burnishing gives the deep, brilliant polish peculiar to steel. The operation is the same as in polishing, except that oil and emery are replaced by charcoal-dust.

262. **Inspection of blades.** The dimensions are compared with the model, and verified by appropriate gauges and patterns.

263. **Proof of blade.** The blade is proved: 1st. By confining the point by a staple, and bending the blade over a cylindrical block, the curvature of which is that of a circle 35 inches diameter, the curvature of the part next the tang being reduced by inserting a wedge 0.7 inches thick at the head, and 14 inches long; 2d. It is struck twice on each of the flat sides on a block of oak wood, the curvature of which is the same as the above; 3d. It is struck on the edge, and twice on the back across an oak block one foot in diameter; 4th. The point is placed on the floor, and the blade bent until it

describes an arc having a certain versed sine. After these trials, the blade is examined to see that it is free from flaws, cracks, or other imperfections, and that it is not *set*, that is to say, does not remain bent.

The scabbard is proved by letting a two-pound weight fall upon it, from the height of 18 inches. The weight should be only one inch square at the base; the scabbard should not be indented.

PORTABLE FIRE-ARMS.

264. **Principal parts.** The essential parts of all portable fire-arms are the *barrel*, the *lock*, the *stock*, the *sights*, and the *mountings*. The nature of the remaining parts will be determined by the manner of loading and discharging, as in *muzzle*-loading and *breech*-loading fire-arms. Portable fire-arms will be treated in the same order as cannon, viz.: 1st. The general principles common to the various kinds; 2d. The peculiarities arising from the uses to which they are applied; 3d. The *manufacture*, *inspection*, and *treatment in service.*

265. **Barrel.** The barrel is the most important part of a fire-arm, its office being to concentrate the force of a charge of powder on a projectile, and give it proper initial velocity and direction; for these purposes, and for the safety of the firer, it should be made of the best materials, and with the greatest care.

Exterior. In determining the exterior form of a barrel, it is not only necessary to give such thickness to the different parts as will best resist the explosive effort of the charge, but such as will prevent it from being bent when used as a pike, or subjected to the rough usage of the service.

Weight, to a certain extent, is necessary to give steadiness to the barrel in aiming, and to prevent it from "springing" in firing. The latter defect generally arises from bad workmanship, whereby there is a greater thickness of metal, and, consequently, less expansion, on one side of the bore than the other. In certain sporting rifles, where great accuracy is sought to be obtained, the barrel is made to weigh from 12 to 15 lbs.; but in the military service, where it is carried by the soldier, it seldom weighs more than 5 lbs.

The thickness of metal of the rifle-musket barrel at the breech is 0.28 inch; from this point it gradually diminishes (the exterior element being a slightly re-entering curve) to the muzzle, which is 0.1 inch.

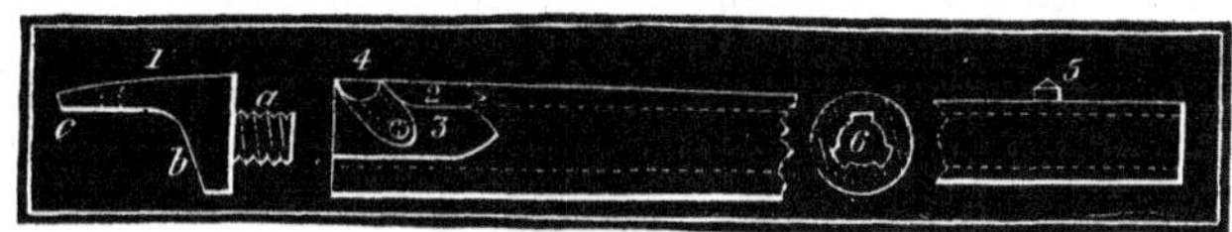

Fig. 92.

Nomenclature. The principal parts of a barrel (see fig. 92) are the *breech*, the *breech-screw* (1); the *flats* (3), *bevels* (2), and *oval*; the *cone*, and *cone-seat* (4); the *bayonet-stud* and *front-sights* (5); the *bore*, the *grooves*, and the *lands* (6).

The *breech-screw* is composed of the *body* (*a*), *tenon* (*b*), and *tang* (*c*). The object of the breech-screw is to close the bottom of the bore; the tenon fits into a mortise cut in the stock, and prevents the barrel from turning in its bed; the tang is the part by which the breech of the barrel is secured to the stock, and, for this purpose, it is pierced with a hole for the *tang-screw*, which passes through the stock, and screws into the guard-plate.

The flats are two vertical plane surfaces, situated at equal distances from the axis of the bore. They serve to prevent the barrel from turning in the jaws of the vice when the breech-screw is taken out; the flat, on the right side of the barrel, also presents a surface of contact for the lock-plate, which prevents the hammer and cone from changing their relative position.

The corners of the flats are bevelled to prevent the barrel from being marred, and to improve its finish.

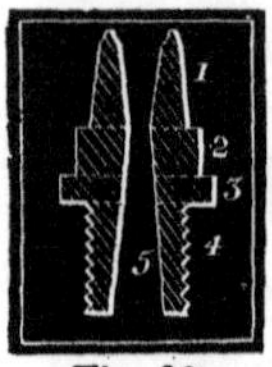

Fig. 93.

The cone. The functions of the cone are to support the percussion-cap when exploded by the hammer, and to conduct the flame to the vent of the piece. The parts (see fig. 93) are the *nipple* (1), upon which the cap is placed; the *square* (2), the part to which the wrench is applied; the *shoulder* (3); the *screw-thread* (4); and the *vent* (5).

To increase the effect of the hammer on the cap, the upper surface of the cone is diminished by chamfering the corners.

The *cone-seat* is a projecting piece of iron welded to the barrel, near the breech, for the purpose of sustaining the cone. The principal parts are the *female screw*, the *vent*, and the *rim;* the latter prevents the flame from penetrating between the lock and barrel. The position of the cone-seat should be such that the vent will have a direct communication with the bore. In the present barrel, however, the vent makes a bend at right angles with the axis of the cone, on account of the peculiar construction of the new self-priming lock. To prevent the blow of the hammer from straining the cone, and breaking it off in the cone-seat, the plane of the face of the hammer should pass through the axis of motion.

Calibre. Three important points are to be considered in determining the calibre of small-arms: 1st. The calibre should be as small as possible, to enable the soldier to carry the greatest number of cartridges; with the present calibre, the number of musket-cartridges is limited to 40; the total weight of which is about $3\frac{1}{2}$ lbs. 2d. To diminish the amount of ammunition required to supply the wants of an army, and to prevent the confusion that is liable to arise from a variety of calibres, there should not be more than two, for all arms of the same service, viz., one for the musket and one for the pistol. 3d. This point relates to the force and accuracy of the projectile. The introduction of elongated projectiles affords the means of increasing the accuracy and range of fire-arms, without increasing the weight of the projectile, simply by reducing the calibre, which diminishes the surface opposed to the air. Too great reduction of calibre, however, gives a very long and weak projectile; and besides, the effect of a projectile on an animate object, depends not only on its penetration, but on the shock communicated by it to the nervous system, or upon the surface of contact. A projectile of very small calibre, having but little inertia, does not expand well into the grooves of the bore, by the action of the powder; it is not, therefore, suited to the present method of loading, at the muzzle.

The foregoing considerations led to a general reduction of calibre on the introduction of rifles for military purposes. The present rifle calibre is .58 inch; that of the pistol (navy size) is .37 inch.

Length of barrel. The length of a gun-barrel is determined by the nature of the service to which it is applied, rather than by the effect which it exerts on the force of the charge. It has been shown by experiment, that the velocity of a projectile, in a smooth-bored musket, increases with the length of the bore, up to 108 calibres, at least; but a musket with this length of barrel, would be too heavy as a fire-arm, and too unwieldy as a pike. The length of the present musket-barrel is 70 calibres, or 40 inches.

Grooves. The principal cause of the deviation of a projectile from its true line of flight is the rotary motion which it receives in the bore of the piece, combined with the resistance of the air. In a smooth-bored barrel, variable causes conspire to produce rotation, consequently the deviation which results from it, is variable and uncertain. In addition, therefore, to giving a projectile the requisite initial velocity and direction, a gun-barrel should be constructed to give it a certain rotary motion that shall continue throughout its flight. This rotary motion, for reasons stated in discussing rifle-cannon, takes place around an axis coinciding with that of the barrel, and is produced by spiral grooves cut on the surface of the bore.

The points to be observed in constructing rifle-grooves for military arms, are range, accuracy of fire, endurance, and facility of loading and cleaning the bore. For expanding projectiles, experiment shows that these points are best attained by making the grooves broad and shallow, and with a moderate twist.

The following is a description of the grooves adopted by the United States' government, viz.:

Number. Three.

Width. Equal to the lands, or one-sixth the circumference of the bore.

Depth. Uniformly decreasing from the breech, where it is .015 in., to the muzzle, where it is .005 inch.

Twist. Uniform, one turn in six feet for long or musket, and one turn in four feet for short or carbine barrels.

The effect of decreasing the depth of rifle-grooves is to increase the accuracy, but diminish the range. The increase of accuracy, undoubtedly, arises from the fact that the projectile is held more firmly by the grooves, as it passes along the bore; while the diminution of range arises from an increase of friction between the projectile and the grooves.

The twist is dependent on the length, diameter, and initial velocity of the projectile; in other words, it should be increased in a certain proportion to the length of the projectile;* and for the same weight of projectile, it should be increased in a certain proportion as the length of the bore is diminished. Experiment, however, is the surest way of determining the most suitable twist for any projectile.

266. **Lock.** The lock is the machine by which the charge in the barrel is ignited. Nearly all the locks of the present day belong to the percussion class, in which fire is produced by the blow of a hammer upon a small charge of percussion powder contained in a copper or paper cap.

* See section on rifle-cannon.

The conditions to be fulfilled in the construction of a military lock, are—

1st. The production of fire, and its communication with the charge, should be certain, and under the perfect control of the soldier.

2d. The cap should be placed upon the cone with facility, and it should not be displaced in handling the piece.

3d. Fragments of the cap should not incommode persons near by, nor should the gas generated by the explosion of the cap corrode or injure the cone, barrel, or stock.

4th. There should be no danger of accidental explosions.

Nomenclature. The ordinary percussion lock is composed (see fig. 94) of the *lock-plate* (1), to which the several parts are attached, and by which the lock is fastened to the stock; the *hammer* (2), which strikes upon the cap, and explodes the composition; the *main-spring* (3), which sets the hammer in motion; the *tumbler* (4), or axle, by which the power of the main-spring is communicated to the hammer; the *sear* (5), or lever, the point of which fits into the notches of the tumbler, and holds the hammer in the required position; the notches are designated as the *full-cock notch*, and *safety-notch;* the *sear-spring* (6), which presses the point of the sear into the tumbler notch; the *bridle* (omitted in the figure), which is pierced with two holes for the inner pivots of the sear and tumbler; the *swivel* (7), which joins the main-spring and tumbler.

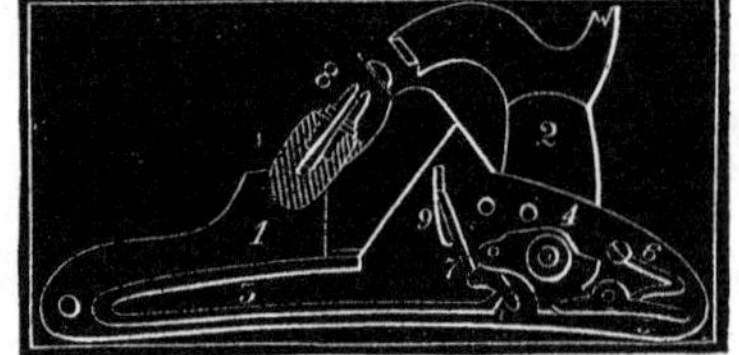
Fig. 94.

Self-priming. The foregoing constitute the essential parts of an ordinary percussion-lock; in addition to these, the new service lock is supplied with Maynard's self-priming apparatus.* The primer used in this apparatus, is a long strip of paper containing about 60 charges of percussion-powder, distributed at uniform intervals. The strip is wound up in the form of a coil, and inserted in a cavity cut into the exterior surface of the lock-plate, called the magazine. One end of the coil protrudes through an opening in the *magazine* (8), so that the centre of the first charge of percussion-powder is directly over, but not in contact with, the top of the cone. When the lock is sprung, the primer is cut off by a knife-edge on the lower side of the face of the hammer, carried forward and exploded on the top of the cone. A *feeding-finger* (9), connected with the tumbler, pushes out another primer, when the hammer is brought to the position of "full-cock."

Other methods are used for self-priming, in some of which the primer is enclosed in the cartridge itself; but few are found, under all circumstances, to be as reliable as the common percussion lock.

Back-action. In the back-action lock, the main-spring is placed in rear of the tumbler, and the sear-spring, as a separate part, is dispensed with. The mortise, which forms a bed for this lock, seriously affects the strength of the stock at the handle; and, for this reason, the front-action lock is generally preferred for military arms.

Accidents. If the head of the hammer be allowed to

* In 1861, the self-priming apparatus was omitted in all arms of the U. S. service, as it was not found to work well in practice.

rest on the cap, an explosion will be liable to follow an accidental blow on the hammer.

267. **Stock.** The stock is the wooden part of a fire-arm, to which all the parts are assembled.

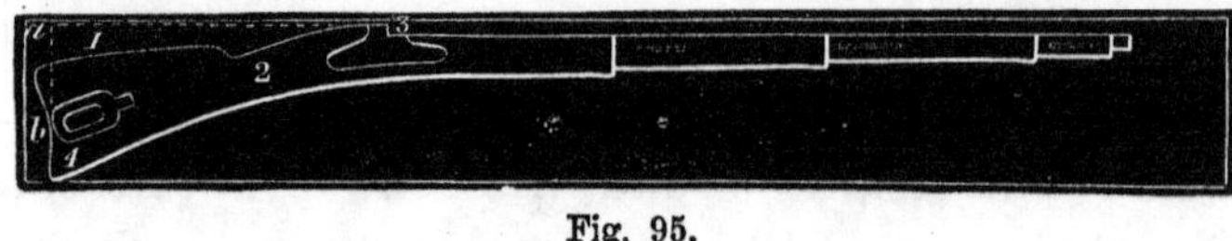

Fig. 95.

The most important portions of the stock (see fig. 95) are the *butt* (1), the *handle* (2), the *head* (3), the *grease-box** (4), the *beds* for the barrel, lock, band-springs, guard-plate, butt-plate; the *shoulders* for the tip and bands, and the *ramrod-groove*.

The material of the stock should be light and strong. Well-seasoned black walnut is generally used for military small-arms.

The butt is intended to rest against the shoulder, and support the recoil of the piece; it should be of such length and shape as will enable it to transmit the recoil with the least inconvenience to the soldier. The longer it is, to a certain extent, the more firmly will it be pressed against the shoulder, and the effect of the recoil will be a *push* rather than a *blow*. The stock is crooked at the handle, for convenience in aiming, and for the purpose of diminishing the direct action of the recoil. Changing the direction of the recoil, in this manner, causes the piece to rotate around the shoulder with an intensity proportional to the lever arm $a\ b$; whence it follows that, if the stock be made too crooked, the butt will be liable to fly up and strike the soldier's face.

268. **Sights.** The sights are guides by which the piece

* Omitted in 1861.

is given the elevation and direction necessary to hit the object. There are two, called *front* and *rear sights*.

The *front sight* is fixed; in the rifle-musket it is formed by sharpening the top of the bayonet-stud, so that its edge shall present a point to the eye of the marksman. The *fineness* of this point is regulated by the length of the barrel, or distance from the eye, and the size and distance of the objects generally aimed at; it is made coarser in military than in sporting arms, to prevent injury.

The *rear sight* is composed of a base, which is firmly secured to the barrel at a short distance from the breech, and a movable part capable of being adjusted for different elevations of the barrel. The sight originally affixed to the rifle-musket had a single leaf, to which was attached a slide, containing the sight notch, which could be adjusted for all distances between 100 and 1,000 yards. By a late order of the war department, this has been replaced by a sight which has three movable leaves, turning on a common axis, and bearing notches adjusted to 100, 300, and 500 yards, respectively.

Aiming a fire-arm consists in bringing the top of the front sight, and the bottom of the notch of the rear sight, into the line, joining the eye and the object. A sight for a military arm should satisfy the following conditions, viz.: 1st. It should be easily adjusted for all distances within effective range; 2d. The form of the notch should permit the eye to catch the object quickly; 3d. It should not be easily deranged by the accidents of the service.

The *globe* and *telescopic* sights are used for very accurate sporting-arms, but they are too delicate in their structure, and too slow in their operation, for general purposes.

In the *absence* of a proper rear sight, the soldier of the line may be taught to point his piece by aiming over the centre of the knuckle of his left thumb; the position of the thumb along the barrel determines the elevation of the piece. This method is practised by certain French troops of the line, for distances less than 400 yards.

269. **Mountings.** The mountings comprise the *butt-plate*, the *guard-plate*, the *bands*, *springs*, and *tip*.

The *butt-plate* protects the end of the stock from injury by contact with the ground; it is curved to fit the shoulder in firing, and is secured in its place by two wood-screws.

The *guard-plate* strengthens the handle of the stock, and serves as a fulcrum for the trigger. It is secured by the tang-screw and two wood-screws.

The *trigger* is a lever used to disengage the point of the sear from the notch of the tumbler, which sets the lock in motion. The force required to set off the trigger, if very great, may disturb the accuracy of the aim; if it be slight, the piece will be liable to accidental discharges, as in the case of the *hair-trigger* used in target-pieces.

The *guard-bow* protects the finger-piece of the trigger from injury, and from accidental blows that might produce explosions.

The *bands* secure the barrel to the stock, and the *springs* keep the bands in their places. If the piece be intended to be carried upon the soldier's back, it is provided with two *swivels*, one of which is fastened to the guard-bow, and the other to a band.

270. **Ramrod.** The ramrod is the long, slender piece employed in muzzle-loading arms, to push the charge to its proper place, and to wipe out the barrel. It is carried in a groove cut into the under side of the stock, and it is kept in its place by the pressure of the *swell* against the tip of the stock. The *head* of the rod is countersunk to fit the point of the projectile; and the *point* has a screw to receive the *wiper* and *ball-screw*—implements that are used to clean and remove obstructions from the bore.

BREECH-LOADING ARMS.

271. **General description.** The term "breech-loading" applies to those arms in which the charge is inserted into the bore through an opening in the breech; and, as far as loading is concerned, the ramrod is dispensed with.

The *interior* of the barrel of a breech-loading arm is divided into two distinct parts, viz., the *bore* proper, or space through which the projectile moves under the influence of the powder; and the *chamber* in which the charge is deposited. The diameter of the chamber is usually made a little larger, and that of the bore a little smaller, than that of the projectile; this arrangement facilitates the insertion of the charge, and causes the projectile to be compressed, and held firmly by the lands in its passage through the bore. As before stated, this operation is called *slugging* the projectile. The bottom of the grooves and the surface of the chamber are generally continuous.

If the chamber be made in the piece which closes the breech, commonly called the *breech-block*, the arm is said to have a *movable* chamber; if it be formed by counter-boring the barrel, it is said to be a *fixed* chamber. The fixed chamber has great advantages, and is now generally used, especially for the metallic case cartridge.

Formerly the most serious defect of breech-loading arms was the escape of the flame through the joint, which not only incommoded the soldier, but by fouling the machinery, seriously interfered with its operation. At present this is entirely overcome by "packing the joint" with a thin steel ring, which is a part of the gun and called the "gas-check," or by the case of the cartridge itself. The elasticity of the case, if made of copper or brass, is such that it may be used repeatedly if any provision is made for re-priming it with fulminate, as in sporting guns.

Systems. At the present time breech-loading arms may be divided into *old* and *new* systems. The old system is characterized by the use of the percussion-cap; the new system by the self-primed metallic case cartridge. About the only breech-loading arms of the old system now found in our service are Sharp's.

272. **Sharp's fire-arms.** In Sharp's fire-arms the chamber is fixed (see fig. 96) and the barrel closed by a sliding breech-piece (a), which moves nearly at right angles to the axis of the piece. Formerly there was no attempt to prevent the escape of gas through the joint, or opening between

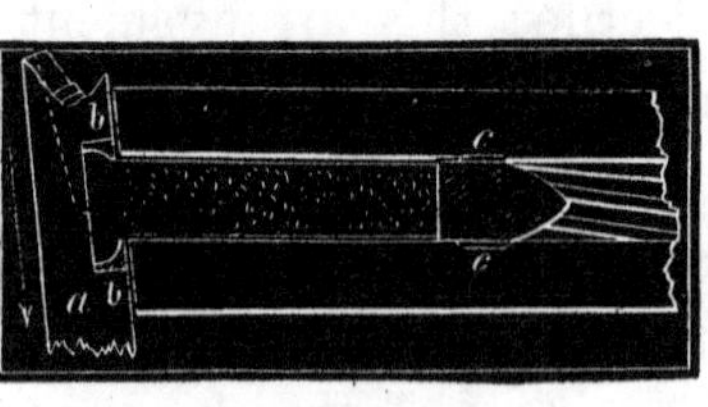

Fig. 96.

the slide and barrel, and after a few discharges in dry weather the movement of the slide became clogged with burnt powder. By boring a recess into the face of the slide, opposite to the chamber, and inserting a tightly-fitting steel ring (*b b*) into it, in such a manner that the inner rim is pressed against the end of the barrel by the force of the powder, the escape of gas is prevented. This fire-arm is loaded by depressing the lever, or trigger-guard, which withdraws the slide and opens the breech for the insertion of the cartridge. The cartridge is composed of a linen cloth cylinder to contain the powder, one end of which overlaps and is gummed to the base of the bullet; the other is closed with a layer of thin bank-note paper. The flame of the percussion-cap penetrates through this paper and ignites the powder. The linen case is carried out with the bullet, and drops to the ground a short distance in front of the piece.

273. **Metallic cartridge.** One of the most marked improvements in small-arms, developed by the late war in this country, is the use of the self-primed metallic case cartridge for breech-loading arms. This cartridge (fig. 97) is composed of a thin copper case (*a*), with a projecting rim (*b*). The case is formed by cutting a circular disk out of a sheet of No. 22 (wire gauge) copper, and punching it into the shape of a cup, or short cylinder closed at one end. To give this cup the proper length, its sides are drawn out by forcing it in succession, with punches, through four sets of dies, each operation making the sides longer and thinner than the preceding.

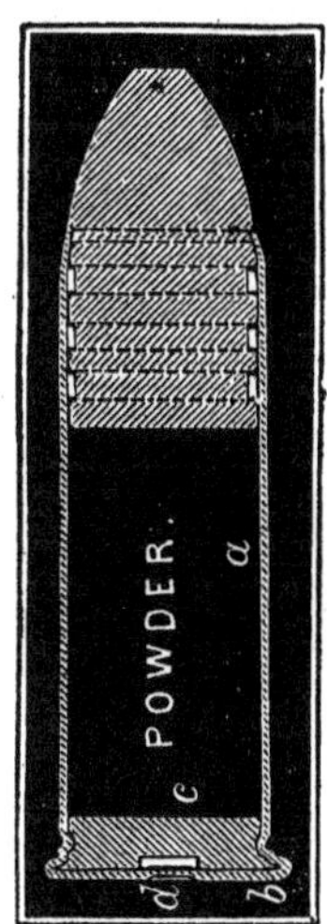

Fig. 97.

In these drawing operations the cases are moistened with soap-suds, and the copper is annealed one or more times (depending on its quality), to prevent cracking. The open end is trimmed in a lathe; and the metal at the head is crowded up so as to overlap the sides and form a projecting rim. This rim furnishes a hold for the extractor, which withdraws the empty case from the gun after firing. In some cartridges the fulminate is placed in this rim, and it is exploded by the firing-pin pressing the rim against the counterbore of the barrel; these are known as *side fire* cartridges. In the cartridge lately adopted for the altered Springfield rifle-musket, the fulminate is placed in the cavity of an iron bar, or *anvil* (*c*), and in contact with the centre of the head of the case, as shown at (*d*) in the figure. The anvil, which has to sustain the blow of the hammer communicated through the firing-pin, is held in place by crimping the sides of the case into the groove at each end. To prevent corrosion, the anvil is tinned. The case is tapered from the head to the open end .02 inch at the same time that it is crimped; this taper facilitates the withdrawal of the case from the chamber of the gun.

The lubricant, which is composed of two parts of *parafine* and one of *bayberry* tallow, is protected in the grooves of the bullet by the overlapping of the case, as shown in the figure. The figure represents the parts of the cartridge of the natural size. The bullet is solid and weighs 450 grs., and the powder 70 grs.

Advantages. The advantages of this kind of cartridge are :—1st. It is self-primed, and the operation of loading is simplified; 2d. The case is strong and not easily disfigured, and the powder and priming within it are protected from injury; 3d. The certainty of putting the proper amount of powder into the piece in loading; 4th. The axis of the bullet is made coincident with that of the bore when the piece is loaded, and an important cause of inaccuracy of fire is obviated; 5th. The cartridge case is a perfect gas-check, which prevents the machinery of the gun from being fouled and worn by the flame of the powder.

When the loss by wear and tear of the ordinary cartridge is taken into consideration, it is thought that it is little, if any, cheaper than those with metallic cases. Experience shows that the latter can be handled and transported with safety, and that the explosion of one or more will not extend to others in the same box. Such cartridges have been kept under water more than two months without injury; a portion of the time the water was frozen.

274. **Principles of breech-loading arms.** Breech-loading arms of the new system may be divided into *simple breech-loaders* and *repeaters*. The principal parts peculiar to the former are: 1st. The *movable breech-block*, by which the chamber is opened and closed; 2d. The *breech frame* upon which the breech-block is mounted and united to the barrel; 3d. The *chamber*, with its counterbored recess, to receive the rim of the cartridge; 4th. The *firing-pin*, which transmits the blow of the hammer to the priming of the cartridge; 5th. The *extractor*, by which the empty case is removed after firing.

In addition to these parts, the repeater has a magazine attached to it to contain a certain number of cartridges, which are successively brought into the chamber by peculiar mechanism.

The foregoing named parts may be said to be essential to all breech-loading arms in which the metallic cartridge is used; the different ways in which they are combined mark the different systems now in vogue. These combinations have reference chiefly to the modes of operating and locking the breech-block. The breech-block may be operated in two ways: 1st, by *rotation*, where it swings on a hinge; 2d, by *sliding*, where it moves in grooves. The former mode is generally to be preferred, as the rubbing surfaces are small, and the power is applied at the end of a lever; in the same way that it is easier to close a hinged door than a sliding one.

In the rotating breech-block, the position of the hinge has an important influence on the facility of operating the block, inserting the cartridge and withdrawing the empty shell, and the most suitable position is deemed to be in front of the centre of the block. In this case the motion of opening and closing the block is natural and easy, the cartridge is pushed into its place by the block, and a very simple retractor serves to withdraw the empty shell after firing. The Allin (Springfield altered), Berdan, Millbank, Lamson, Remington, Laidley, and others belong to this class. In the last two named the hinge is below, in the others it is above the axis of the block. In the Snider (Enfield altered), Warner, Maynard and others the hinge is on the side of the block; while in the Peabody, Roberts and others it is in rear. In the Prussian-Needle, the French Chassepot, Root, Meigs, Sharp and other guns the breech-block is made to slide. There is another system, however, in which the breech is opened by moving the barrel. This system is better adapted to sporting than to military guns.

The following are among the more important conditions to be fulfilled in constructing a breech-loading gun of the new system, viz.: 1st. The strength and union of the parts should be such as not only to resist repeated discharges, but the bursting of a cartridge case, which sometimes occurs from defective material or workmanship. 2d. The locking of the breech-block should not only be secure, but all the parts by which it is effected should work freely—without sticking. 3d. The parts should be so arranged that the hammer cannot strike the firing-pin until the breech-block is properly locked. 4th. The piece should not be carried loaded with the hammer resting on the firing-pin. 5th. Avoid, if possible, the necessity of bringing the hammer to the full cock in order to unlock the breech-block. 6th. The working parts should, as far as possible, be covered from dust and water. 7th. The extractor should be so arranged as to require no cuts or openings in that part of the chamber which surrounds the body of the cartridge-case.

275. **Altered Springfield musket.** There being a large number of serviceable Springfield rifle-muskets on hand, the War Department has decided to reduce the calibre of a portion of them and change them into breech-loaders for issue to the army. By these alterations, the range, accuracy and rapidity of fire are increased, and the weight of the ammunition diminished.

Reduction of calibre. The bore is reduced from its original size—.58 inch—to its present size—.50 inch—by reaming out the grooves and inserting a tube of wrought-iron. To give a proper bearing to the tube, the bore and exterior of the tube are slightly tapered from the breech to the muzzle. The tube is then introduced through the breech of the barrel, and driven with a hammer until the two are in close contact throughout. Both ends are then made solid by brazing under a blow-pipe, after which the barrel is re-proved, finish-bored, and rifled with three broad grooves with a twist of 42 inches, and a depth of .0075 inch.

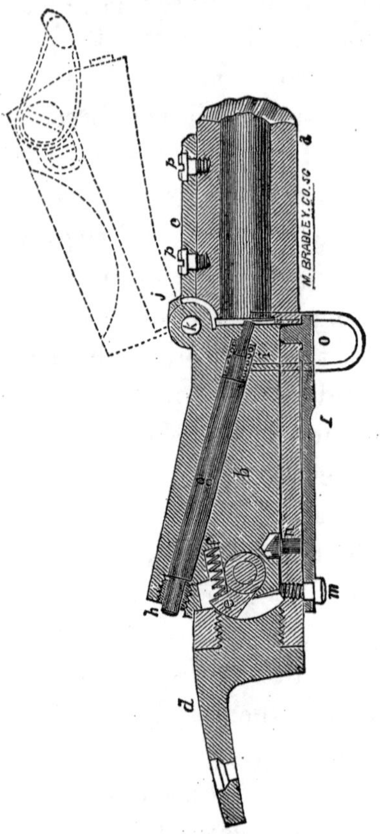

Fig. 98.

Breech-loading alteration. In general terms, the breech-loading alteration is made by cutting an opening into the top of the barrel for about 3 inches from the breech-screw, and inserting into this opening a breech-block, which is hinged at its forward end to the barrel. Its rear end is locked by a cam, which fits into a circular recess cut in the end of the breech-pin. Fig. 98 represents a vertical section of the musket thus altered.

(*a*) is a portion of the barrel around the *chamber;* (*b*) is the *breech-block;* (*c*) the *hinge-strap* to which this block is attached; (*d*) is the breech-pin with its *circular recess;* (*e*) is the *cam-latch* which fits into this recess, thus locking the breech-block in place; (*f*) is a *cam-latch spring*, to press the cam-latch into the recess; (*g*) is the *firing-pin* which transmits the blow of the hammer to the priming of the cartridge; (*i*) is the *firing-pin spring*, to press the pin back when the hammer is raised; (*j*) is a projection on the breech-block, called the *extractor hook*, to loosen the empty cartridge-case when the breech-block is thrown forward into the position of loading shown in the broken lines of the figure; (*o*) is the *ejector spring*, one end of which is fastened to the barrel and the other projects through the barrel and presses on the inner portion of the rim to eject the case from the gun; (*l*) is the *friction spring*, the point of which presses against the rim of the case and holds it against the effort of the ejector spring, until the hook starts it; this will not take place until the chamber has been uncovered by the breech-block; (*n*) is the *ejector stud* against which the case strikes and is deflected from the piece; (*m*) is the screw which fastens the ejector spring to the barrel; (*p p*) are screws which attach the hinge-strap to the barrel, and (*k*) is the pivot screw about which the breech-block turns. The cam-latch is raised and lowered by a thumb-piece attached to the *cam-latch shaft* which projects through the old cone-seat of the barrel. To the thumb-piece is attached a branch which covers the head of the firing-pin, and prevents an accidental discharge of the gun until the cam-latch is in place, and the breech-block is locked.

276. **Spencer repeating arms.** The magazine of the Spencer repeating fire-arm lies in the butt of the stock —fig. 98 (*a*)—and is capable of holding seven copper case cartridges. Weight of powder 45 grs., bullet 350 grs.

A *follower* (*a*) impelled by a spiral spring pushes the

line of cartridges towards the chamber of the barrel. When the chamber is closed, the point of the foremost

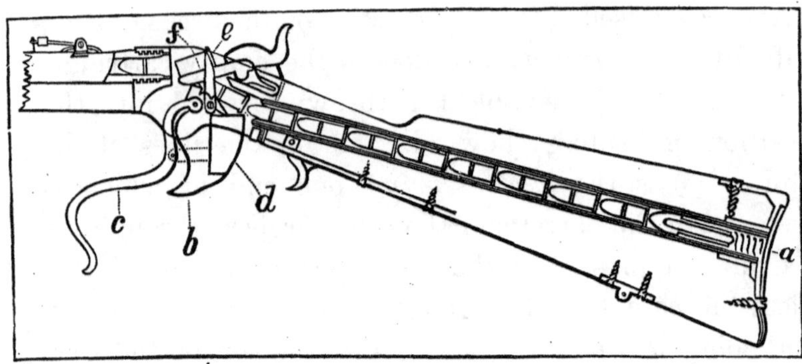

Fig. 98 (*a*).

cartridge rests against the *carrier-block* (*b*). When it is opened, which is done by depressing the *lever guard* (*c*), this cartridge is pushed forward into the position shown in the figure. By raising the lever guard the cartridge is carried around and pushed into the mouth of the chamber, which is firmly closed by the *breech-block* (*d*). The *extractor* (*e*) is a flat lever attached to the left side of the carrier-block, and withdraws the empty case by pressing against the under side of the rim. Another small lever, called the *guide* (*f*), falls into the space occupied by the carrier-block, and forms an inclined plane, up which the empty case moves to clear the piece. A *key* (not shown in the figure) has lately been introduced into this arm, by which the supply of cartridges can be cut off or let on at pleasure, and enables the soldier to reserve the cartridges in the magazine for an emergency. When the magazine is locked, the piece can be loaded directly from the cartridge-box, as a simple breech-loader. The operation of this key is simply to prevent the carrier-block from falling so far as to uncover the magazine, at the same time it falls far enough to uncover the chamber for the insertion of a cartridge by hand.

Other ingenious repeating arms have been devised, but none have been so generally used and approved as the Spencer arm.

277. **Advantages, &c.** The advantages of breech-loading over muzzle-loading arms are: 1st. Greater certainty and rapidity of fire; 2d. Greater security from accidents in loading; 3d. The impossibility of getting more than one cartridge into the piece at the same time; 4th. Greater facility of loading under all circumstances, and particularly when the soldier is mounted, or lying upon the ground, or firing from behind a cover; 5th. The greater security with which the charge is kept in place when the piece is carried on horseback with the muzzle down.

The results of the late wars in this country and Germany have led to the introduction of breech-loading small-arms for all branches of military service.

SMALL-ARM PROJECTILES.

278. **Forcing.** "Forcing," as applied to a projectile, is the operation by which it is made to take hold of the grooves of a rifled barrel, and follow them in its passage through the bore. It may be accomplished in various ways, most of which depend upon the soft and yielding

nature of lead, the material of which small-arm projectiles are made, viz.:

1st. By the action of the ramrod.

2d. By the action of the powder.

3d. By the action of ramrod and powder combined.

4th. By the form of the bore or projectile, as in breech-loading arms, &c.

279. **By the action of the ramrod.** When rifles were first made, forcing was effected by making the projectile a little larger than the bore, and driving it down with a mallet applied to the point of the ramrod; although this caused the lead to fill the grooves completely, converting the projectile into a screw, whereof the barrel was the nut; the operation was slow and laborious, and the accuracy of the projectile was impaired by the consequent disfiguration.

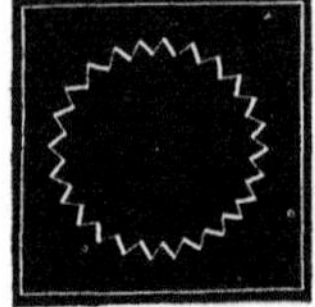

Fig. 99.

The form of the grooves then used is shown in fig. 99. They were liable to be injured by the ramrod, and were difficult to clean.

280. **Patch.** The foregoing plan was improved by making a projectile a little smaller than the bore, and wrapping it with a patch of cloth, greased, to diminish friction in loading. The thickness of the cloth was greater than the windage; this caused the patch to press upon the projectile with so much force as to compel it to follow the winding of the grooves without materially altering its shape. The patch is still used in sporting rifles, and gives excellent results; but the loading is too slow and difficult for a military arm.

281. **Delvigne's plan.** M. Delvigne, an officer of the French infantry, appears to have been the first person

who overcame the difficulty of loading rifles, thereby removing the principal obstacle to their introduction into the military service. The plan proposed by him, in 1827, was to make the projectile small enough to enter the bore easily, and to attach it to a *sabot*, or block of wood (*a*, fig. 100), which, when in position, rested upon the the shoulders of a cylindrical chamber (*b*), formed at the bottom of the bore, to contain the powder. In this position, the projectile was struck two or three times with the ramrod, which expanded the lead into the grooves of the barrel. To the bottom of the sabot was attached a piece of greased serge, which served to soften the residuum of the powder and facilitate the loading. By this plan the accuracy of the round projectile was increased, but its range was diminished.

Fig. 100.

Elongated projectile. In 1742, Robins pointed out the superiority of the oval, or elongated form of projectile, and since this many attempts have been made to employ it in rifled arms, especially in this country, but it remained for M. Delvigne, followed by MM. Thouvenin and Minié, of the French service, to apply it successfully to the military service.

The form of projectile proposed by these officers was composed of a cylinder and conoid. The cylinder served as the base of the projectile, and gave it stability in the bore of the piece; the conoidal surface, which formed the point, was well adapted to diminish the effect of the air, by increasing the penetrating power of the projectile. A single groove was formed around the cylinder, to contain a greased woollen thread, in place of the woollen patch of Delvigne.

It was shown by the trials which followed, that the presence of this groove improved the accuracy of the projectile—a fact which gave a new turn to the investigations, and led to the adoption of two additional grooves. The theory advanced in explanation of the action of these grooves was, that they oppose a resistance to the air, which, acting on the rear portion of the projectile, tends to keep the point foremost in flight, thereby rendering the resistance of the air uniform, and at the same time a minimum.

The correctness of this theory may be well questioned; but that the grooves exert a beneficial effect, by diminishing adhesion to the surface of the bore, and by facilitating expansion, can scarcely admit of a doubt.

282. **Tige, or spindle.** Colonel Thouvenin proposed to replace the chamber of Delvigne by a spindle of iron, screwed into the centre of the breech-screw (see *a*, fig. 101). This was found to be an excellent point of support for the base of the elongated bullet when forced by the blows of the ramrod. The expansion of the lead into the grooves secured the bullet in place, and protected the powder from moisture.

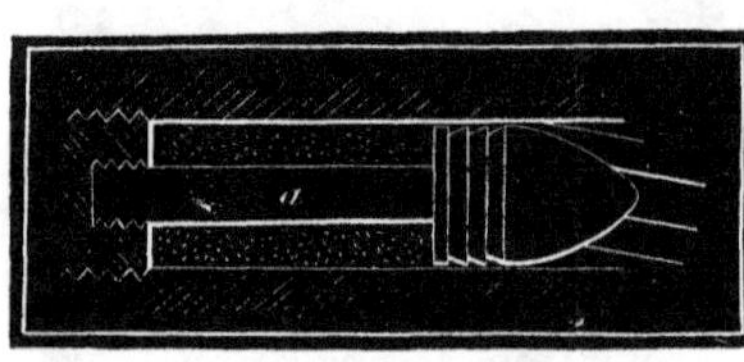

Fig. 101.

Considerable difficulty, however, was experienced in cleaning the space around the spindle; and, like all plans of forcing by the ramrod, it is subject to variation, arising from the particular care and strength exercised by the soldier.

283. **By form of projectile.** This method of forcing is illustrated in the Whitworth rifle. The form of the bore, as in the cannon, is a twisted hexagonal prism, making a complete turn in 20 inches. The projectile (fig. 102) is made nearly of the exact form and size of the bore, and is about three diameters in length. To prevent disfiguration and *stripping*,* which are very liable to occur in bullets of this length, fired with high velocities, the lead is hardened by alloying it with tin and manganese; and to obviate fouling, a greased wad is placed between the powder and bullet. As might be expected from the length of the bullet, the amount of twist, and the extreme accuracy with which the bullet fits the bore, the results obtained with this arm are much superior to those obtained with service-arms.

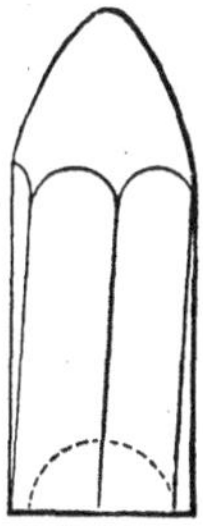

Fig. 102.

284. **By the powder.** It appears that the first attempt to force a projectile by the action of the powder was made by Mr. Greener, an English gunsmith, in 1836. The plan which he tried consisted in forming a cavity at the base of an oblong bullet, and partially inserting in it a conical pewter wedge, which was driven in by the force of the powder in such manner as to expand the outer part of the bullet into the grooves of the barrel.

Fig. 103.

Some years after this, Colonel Minié produced a projectile constructed on the same principle, but instead of a solid wedge, he used a cup of sheet iron, which was inserted into a conical cavity (fig. 103) at the base

* Stripping is the tearing away of the metal when the projectile passes out of the bore without following the grooves.

of the bullet. The point of the ball was cut off to prevent disfiguration by the flat head of the ramrod. This projectile, when fired from a rifle of service calibre, generally possessed great range and accuracy; but it had certain defects which prevented it from being extensively used in military service, viz.: it was compound in its structure; the cup was sometimes forced in obliquely, producing unequal expansion; and, from the large size of the cavity, the top was occasionally blown off, leaving the cylindrical portion adhering to the sides of the bore.

285. **Present methods.** Not long after the introduction of the Minié bullet, it was discovered that, by giving a suitable size and shape to the cavity, the wedge could be dispensed with. The projectile thus obtained was simple in its structure, and gave better and more reliable results than the one from which it was derived.

The particular form and mode of expanding bullets, varies in most military services; in general terms, however, all modern small-arm projectiles are *cylindro-conoidal* in shape, and a majority of them are forced by the action of the powder. The effect of the powder may be direct, as in the case where it acts in the cavity of a bullet; or it may be indirect, as when it compresses the bullet lengthwise, or, technically, "*upsets*" it.

286. **United States.** The bullet used in the United States service, is derived from that of the *carabine à tige*, chiefly, by making a conical cavity in its base. (See fig. 104.) The shape of the first cavity employed, was that of a frustum of a cone; but this was found defective when used in the rifle-musket, inas-

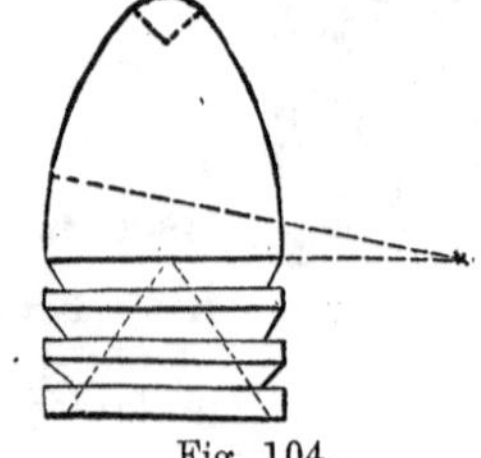

Fig. 104.

much as it rendered the bullet too weak at the juncture of the two exterior surfaces. For arms with reduced charges of powder, as in the carbine and pistol, the large cavity is most suitable.

A description of the musket-bullet has been given in chapter II. A distinguishing feature of this bullet is, that no patch of any kind is used in loading; in nearly all other modern bullets a greased patch of cloth, or paper, envelops them when placed in the bore.

287. **England.** The British bullet (sometimes known as the Pritchett bullet) has a perfectly smooth exterior. (fig. 105.) A conical plug of box-wood is inserted into the opening of the cavity, it is said, more for the purpose of preserving the form of the bullet in transportation than aiding in the expansion. The diameter and weight of this bullet are nearly the same as in the United States bullet.

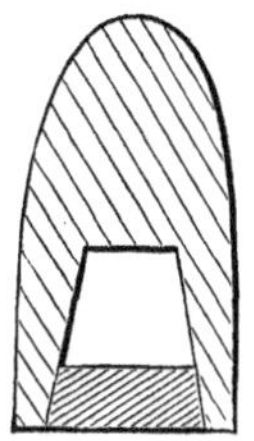

Fig. 105.

288. **France.** Two distinct bullets are employed in the French army. The first is shown in fig. 101; it is heavy, and is intended to have great force and accuracy at long distances. It is used by troops armed with the *carabine à tige*, as the Chasseurs and Zouaves. The second bullet is shown in fig. 106; it is light, and without much accuracy, describes a flattened trajectory, which increases the chances of hitting a line of men at the usual fighting distance. This bullet is used by troops of the line, who are not supposed to be skilful marksmen.

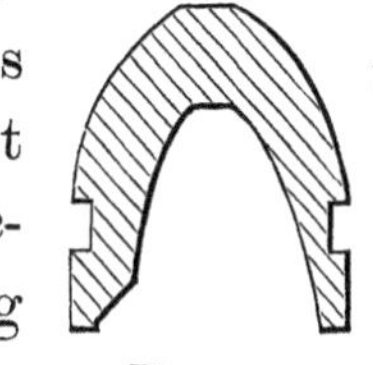

Fig. 106.

289. **Austria.** The Austrian bullet belongs to the class of solid expanding projectiles. In this particular

case, expansion is effected by the crowding up of the disks, formed by cutting two deep grooves around the cylinder. (Fig. 107.) A portion of the Austrian rifles (those carried by the non-commissioned officers, and men of the third rank, who act as skirmishers) have a spindle attached to the breech-screw; the object of which is, not to aid in expanding the bullet, but to give it an invariable position with reference to the powder, and thereby secure uniformity of action.

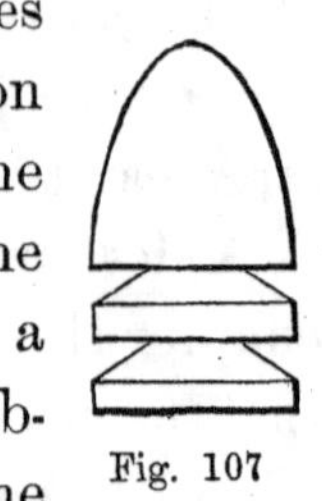

Fig. 107

290. **Switzerland.** Fig. 108 shows the form of the bullet used in the Swiss service. It is solid, and is forced by a cloth patch tied around the grooves. The position of the bullet with reference to the powder is constant; this is determined by a notch on the ramrod—the notch being so arranged as to leave an interval between the powder and the bullet.

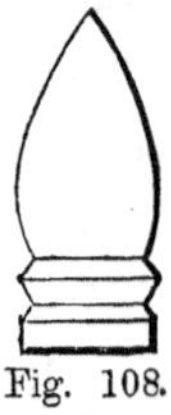

Fig. 108.

The diameter of this bullet is much less than that of any other service; and, in consequence of its lightness, it is fired with a larger proportional charge of powder. Within the usual range of small arms, it is said to have a flatter trajectory, and greater accuracy, than any other small-arm projectile; but at extreme ranges it loses its velocity very rapidly.

CHARGE OF POWDER.

291. **Conditions.** The proper charge of powder, for a small-arm, depends on the calibre, windage, length of barrel, weight of the piece, and character of the projectile. The charge of the old smooth-bored musket was

not far from one-third the weight of the projectile; this was necessary to make up for the loss of force by great windage, and to give the round bullet the necessary momentum. When the elongated bullet was introduced, it became necessary to reduce the charge to prevent too severe recoil; besides, the mass of the bullet being increased, a diminished velocity sufficed to produce the same effect.

In the case of expanding bullets, too small a charge will be insufficient to force the lead into the grooves of the barrel; at the same time, it is shown by experience that, if the charge be increased beyond a certain point, the bullet is liable to be disfigured by *upsetting*, and its accuracy is diminished. *The proper charge for elongated expanding bullets varies from one-tenth to one-seventh the weight of the projectile.*

LUBRICANT.

292. **Necessity.** After a fire-arm has been discharged several times, the residuum of the burnt powder collects on the surface of the bore, forming a hard substance which *seriously* obstructs loading; and unless the windage be very great, it becomes necessary to wipe out the bore, or apply some lubricating substance to the projectile.

293. **Qualities.** A proper lubricating substance for small-arms should be unaffected by changes of climate; *i. e.*, it should not be melted by hot, nor rendered too hard by cold, weather; it should not corrode the projectile, nor weaken the paper of the cartridge, even when

kept in store (as all ammunition is liable to be) for a considerable length of time.

294. **Methods.** The insertion of a few drops of water, or oil, in the bore, has been tried with some success, but the most common lubricating substance is *beeswax*, or *beeswax* and *tallow*, applied to the projectile, or its patch. Beeswax answers well in a hot climate, and, if it be free from acid, does not act on the bullet, nor the patch; tallow alone lubricates the bore well in all climates, but it corrodes the lead of the projectile, and, in the course of time, dries away.

The proportion used in the United States service is *eight* of beeswax to *one* of tallow, applied by dipping the bullet into the melted substance, and immediately withdrawing to cool. The bullet should be previously warmed, to prevent the substance from peeling off by too rapid cooling. The rifle-musket can be thus fired 200 times, at least, without inconvenience.

DIFFERENT KINDS OF SMALL-ARMS.

The small-arms of the United States service are the *rifle-musket*, *rifle*, *carbine*, and *pistol*.

295. **Rifle-musket.** The present rifle-musket was adopted in 1855, with a view of combining in one piece the range and accuracy of the rifle with the advantages of the musket, as regards lightness, quickness of loading, and facility of handling as a pike. It is, therefore, the appropriate arm of troops acting on foot, and in line.

Length of barrel, ...	40.00 in.
Length of arm with bayonet, . .	74.00 in.
Weight of barrel, ...	4.25 lbs.

Weight of arm complete, . .	9.90 lbs.
Weight of projectile, . . .	550.00 grs.
Weight of powder, . . .	60.00 grs.
Initial velocity, ...	960.00 feet.

The cadet-musket only differs from the foregoing in the length of its barrel and bayonet—the former being 38 in., and the latter 16 in. It would make a suitable arm for light troops.

296. **Rifle.** The rifle differs from the rifle-musket in having a shorter and stouter barrel, a sword-bayonet, in the mountings, which are made of brass instead of iron, and in having its barrel browned.

Length of barrel, . . .	33.00 in.
Length of arm with bayonet, . .	72.00 in.
Weight of barrel, . . .	4.80 lbs.
Weight of arm complete, . .	13.00 lbs.
Charge (projectile and powder), same as in rifle-musket.	
Initial velocity, ...	910.00 feet.

297. **Carbine.** The term carbine is applied to an arm used by mounted troops, and intermediate in weight and length between the rifle and pistol. Both breech and muzzle loading carbines are employed, but the former are generally preferred. The ramrod of the muzzle-loading carbine is attached to the barrel by a swivel, which permits it to be handled freely, but at the same time prevents it from falling to the ground. The carbine is secured to the person of the soldier by a sling, which hooks on to a ring, moving on a swivel-bar attached to the left side of the carbine, thereby affording a play to the piece in loading and firing.

Length of barrel, ...	21.00 in.
Weight of piece, ...	7.50 lbs.
Weight of projectile, ...	450.00 grs.
Weight of powder, ...	55.00 grs.
Initial velocity, ...	820.00 feet.

298. **Pistol-carbine.** The pistol-carbine is a muzzle-loading pistol, with a false butt, which permits it to be used either as a pistol or carbine. It is particularly suited to the service of light artillery.

Length of barrel, ...	12.00 in.
Weight complete, ...	5.00 lbs.
Weight of projectile, ...	450.00 grs.
Weight of powder, ...	40.00 grs.
Initial velocity, ...	603.00 feet.

299. **Colt's pistol.** Colt's pistol is constructed on the revolving principle, and is composed of a *cylinder* (containing six charges), a *rifled barrel*, and a *handle* or stock. By cocking the hammer, the cylinder is made to rotate around a spindle in such a way that a new charge is presented to the breech of the barrel every time the piece is cocked. The principal defects of revolving pistols are, that more than one charge is liable to go off at a time; that the fragments of the cap are liable to clog the cylinder; and that there is an escape of gas through the opening in front of the cylinder.

The advantage is rapidity of fire for six discharges. Colt's pistol is considered a very reliable weapon, particularly in partisan warfare.

Length of bore (navy), ...	9.00 in.
Weight of do.	2.40 lbs.
Weight of projectile, . . .	125.00 grs.

Weight of powder, . . .	14.00 grs.
Initial velocity, . . .	760.00 feet.

300. **Sporting rifle.** American sporting rifles have long enjoyed a reputation for extreme accuracy of fire. This has been attained by introducing into their construction many refinements which, though ingenious and effective, are incompatible with the strength, safety, and rapidity of fire of a military arm. To give stiffness and steadiness to the barrel, it is made very heavy in proportion to the charge; to prevent the bullet from being disfigured by a heavy proportional charge of powder, the calibre is made as small as the range will permit; to render friction in the bore uniform, the surface is carefully wiped after each discharge; to prevent disfiguring the corners of the muzzle, the bullet is inserted into the bore through a *false muzzle;* to *centre* the bullet properly in the bore, it is started with an instrument called the *straight starter;* and, finally, the piece is aimed with a globe, or telescopic sight, and fired with a hair-trigger.

The dimensions, &c., of a James's rifle, of this class, belonging to the museum of the Academy, are as follows, viz.:

Length of barrel, ...	32.50 in.
Weight of do.	16.50 lbs.
Calibre, ...	00.45 in.
Weight of bullet, . . .	217.00 grs.
Weight of powder, ...	100.00 grs.
Initial velocity, ...	1,900.00 feet.

MANUFACTURE OF SMALL–ARMS.

301. **Where made.** With the exception of swords and patent-arms, all small-arms for the United States army and militia are made at the national armories, situated at Springfield, Mass., and Harper's Ferry,* Va. These armories are under the general charge of the chief of ordnance, who, by the authority of the war department, furnishes the models, and prescribes the kind and quantity of work to be done; the operations are conducted by civilians.

302. **How made.** A principal requisite, in the manufacture of small-arms, is, that similar parts of the same kind of arm, or *model*, shall be capable of interchange. This demands a higher degree of accuracy in the workmanship than can be attained by hand-labor, without great cost, and the consequence is, that machinery is now very generally employed in this branch of manufacture.

303. **Operations.** The principal operations of manufacturing arms are *welding*, *swaging*, *boring*, *turning*, *drilling*, *tapping*, *milling*, *cutting and filing*, *grinding*, *case-hardening*, *tempering*, and *polishing*. Welding and swaging are performed by blacksmiths; the other operations, by armorers or finishers.

304. **Welding.** Welding is the process of uniting certain metals by means of heat and pressure. To bring the heated substances into perfect contact, the joining surfaces should be freed from the coating of

* Since the first edition of this work was published, the Harper's Ferry armory has been destroyed, and is no longer used for government purposes.

oxide which generally covers them; and this is done by applying a composition of *ten* parts of borax to *one* of sal-ammoniac.

The most important welds in the musket are those of the barrel, and the blade and socket of the bayonet. The first was formerly done under the trip-hammer; it is now better and more economically performed by rollers.

305. **Rolling barrels.** The material from which a musket-barrel is made is a flat bar of wrought iron, 14 inches long, $5\frac{3}{8}$ inches wide, and $\frac{9}{16}$ inches thick; the edges are bevelled so that they will make a perfect lap-joint when united as a tube. The several processes of welding are *curving*, *welding*, and *straightening*.

Curving. The *plate* is heated in a reverberatory furnace, to a red heat, and then passed between the grooves of the curving-rolls (fig. 109), to bring the bevelled edges in contact. There are *five* grooves, *two* being open grooves, and *three* have tongues upon the upper roll to bend the plate down into the lower groove. The grooves also differ in size. The first one gives the plate the shape of a trough; the second and third gradually contract it, without changing its form; the fourth and fifth are parallel grooves, which bring the edges of the plate in contact. The object of so many grooves is, to bend the plate gradually, and prevent it from being split open, in case the iron is brittle. In this way, 450 plates can be bent by one set of rolls in a day.

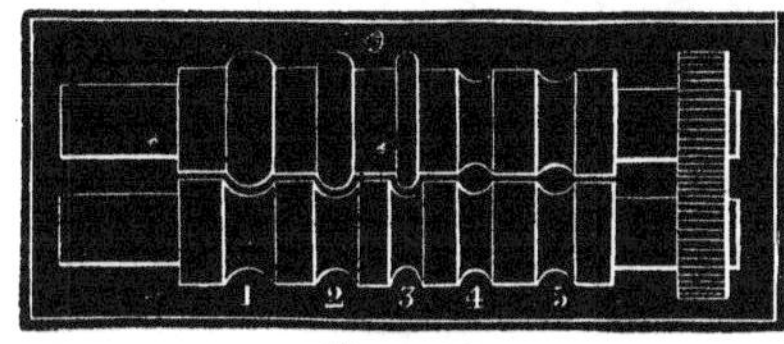

Fig. 109.

Welding. The plates thus bent, or "cylinders" are

replaced in the furnace to prepare them for the welding-rolls. The workmen are supplied with eight steel *mandrels*, or rods terminated at the point with an egg-shaped bulb; the bulbs vary from .71 in. to .46 in. in diameter. When one of the cylinders is brought to a white, or welding heat, a workman thrusts the largest mandrel through it, whilst it is in the furnace. He then carries it to the rolls (only one of which is shown in fig. 110), and placing the mandrel through the frame, he introduces the end of the cylinder into the first groove; the action of the rolls is to slip the cylinder over the mandrel, the centre of the bulb being placed and held in the plane of the axis of the two rolls. The cylinder is then straightened by striking it on a flat iron table, and placed in the furnace to be reheated. The second size mandrel is then inserted, and the cylinder is passed through the second groove in the rolls, which is smaller than the first, and the welding is completed.

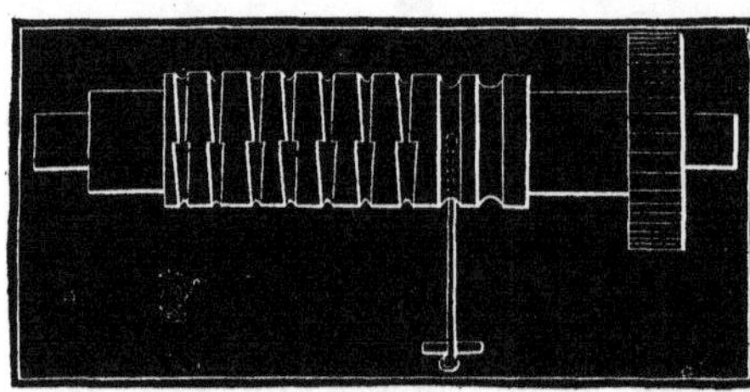

Fig. 110.

The object of the remaining grooves is, to give the proper form, or taper, to the cylinder, and for this purpose they are made of the same shape as the required barrel. As each groove makes a single circuit of the rolls, and as the rolls are continually in motion, it requires some dexterity on the part of the workman to insert the end of the cylinder at the right moment.

In this way the cylinder is passed, breech foremost, through five of the taper grooves; it is then passed twice through the last groove, without a mandrel, to make it smooth.

In passing through each taper groove, the barrel is reheated (to a red heat), as the bulb of the mandrel chills the interior surface.

Straightening. The welding process being completed, a workman places the barrel in the straightening machine, which is composed of two dies, each of the length and shape of the half-barrel, and which close upon each other as the workman turns the barrel around its axis with a pair of tongs.

In this way about 75 barrels can be finished by one set of workmen in a day.

306. **Swaging.** Swaging is the operation by which the rough iron or steel is converted into a piece of suitable size and shape for the finisher. It is done by forcing the piece of heated metal into a die by means of a heavy drop-weight; the machine is called a *drop-hammer.*

307. **Boring.** Boring is the operation of forming the bore of the barrel. The manner of performing it, and the character of the tools used, depend on the metal employed. If it be steel, the piece to be bored is formed into a solid bar, of homogeneous texture; if of wrought iron, it is formed into a tube, some portions of which are liable to be harder and more difficult to cut than others. In the first case, a stationary drill is driven through the piece which revolves; in the second case, the boring instrument revolves rapidly, and, at the same time, is drawn through the hole left by the welding mandrel.

308. **Drilling and tapping.** The object of *drilling* is to form holes for the screws, rivets, &c.; and that of

tapping, is to convert the surface of the hole into a female screw. The former operation is performed by the *drill-press;* the latter by an instrument called a *tap*, which is made of a piece of steel, of a pyramidal form, and on the edges of which are segments of screw threads. In all operations of cutting and drilling wrought iron, it is necessary to use oil or water to preserve the temper of the tools. In working cast iron, no cooling substance is required.

309. **Turning.** The object of *turning* is to give shape and smoothness to the exterior of a body, and is accomplished in a machine called a *lathe.* The body is generally made to revolve around a fixed axis, and a cutter, which has a motion parallel to this axis, is made to press against its surface; the combination of these two motions cuts away a spiral chip, and leaves a new surface concentric with the axis. It will be easily seen that if the cutter has, in addition to its motion parallel to the axis of rotation, another perpendicular to it, that the resulting figure will be no longer round, but irregular.

This constitutes the principle of *eccentric turning*, and affords the means of turning an almost infinite variety of shapes, simply by regulating the motion of the cutter by a pattern, or model of hardened steel. In this way gun-stocks, and other irregular figures, are formed by machinery; the principle has even been used in copying statuary.

310. **Milling.** Pieces of metal which are not suited to the turning-lathe, may be reduced to their proper shape by milling, an operation adapted to nearly all surfaces which have right-line elements.

It is performed by a revolving cutter, armed with

saw-teeth, while the piece to be cut is fastened on a carriage, which moves steadily under the cutter, and along a plane director.

The shape of the cut surface depends on the shape of the profile of the cutter; for instance, to dress the sides of a lock-plate, a cylindrical cutter would be used; to trim the edges, a curved one. By combining different shaped cutters on the same arbor, or shaft, a great variety of surfaces can be formed.

311. **Cutting and filing.** Cutting and filing are done by the hand—the former with a cold-chisel, and the latter by a file. They are employed to finish such parts as are not well adapted to machinery. To guide the workman in giving the proper form, the piece is placed in a hardened steel frame, called a *jeg.*

312. **Grinding and polishing.** Grinding is done with rapidly-revolving grindstones, and is principally confined to finishing the bayonet, and exterior of the barrel. Polishing the surface of finished parts is done with emery-wheels, which revolve with great rapidity. The wheels are made of wood, and the circumference is covered with buff leather, to which is glued a coating of emery.

313. **Case-hardening.** Case-hardening is the conversion of the surface of wrought iron into steel, to enable it to receive a polish, or bear friction. The process consists in heating the iron to a cherry red, in a close vessel, in contact with carbonaceous matter, and then plunging it into cold water. Old shoes are generally employed for this purpose at the armories, although bones, hoofs, soot, &c., will answer. The materials should be first burnt, and then pulverized.

314. **Hardening and tempering steel.** Hardening is effected by heating the steel to a cherry red, or until the scales of oxide are loosened on its surface, and plunging it into a liquid, as water, oil, &c., or placing it in contact with some cooling solid; the degree of hardness depends on the heat, and the rapidity of cooling. Steel is thus rendered so hard as to resist the hardest file; and it becomes at the same time extremely brittle.

Tempering. In its hardest state steel is too brittle for most purposes; the requisite strength and elasticity are obtained by tempering, which is done by heating the hardened steel to a certain degree, and plunging it into cold water.

The requisite heat is usually ascertained by the color which the surface of the steel presents, due to the film of oxide formed on it:

At 450° Fahr., a pale straw color.	Suitable for hard instruments, as the faces of hammers, &c.
At 600° Fahr., a greyish blue.	Gives a spring temper, or one that will bend before breaking: suitable for saws, sword-blades, &c.

Shades of colors between these extremes, give intermediate degrees of hardness. If steel be heated above 600°, the effect of the hardening process is destroyed. The parts of small arms are tempered by dipping them in oil, then heating them until the oil is burned off, when they are again plunged into cold water.

315. **Blueing.** A blue color may be given to the surface of iron and steel parts, by subjecting them to a

certain degree of heat. As soon as the proper shade of blue makes its appearance, the piece is removed and allowed to cool, when the color becomes fast.

316. **Browning.** Browning is the coating given to a gun-barrel to protect it from the action of the atmosphere, and to prevent the surface from reflecting the sunlight.

The process consists in forming a coat of rust, with a mixture of such materials as *spirits of wine*, *blue vitriol*, *tincture of steel*, *nitric acid*, *&c.* (see Ord. Manl.), on the clean surface of the barrel, and then rubbing it well with a *steel scratch-card* until it has a metallic lustre. This operation is repeated about a dozen times, until the coating has a deep brown color. The barrel is then washed with boiling water, to dissolve away any of the corroding mixture that may remain, and, when cold, is covered with sperm oil.

When the browning has been worn away in places, it may be entirely removed—first, by boiling in lime-water, to remove the varnish or grease, and then soaking in vinegar, which loosens the browning so that it can be wiped away with a rag.

INSPECTION OF SMALL-ARMS.

317. **Object.** The objects of inspecting small-arms are, to verify the dimensions, the workmanship, and the quality of the materials of the various parts.

Inspections at the armories are made by the foremen of the several departments of work, under the direction of the master armorer. To secure uniformity in all service-arms, comparative inspections are occasionally made of the work from the different armories; the parts of one set are required to interchange freely with those of another. Partial inspections are made at the different stages of manufacture, to prevent unnecessary labor from being expended on defective pieces. Contract-arms are inspected by an officer of the ordnance department, and by sworn assistants taken from one of the armories.

The following regulations for the inspection of finished arms, the care and preservation of arms in service, &c., are taken from the Ordnance Manual.

318. **Finished arm.** The inspector will examine the finished arm on every side, to see that the parts are well fitted together; he will also verify the principal dimensions and forms, by means of appropriate gauges and patterns.

319. **Barrel.** The diameter of the bore should be verified with the standard and limit gauges. The standard gauge is a cylinder of the diameter of the bore (.58 in.), and the limit gauge is .0025 inch greater. The former should pass freely through the bore, and the latter should not enter it. The barrel should enter the groove of the stock, one-half of its diameter, and should bear uniformly throughout, particularly at the breech. The *vent* should be accurate in its dimension, position, and direction, and a wire should be passed through it, to see that it is free. The *cone* should be sound. The shoulders of the *breech-screw* should fit close to the end of the barrel, and it should be free from cracks or flaws about the tang-screw hole. The *straightness* of the barrel may be ascertained by turning out the breech-

screw, and holding the barrel up to the light, and reflecting the image of a straight-edge from the surface of the bore. If the barrel be straight, the reflected image will be straight in all positions of the barrel. The bore should be free from hammer-marks, ring-bores, cinder-holes, flaws, cracks, &c., and the bayonet-stud and sight notches, should not be cut too deep.

320. **Ramrod.** The temper of the ramrod may be tested by springing it in four directions, with the point resting on the floor. When the musket-rod is bent six inches out of line, it should spring back perfectly straight without setting. Its soundness may be tested by striking it with a piece of metal, or by bending it over the edge of a block of wood; in the first case the sound emitted should be clear, and in the second case the flaws or cracks will be opened. The screw on the point of the rod should be properly cut; it should bear properly in its groove, neither too light, nor too loose. The point should rest on the stop.

321. **Bayonet.** The form and dimensions of the bayonet are verified with the proper gauges; the temper is tried by resting the point against the floor, and springing the blade smartly in four directions—toward the back, face, and two edges—grasping the butt of the stock with the right hand, and the middle of the barrel with the left. After this, inspect for cracks and flaws. To test the welding of the blade to the socket, strike the elbow smartly on the work-bench.

322. **Stocks.** The wood should be straight-grained, well-seasoned, and free from sap and worm-holes. The effect of unseasoned wood will be to rust the lock and barrel. It may be detected by the odor of a fresh cut,

or by the crumbling of a chip when pressed in the fingers. The edges should be sharp and clear, and free from splits. The dimensions, which concern the fitting of the parts, should be carefully verified.

323. **Lock.** All parts of the lock should be sound, well filed, and of proper form and dimensions. The temper of the hardened parts should be tried with a fine-cut file. See that the main and sear springs have the requisite power.

Examine carefully the action of the lock; see that the movable parts are *free*, *i.e.*, do not rub against other parts when in motion. Snap on the cone, and see that it fits its seat properly. Let the hammer down several times, to judge of the working of the parts. See that the interior parts are not wood-bound; that it does not go off at half-cock when the trigger is pulled hard, and that it goes neither too hard nor too easy when cocked.

324. **Mountings.** The trigger should work freely, but should have no lateral motion in the guard-plate. The guard-plate should not be screwed up too hard, lest the trigger be brought too close to the sear. The bands should be close to the stock, but not so tight as to require much force to move them. The band-springs should spring back freely when pressed down.

The sights are aligned by the flats of the barrel, which are equidistant from the axis of the bore, by construction. The alignment can only be verified by firing at a target.

PACKING AND STORAGE OF ARMS.

325. **Boxes.** Packing-boxes for muskets are made of

well-seasoned pine boards. Each box contains twenty muskets, in two rows of ten each. The pieces are kept from jostling and injuring each other by grooved clamps. The bayonets are unfixed, and placed securely on the bottom of the box, and the appendages are placed in a small apartment at the end.

When the regular packing-box cannot be had, arms may be packed in boxes with straw that is dry and free from dust, by forming it into a rope, and wrapping it around them; hay will not answer. They are then placed in rows, the lower row resting on three cushions of straw placed on the bottom. The butts are kept apart by wedges of straw; and the top row is covered with straw, pressed in by the cover, which is fastened by two hoops.

326. **Storage.** Arms are kept at the arsenals either in the boxes in which they are received from the armories, or in racks.

Each kind is kept separate, and arranged according to model, the place and year of construction, and the time when they were last cleaned.

New arms are kept distinct from those which have been repaired. Arms of peculiar kinds, arms to be repaired, and unserviceable or condemned arms, are kept separate.

Limbs and spare parts, intended for repairs of arms, should be kept in store by themselves, in a dry place, classed according to the kind of arms, and to the model and year of fabrication, and labelled accordingly.

All arms in store should be frequently examined, to see that they are not rusty. Those which are rusty should be immediately cleaned and oiled with *sperm oil.* If browned arms are affected with specks of rust, they should be rubbed with linseed oil; and if the acid be not neutralized, proper authority should be obtained to remove and renew the browning. Empty packing-boxes, from which the arms in racks are taken, should be kept, with the necessary parts, in the attics, or other dry situations. Storehouses for arms should be aired in clear, dry weather.

PRESERVATION AND CARE OF ARMS IN SERVICE.

327. **Instruction.** The officers, non-commissioned officers, and soldiers, should be instructed and practised in the nomenclature of the arms, and the manner of dismounting and mounting them, and the precautions and care required for their preservation.

Each soldier should have a *screw-driver* and a *wiper*, and each non-commissioned officer a *wire tumbler-punch* and a *spring-vice.* No other implements should be used in taking arms apart, or in setting them up.

In the inspection of arms, officers should attend to the qualities essential to service, rather than to a bright polish on the exterior. Arms should be inspected in the quarters, at least once a month, with the barrel and lock separated from the stock.

328. **Dismounting by a soldier.** The rifle-musket should be dismounted in the following order, viz.:—1st. Unfix the bayonet; 2d. Insert the tompion; 3d. Draw the ramrod; 4th. Turn out the tang-screw; 5th. Take off the lock; to do this, put the hammer at half-cock, and partially unscrew the side-screws, then, with a slight

tap on the head of each screw with a wooden instrument, loosen the lock from its bed in the stock; turn out the side-screws, and remove the lock with the left hand; 6th. Remove the side-screws without disturbing the washers; 7th. Take off the bands in order, commencing with the uppermost; 8th. Take out the barrel. In doing this, turn the musket horizontally, with the barrel downward, holding it loosely, with the left hand below the rear sight, and the right hand grasping the stock by the handle; tap the muzzle on the ground, if necessary, to loosen the breech. If an attempt were made to pull the barrel out by the muzzle, it would, in case it were wood-bound, be liable to split at the head of the stock.

The foregoing parts of the rifle-musket are all that should usually be taken off, or dismounted by the soldier. The breech-screw should be taken out only by an armorer, and never in ordinary cleaning. The mountings, cone, and cone-seat screw, should not be taken off, nor should the lock be taken apart, except by permission of an officer.

329. **To clean the barrel.** 1st. Stop the vent with a peg of soft wood, or piece of rag or soft leather pressed down by the hammer; pour a gill of water (warm, if it can be had) into the muzzle; let it stand a short time, to soften the deposit of powder; put a plug of soft wood into the muzzle, and shake the water up and down the barrel; pour it out, and repeat the washing until the water comes out clear; remove the peg from the cone, and stand the barrel, muzzle downward, to drain, for a few moments.

2d. Screw the wiper on the end of the ramrod, and

put a piece of *dry cloth*, or *tow*, round it, sufficient to prevent it from chafing the grooves of the barrel; wipe the barrel dry, changing the cloth two or three times.

3d. Put no oil into the vent, as it will clog the passage, and cause the first primer to miss fire; but, with a slightly oiled rag on the wiper, rub the bore of the barrel, and the face of the breech-screw, and immediately insert the tompion into the muzzle.

4th. To clean the exterior of the barrel, lay it flat on a bench or board, to avoid bending it. The practice of supporting the barrel at each end, and rubbing it with a strap, buffstick, ramrod, or any other instrument, to *burnish* it, is pernicious, and should be strictly forbidden.

5th. After firing, the barrel should always be washed as soon as practicable; when the water comes off clear, wipe the barrel dry, and pass into it an oiled rag. Fine *flour of emery cloth* is the best article to clean the exterior of the barrel.

330. **To clean the lock.** Wipe every part with a moist rag, and then a dry one; if any part of the interior shows rust, put a drop of oil on the point or end of a piece of soft wood dipped into flour of emery; rub out the rust, and wipe the surface dry; then rub every part with a slightly oiled rag.

331. **To clean the mountings.** For iron and steel parts, use fine emery moistened with oil, or emery cloth. For brass parts, use rotten-stone moistened with vinegar or water, applied with a rag, brush, or stick; oil or grease should be avoided. The dirt may be removed from the screw-holes by screwing a piece of soft wood into them. Wipe all parts with a linen rag, and leave the parts slightly oiled.

332. **Dismounting by an armorer.** The parts which are specially assigned to be dismounted by an experienced armorer will be stated in their regular order, following No. 8, viz.:

9th. Unscrew cone; 10th. Take out cone-seat screw; 11th. Take out band-springs, using a wire punch; 12th. Take out the guard-screws. Be careful that the screw-driver does not slip, and mar the stock; 13th. Remove the guard without injuring the wood at either end of the plate; 14th. Remove the side-screw washers with a drift-punch; 15th. Remove the butt-plate; 16th. Remove the rear-sight; 17th. Turn out the breech-screw by means of a "breech-screw wrench" suited to the tenon of the screw. No other wrench should ever be used for this purpose, and the barrel should be held in clamps, neatly fitting the breech.

333. **Lock.** To take the lock apart:—1st. Cock the piece, and apply the spring-vise to the mainspring; give the thumb-screw a turn sufficient to liberate the spring from the swivel and mainspring notch; remove the spring; 2d. The sear-spring screw; 3d. The sear-screw and sear; 4th. The bridle-screw and bridle; 5th. The tumbler-screw; 6th. The tumbler. This is driven out with a punch, inserted in the screw-hole, which at the same time liberates the hammer; 7th. Detach the mainspring swivel from the tumbler with a drift-punch; 8th. Take out the feed-finger and spring; 9th. The catch-spring and screw.

As a general rule, all parts of the musket are assembled in the inverse order in which they are dismounted. Before replacing screws, oil them slightly with good sperm oil (inferior oil is converted into a gum which clogs the operation of the parts). Screws should not be turned in so hard as to make the parts bind. When a lock has, from any cause, become gummed with oil and dirt, it may be cleaned by boiling in soap-suds or in pearlash or soda water; heat should never be applied in any other way.

334. **Precautions in using.** In ordering arms on parade, let the butt be brought gently to the ground, especially if the ground be hard. This will save the mechanism of the lock from shocks, which are very injurious to it, and which tend to loosen and mar the screws, and split the woodwork.

The ramrod should not be "sprung" with unnecessary force, for fear of injuring the corners of the grooves; and, in stacking arms, care should be taken not to injure the bayonets by forcibly straining the edges against each other.

No cutting, marking, or scraping the wood or iron should be allowed; and no part of the gun should be touched with a file. Take every possible care to prevent water from getting between the lock, or barrel, and stock. If any should get there, dismount the gun as soon as possible, clean and oil the parts as directed, and see that they are perfectly dry before assembling them.

INSPECTION OF ARMS IN SERVICE, &c.

335. **Gauges.** The inspecting instruments are the standard and limit gauges of the bore and exterior of the barrel, and a screw-plate with taps for the holes of the lock-plate.

336. **Inspection.** The following are the principal points to be attended to in the inspection:

Barrel. Defects for which the barrel must be condemned as unfit for service. The large gauge entering the whole length of the barrel. The small or standard gauge not entering, unless the diminution of the bore is caused by the barrel being indented or bent, defects which may be remedied. A diminution of the exterior diameter at the breech, or at the muzzle, so as to enter the small receiving gauges; this diminution is 0.1 inch at the breech; 0.03 inch at the muzzle, for arms with bayonets; 0.045 inch for arms without bayonets. A diminution of 0.5 in length of the barrel, splits, cross-cracks, and other serious defects, caused either by bad workmanship, or by use.

See that the bayonet-stud is not too much worn or broken; that the cone-seat is perfect, and the vent unobstructed. See that the breech-screw is tight after entering five or six threads; that the threads are sharp and sound; that the body fills the bore of the female screw; that the tang is not broken or cracked at the screw-hole, and that it is even with the upper surface of the barrel. If it have any of these defects, replace it with a new one.

Cone. See that the chamfered end is not broken or bruised, and that the thread and vent are in proper condition.

Bayonet. A bayonet is considered unserviceable if the blade is one inch too short. See that it is sound and perfect in all its parts; that it fits the barrel; and that the clasp is in good order, and turns freely.

Lock. See if the fixed branches of the *springs* fit

closely to the lock-plate, if the movable branches are clear of it, and if any of the parts are wood-bound. Renew the springs and the bridle of the tumbler when their pivots are broken. If the *sear* rubs on the plate, have it adjusted. The friction of the tumbler may be caused by the bridle being badly pierced, in which case renew the bridle. If the *hammer* rubs on one side, have it adjusted; if it rubs everywhere, the arbor of the tumbler does not project sufficiently, and the tumbler should be renewed. If the *notches of the tumbler* are broken, or the edges blunt, have them dressed; if the hook of the tumbler projects beyond the edge of the lock-plate when the hammer is let down, the tumbler should be renewed. The arbor and pivot of the tumbler should fit well in their holes. Examine the *sear* closely, and have it renewed when the nose is too thin, or is worn on the side next the lock-plate, although it may be perfect on the exterior.

If the hammer is not steady, the tumbler should be renewed. Try the action of the hammer, to see that it explodes the cap with certainty.

Renew the lock-plate when the screw-holes are too much worn to be dressed over. Renew every limb that is broken or cracked, the screws which are too much worn, or of which the stems are bent, or the slits too much enlarged.

Mountings. See if the parts be complete and sound.

Ramrod. See if it be sound, and have a good thread, and be of the proper length; otherwise replace it.

Stock. Examine carefully the *bed of the lock*, and the holes for the band-springs. Press the thumb against the *facings*, to see if they are split at the holes for the side-screws, and renew the stock if it be split at any other part to an injurious extent.

337. **Classification.** Arms that have been in service may be classified as follows:

1st. *Serviceable arms.*

2d. *Arms requiring repairs.*

3d. *Irreparable arms.*

Arms in the hands of the troops may be repaired by replacing the defective parts by new ones, or by transferring parts from other arms of the same model. Every officer in charge of arms should be supplied with a suitable number of spare parts for making repairs in the field.

Arms are considered *irreparable* when both the barrel and stock are unfit for service; or when they require extensive repairs, and the parts can be used for repairing other arms.

DURABILITY AND STRENGTH OF THE MUSKET-BARREL.

338. **Durability.** Some idea may be formed of the endurance of small-arms generally, by that of the French musket-barrel—the barrel being the most important part of any arm. It has been shown that this barrel will bear 25,000 discharges without becoming unserviceable. In time of war a musket is not fired more than five hundred times a year; with good care, therefore, it ought to last fifty years.

The principal cause of weakness in a barrel is the diminution of the exterior diameter, at the breech, by wear. This diminution is limited to 0.1 inch, although a barrel worn away 0.13 in. has borne the discharge of two cartridges, placed one upon the other.

339. **Strength.** Trials made at Mutzig, in 1829, with arms sent there for repairs, show the following results:

1st. When a musket-barrel is charged with a single cartridge placed in any part of the barrel, or with two, or even three cartridges, inserted regularly, without any interval between them, there is no danger; with four cartridges, inserted regularly over each other, or with two or three cartridges placed over each other with slugged balls, there is danger only in case of some defect of construction, or some deterioration in the barrel; with more than four cartridges inserted regularly over each other, or with two, three, or four cartridges, with intervals between them, it is not safe to fire. Late experiments with the rifle-musket, show that any number of cartridges can be placed one upon the other, and the piece be fired, without injury. In consequence of the expansive nature of the projectile, which cuts off the passage of the flame, but two charges will be inflamed, and their force will be expended through the vent.

2d. No danger of bursting is occasioned by leaving a ball-screw in the barrel. There may be danger from a plug of wood driven tightly into the muzzle when the barrel is loaded with two cartridges; or from a cork rammed into the barrel to a certain distance from the charge, with another cartridge over it.

Snow, clay, and sand, accidentally introduced into the barrel, are not dangerous, if they lie close to the charge; but they are so when there is a space between them and the charge; in this case sand is the most dangerous, then clay and snow. Balls or pieces of iron inserted over the charge, are not attended with danger when placed close to it, even when their weight amounts to $1\frac{1}{4}$ lbs.; but there is danger from a piece of iron 0.5 inch square, weighing $\frac{1}{4}$ lb., if placed 20 inches or more from the breech.

3d. A barrel, with a defect which might have escaped the inspector, bore the explosion of three cartridges, regularly inserted. In these trials, barrels originally 0.272 inch thick at the breech did not burst when loaded with two cartridges, until the thickness was reduced to 0.169 inch, and with one cartridge to 0.091 inch.

Of the 27,574 muskets picked up on the battle field of Gettysburgh, and turned into the Washington Arsenal, at least 24,000 were loaded. About one-half of this number contained *two* charges each, one-fourth contained from *three* to *ten* charges each, and the balance *one* charge. The largest number of cartridges found in any one piece was twenty-three. In some cases the paper of the cartridges was unbroken, and in others the powder was uppermost. The experience of the war goes to show that the principal causes of the bursting of the barrels of small arms are the clogging of the muzzle with mud, and thoughtlessly allowing the tompion to remain in, in firing.

Experience shows that the average duration of arms in active service is about as follows, viz.: seven years for infantry muskets, five for cavalry carbines, and four years for cavalry sabres and revolvers. Four and six years are found to be the average wear of cavalry and infantry accoutrements respectively.

CHAPTER VII.

PYROTECHNY.

340. **Definition.** Pyrotechny is the art of preparing ammunition and fireworks for military and ornamental purposes. It will be treated under the head of *buildings*, *materials*, *ammunition*, *military fireworks*, and *ornamental fireworks.*

BUILDINGS, &c.

341. **How arranged.** To conduct the operations of a military laboratory with safety and convenience, the following rooms are necessary, viz.:—

1st. *Furnace-room*, for operations requiring the use of fire.

2d. *Cartridge-room*, for making all kinds of cartridges.

3d. *Filling-room*, for filling cartridges with powder.

4th. *Composition-room*, for mixing compositions.

5th. *Driving-room*, for driving rockets, fuzes, &c.

6th. *Packing-room*, for putting up articles for transportation.

7th. *Carpenter's and tinner's shop.*

8th. *Magazine*, for storing powder and ammunition.

A laboratory, like a powder-mill, should be situated apart from inhabited buildings; and, for convenience of communication, the rooms, with the exception of the furnace-room, carpenter's shop, and magazine, should be situated under one roof.

342. **Furnaces.** A furnace is composed of a cast-iron kettle, 2 feet in diameter, set in a fire-place of brick. In the field, sods may replace the brick, if the latter cannot be obtained.

Two kinds of furnaces are employed in a laboratory; in the first, the flame circulates around both bottom and sides of the kettle; in the second, it only comes in contact with the bottom; the latter is used for compositions in which gunpowder forms a part.

343. **Precautions.** To prevent accidents in the operations of a laboratory, avoid, as much as possible, the use of iron in the construction of the buildings, fixtures, &c.; sink the heads of iron nails, if used, and cover them with putty; cover the floor with oil-cloth, or carpets, and have it frequently swept. Let the workmen in the powder-room wear socks, and take them off when they go out. Keep no more than the requisite quantity of powder in the laboratory, and have the ammunition and finished work taken to the magazine. Let powder-barrels be carried in hand-barrows made with leather, or with slings of rope or canvas, and the ammunition in boxes. Let every thing that is to be moved be lifted, not dragged or rolled on the floor. Never drive rockets, port-fires, &c., in a room where there is any powder or composition, except that used at the time. Never enter the laboratory at night, unless it is indispensable, and then use a close lantern, or wax or oil light well trimmed. Allow no tobacco to be smoked, nor friction matches to be carried in or around the laboratory.

MATERIALS.

344. **Classified.** Laboratory materials may be divided into four classes, viz.:

1st. Those for producing light, heat, and explosion.

2d. Those for coloring flames, and producing brilliant sparks.

3d. Those used in preparing compositions.

4th. Those used in making cartridge-bags, cases, &c.

345. **1st. class.** *Nitre.* For laboratory use, nitre must be reduced to a fine powder, or very minute crystals. It is best pulverized in rolling-barrels at the powder-mills, but it may be pulverized by hand, in the laboratory, with a rolling-barrel, or by pounding in a brass mortar, or by stirring a crystallizing solution.

Chlorate of potassa. Chlorate of potassa is formed by passing a current of chlorine, in excess, through lime-water, and then treating the mixture with the chloride of potassium, or by the carbonate or sulphate of potassa. The chlorate of potassa and chloride of calcium are formed—the former crystallizes, the latter remains in solution. It is soluble in water, but not sensibly so in alcohol. As before stated, it is a more powerful oxydizing agent than nitre; and, when mixed with a combustible body, easily explodes by shock or friction. It is inflamed by simple contact with sulphuric acid, and thus affords a simple means of exploding mines.

A convenient form of apparatus for this purpose, is a glass vessel with two compartments, one containing sulphuric acid, and the other chlorate of potassa and gunpowder. It is placed near the surface of the ground, and, when broken under the feet of the enemy, the two substances are brought in contact, producing fire, which explodes the mine.

Charcoal. For laboratory use, charcoal may be made by charring wood in an iron kettle buried in the ground. It may be pulverized by rolling in a barrel with bronze balls, or by beating in a leather bag with a maul. It should be kept in close barrels, in a dry place.

Sulphur. When melted sulphur is to be used, care must be taken that it does not become thick, which occurs at about 400°. It may be pulverized in a rolling-barrel, or by being pounded in a mortar and sifted. Roll brimstone is better for melting than flowers of sulphur. When flowers of sulphur are to be mixed with chlorate of potassa, it should be washed to remove the free sulphuric acid. Sulphur retards the combustion of compositions to which it is added.

Antimony. Antimony, or *regulus of antimony*, is a grayish white metal, easily reduced to a powder, and, by its combustion with sulphur, produces strong light and heat; the color of the flame is a faint blue.

Sulphuret of antimony. Sulphuret of antimony is mixed with inflammable substances to render them more easily ignited by flame or friction.

Gunpowder. For compositions, gunpowder is pulverized, or *mealed*, by the rolling-barrel, or by grinding with a muller on a mealing-table, or by beating in a leather bag. The simple incorporation of the ingredients of gunpowder does not answer the desired purpose.

Lampblack. Lampblack is the result of the incomplete combustion of resinous substances. It is composed of about 80 parts of carbon, and 20 of impurities. It is employed to quicken the combustion of certain mixtures; but, before it is used, it should be washed with a hot alkaline solution, to remove all traces of empyreumatic oil.

346. **2d class.** *Coloring materials.* A flame is colored by introducing into the composition which produces it, a substance, the particles of which on being interspersed through the flame, and heated to the incandescent state, give it the required color. Coloring substances do not generally take part in the combustion, and their presence, more or less, retards it; it is for this reason that chlorate of potassa, a more powerful oxydizing agent than nitre, is used in lieu of it, in compositions for colored fires.

Colors. There are a great variety of substances which give color to flames, the principal of which are, *nitrate and sulphate of strontia* and *chloride of strontium*, for red; the *nitrate of baryta*, for green; the *bicarbonate of soda*, for yellow: the *sulphate*, *carbonate*, and *acetate of copper*, for blue. *Lampblack* is employed to give a train of rose-colored fire in the air, *powdered flint-glass* for white flames, and *oxide of zinc* for blue flames.

Sparks. Brilliant sparks are produced by introducing into the composition, filings or thin chips of either wrought iron, cast iron, steel, or copper, or by fragments of charcoal; the effect depends on the size of the particles introduced. The particles should be freshly prepared, or should have been well preserved from rust.

347. **3d Class.** *Preparing compositions.* *Turpentine* is the substance which exudes from the freshly-cut sur-

face of a pine tree in warm weather. The first year's running is called *virgin*, or *white turpentine;* after this it becomes more hard and yellow.

Spirits of turpentine. This is the *essential oil* obtained by distilling native turpentine.

Rosin. This substance is sometimes called *colophony*, and is the residuum of the distillation of turpentine.

Tar. Tar is a semi-fluid substance, obtained from the heart of the pine-tree by a smothered combustion, as in charcoal pits.

Pitch. Pitch is obtained by boiling tar down to the requisite consistency, either by itself, or combined with a portion of rosin; it becomes solid on cooling, but is softened by the heat of the hand.

Venice turpentine. Venice turpentine is obtained from the larch; but what is commonly known by that name, is a compound of melted rosin and spirits of turpentine. The foregoing substances are chiefly employed in the preparation of compositions for producing light.

Alcohol, &c. *Alcohol* (spirits of wine), *brandy*, *whiskey*, or *vinegar*, is used for mixing compositions in which nitre enters, because this salt is but slightly soluble in these liquids.

Gum-arabic. Gum-arabic in solution is employed to give body to certain compositions. It retards combustion; and, as the solution is liable to spontaneous decomposition, it should only be prepared as wanted.

Beeswax and mutton tallow are employed chiefly in mixing compositions intended to produce heat and light.

348. **4th Class. Preparing cartridges, &c.** The size and strength of laboratory paper is regulated by the use to which it is applied. It is arranged in seven classes, the

strongest being for cannon-cartridges, and the thinnest for musket-cartridges.

Paste. Ordinary paste is made of rye flour, stirred and boiled in water.

Flannel, *wildbore*, or *serge*, for cartridge-bags, should be made entirely of wool or silk; the fabric should be soft, and closely woven, to prevent the powder from sifting out.

Fabrics of cotton and flax are not used, because the powder sifts through them, and they are more apt to leave fire in the gun than woollen stuffs.

Canvas. Canvas is used for sacks, &c.; it should be strong and closely woven.

Twine, &c. Twine should be strong, smooth, and well twisted.

AMMUNITION FOR SMALL-ARMS.

349. **Bullets.** Bullets, for the military service, are made by pressure. To prepare the lead for the press, it is cast into cylinders, or drawn out into a wire of a diameter somewhat less than that of the bullet. A piece, just sufficient to make a bullet, is then cut off, and transferred by a movable arm to a pair of dies (*a a*, fig. 111), which are firmly closed by two movable wedges (*b b*); as soon as this is done, the punch (*c*) descends upon the lead and forms the cavity at the base of the bullet. To disengage the bullet, the punch rises, but before it has completely cleared the cavity, the dies open, and the bullet falls into its receptacle.

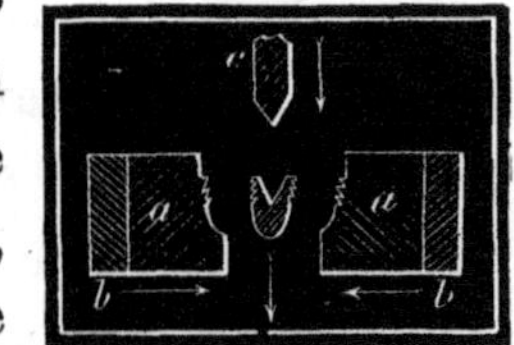

Fig. 111.

One press is capable of making 3,000 bullets in an hour. Bullets may be also cast in moulds, and afterward *swaged* in a die to the proper size and shape.

350. **Cartridge.** After the bullet has been greased (see page 315), it is made up into a cartridge along with its charge of powder. The rifle-musket cartridge (fig. 112) is formed of *three* parts; the *bullet* (*b*), the *cylinder* (*a*), which contains the powder, and the *wrapper*, which unites the cylinder with the bullet. The bottom of the cylinder should be perfectly tight, to prevent the powder from sifting through.

Fig. 112.

To use this cartridge, tear off the fold of the wrapper and pour the powder into the bore, break the cartridge at the junction of the bullet and powder-cylinder, force out the bullet by pressing with the thumb and forefinger, and insert it in the bore. Care should be taken to pour all of the powder into the barrel.

351. **Buckshot cartridge.** The number of projectiles in a buckshot cartridge is twelve, or four layers of three each (fig. 113). The layers are kept in position by passing one half hitch of the choking thread, between every two layers; the thread is secured by passing two half-hitches around the upper layer. For rifled arms, the shot end of the cartridge should be dipped in the composition used for lubricating bullets; with this precaution all leading of the grooves will be avoided. Buckshot cartridges are principally used in Indian warfare, and especially in night-firing.

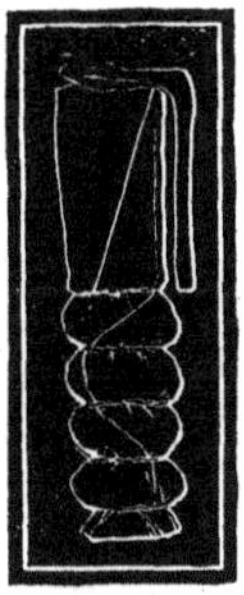

Fig. 113.

352. **Packing.** Small-arm cartridges are wrapped in

bundles of ten each, and packed in boxes of 1,000. The date and place of fabrication are marked on the inside of the cover, and the contents, on the outside of one end of the packing-box.

353. **Pistol cartridge.** The powder-cylinder of Colt's cartridge is made of combustible paper (prepared after the manner of gun-cotton); it is attached to the base of the bullet, and is inserted in the piece entire.

354. **Percussion-caps.** The military percussion-cap is made slightly conical, to fit the cone tightly, and has a rim around the open end for convenience in handling. It is made of sheet copper, which is first cut into the form of an equilateral cross (*a*, fig. 114), and then transferred to a die, where it is punched into the required shape (*b*). It is charged with half a grain of powder, composed of one part of nitre and two of fulminate of mercury; the object of the nitre being to retard the combustion of the composition, and give density to the flame.

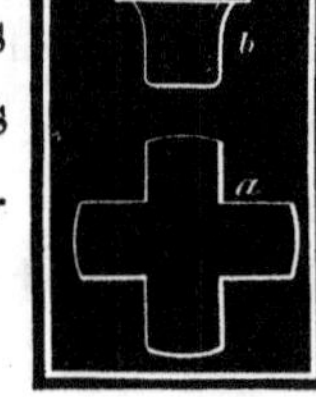

Fig. 114.

The composition is pressed into the cap in a dry state, and covered with a drop of shellac varnish to fix it in its place and protect it from moisture. With the exception of varnishing, all the operations of making percussion-caps are performed by a single machine, at the rate of 50,000 per day. Percussion caps were invented in the United States, in 1817.

355. **Maynard's Primer.** This primer is made by indenting a sheet of paper at regular intervals (*a a*, fig. 115), filling each indentation with a small charge of percussion powder, and

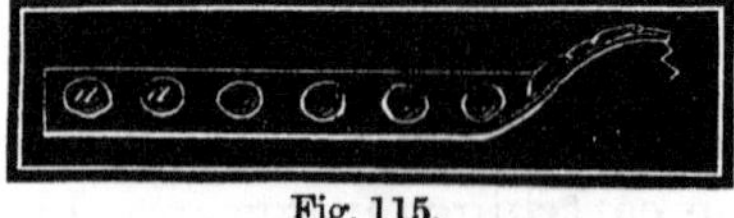

Fig. 115.

covering the whole with another sheet of paper, firmly pasted on. The sheet is then cut into strips, each strip containing 60 primers in a single row, and, to protect it from the moisture, it is covered with a thick coat of shellac varnish.

FIELD AND MOUNTAIN AMMUNITION.

356. **Composition.** Ammunition for the field-service is composed of *solid shot*, *shells*, *spherical-case shot*, and *canister-shot*; in the mountain service, the solid shot are omitted.

For convenience in loading, and safety in transportation, cannon ammunition should be prepared in a peculiar manner, and with great care.

357. **Stand of ammunition.** A *stand* of ammunition is composed of the *projectile* (*a*, fig. 116), the *sabot* (*b*), the *straps* (*c*), the *cartridge-bag* (*d*), and the *cylinder* (*e*) and *cap*. The preparation of the projectile itself, is described in chapter II.

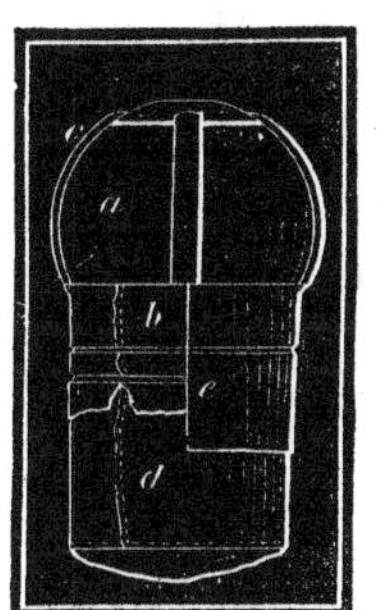

Fig. 116.

Sabot. The sabot is a thick, circular disk of wood, to which the cartridge-bag and projectile are attached.

For a spherical projectile, the sabot has a spherical cavity (*a*, fig. 117), and a circular groove to which the cartridge-bag is tied; in the canister-sabot, the spherical cavity is omitted, and a circular offset (*b*) is added.

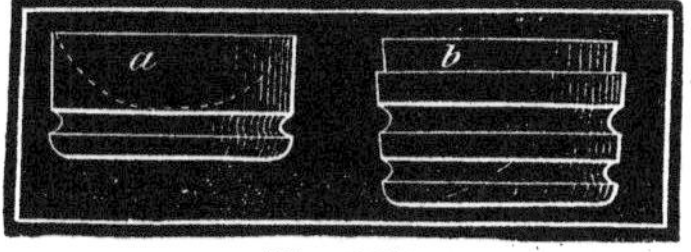

Fig. 117.

The effects of a sabot are:—1st. To prevent the

formation of a *lodgment* in the bore. 2d. To moderate the action of the powder on the projectile; and, 3d. To prevent the projectile from moving from its place. In consequence of the scattering of the fragments, it is dangerous to use the sabot in firing over the heads of one's own men.

Straps. The projectile is secured by two tin straps, fastened at the ends with tacks driven into the sabot. The straps cross each other at right angles; for solid shot, one strap passes through a slit in the other; for hollow projectiles, both straps are fastened to a tin ring which surrounds the fuze-hole.

Cartridge-bag. The materials of which cartridge-bags are made are described on page 348. A cartridge-bag for the field service is made of two pieces—a rectangular piece for the sides, and a circular piece for the bottom. The rectangular piece should be cut in the direction of the warp, to prevent the bag from stretching in the direction of its diameter; the seams should be sewed with *woollen* yarn, 12 stitches to the inch, and the edges should be basted down, to prevent the powder from sifting through. The charge is determined by measurement.

Cylinder and cap. The cylinder and cap are made of stout paper. The cylinder is used to give stiffness to the cartridge at the junction of the sabot and bag; the cap covers the exposed portion of the bag, and is drawn off before loading, and placed over the projectile, or thrown away. The cap is made by cutting off a portion of the cylinder, and choking one end. The cartridge-bag is attached to the projectile by tying it around the grooves of the sabot with twine.

358. **Strapped ammunition.** Ammunition thus prepared is called *fixed ammunition.* In the large field howitzers, it is not convenient to unite the cartridge-bag and projectile, on account of the difficulty of packing them in the ammunition chests; the bag and projectile are therefore carried separately. The projectile is attached to a sabot without grooves; and, to give a proper form to the cartridge-bag the mouth is closed with a *cartridge-block*, which resembles a sabot; hence the name *strapped ammunition.*

359. **Packing, &c.** As soon as ammunition is finished it should be gauged, to see that it is of the proper calibre; it is afterward packed in boxes containing ten *rounds* each, with scraps of paper, or tow, well rammed into the interstices.

Ammunition may be distinguished by the color of the cap; for spherical-case shot, it is *red;* for shells, *black;* and for solid-shot and canister, it is the *natural color* of the paper. The outside of the packing-box is colored *red* for spherical-case shot; *black* for shells; *olive* for shot; and, for canister, the *natural* color of the wood; besides this, it is marked with the number and character of the contents in letters and figures.

SIEGE AND SEA-COAST AMMUNITION.

360. **Cartridge-bags.** On account of the great weight of siege and sea-coast ammunition, the cartridge-bag and projectile are carried separately.

The cartridge-bags for large charges of powder are made of two pieces of woollen stuff (fig. 118), or of a

Fig. 118.

paper tube, with woollen cloth bottom. The former are preferred for rapid firing.

For *sea-coast howitzers*, the bag should fill the chamber; if the piece be fired with a reduced charge, a cartridge-block should be inserted into the bag to give it proper size. For *mortars* the bag is only used to carry the powder, and when the piece is loaded, the powder is poured into the chamber; bags of any suitable size will answer for this service. For *hot-shot* cartridges, bags are made double, by putting one bag within another. Care should be taken to see that the bags are free from holes.

For *ricochet firing*, or other occasions when very small charges are required, a cartridge-bag of inferior calibre may be used; or else, after the charge is poured into the bag, place it on another bag filled with hay, pressing it with the hands to reduce the diameter; after having shaken this bag down, and rolled and flattened the empty parts of the two bags, tie them with woollen yarn, like a bundle of musket-cartridges, placing the knot on top.

361. **Strapping shells, &c.** In the siege and sea-coast services, solid shot are transported and loaded loosely, but hollow projectiles are strapped to sabots to prevent the fuze from coming in contact with the powder of the charge. The sabots are made from thick plank (fig. 119), and the straps are fastened as in the field-service. The fuze-hole is placed in one of the angles of the straps, in such a manner that its axis shall

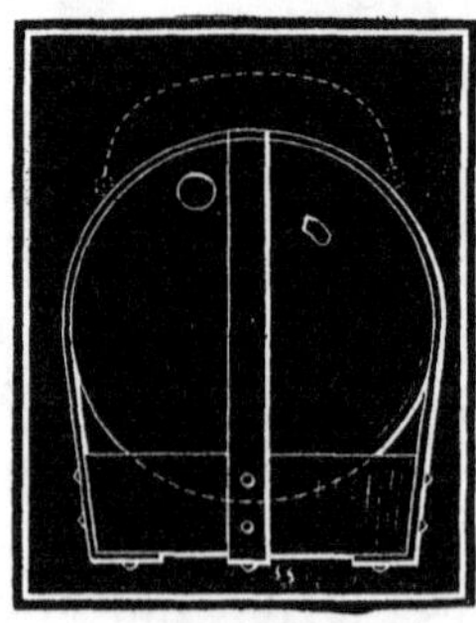
Fig. 119.

make an angle of 45° with that of the sabot. In loading, care should be taken to place the fuze-hole uppermost.

When there are no ears on the projectile, a handle of rope-yarn is attached to two loops soldered to one of the straps, or it is passed through two holes bored in the bottom of the sabot.

362. **Filling shells.** Shells are filled with cannon-powder alone, or with mortar-powder and some kind of incendiary composition. Before filling, the shell should be inspected, to see that it is dry, clean, and free from defects.*

The service bursting-charges of powder for the larger shells are:—

	15-INCH.	13-INCH.	10-INCH.	8-INCH.
Mortar-shell, ...		7 lbs.	5 lbs.	$2\frac{1}{2}$ lbs.
Columbiad-shell, .	17 lbs.	"	3 lbs.	$1\frac{1}{2}$ lbs.

363. **Wads.** Junk wads having been found detrimental in ordinary firing, are only used for proving cannon.

For *firing hot-shot*, a hay wad, soaked in water, is interposed between the powder and shot. The wad is made by twisting the hay into a rope, winding the rope into a coil, which is driven into wooden moulds of the proper size. To preserve the size and form of the wad, it is afterward wrapped tightly with rope-yarn.

Grommets. Grommets, or *ring wads*, are useful in increasing the accuracy of fire, and keeping the projectile in its place when the piece is moved or depressed.

* To find the quantity of powder which a shell will contain, multiply the cube of the interior diameter of the shell in inches by 0.01744, the result is the weight of powder in pounds.

It is made by bending a strand of rope into the form of a circle, and wrapping it with rope-yarn. The size of the ring is the full diameter of the bore, that it may fit tightly. This wad is attached to the front of the projectile with twine; or, it may be inserted after the projectile, like ordinary wads.

MILITARY FIREWORKS.

364. **Comprise what.** Military fireworks comprise preparations for the service of *cannon ammunition*, and for *signal*, *light*, *incendiary*, and *defensive* and *offensive* purposes.

365. **Compositions.** The term composition is applied to all mechanical mixtures which, by combustion, produce the effects sought to be attained in pyrotechny. If these compositions be examined, it will be found that many of them are derived from gunpowder, by an admixture of sulphur and nitre, in proportions to suit the required end; a German writer has even proposed to extend this method to the formation of all the principal compositions, but the simplicity of the plan has never been fully realized in practice.

Preparation. Compositions are prepared in a dry or liquid form; in either case it is necessary that the ingredients should be pure, and thoroughly mixed.

For dry compositions, the ingredients are pulverized separately, on a *mealing*-table, with a wooden *muller;* they are then weighed, and mixed with the hands, and afterward passed three times through a wire sieve of a certain fineness. When a highly oxydizing substance, as the chlorate of potassa, is present, great care must be observed in mixing, to avoid friction or blows, which might lead to an explosion. When coarse charcoal, or metals in grains are used, they should be added after the other ingredients have been mixed and sifted.

For the liquid form. When it becomes necessary to use fire to melt the ingredients, the greatest precaution is necessary to prevent accidents, particularly when gunpowder enters. The dry parts of the composition may be, generally, mixed together first, and put by degrees into the kettle, when the other ingredients are fluid, stirring well all the time. When the dry ingredients are very inflammable, the kettle must not only be taken from the fire, but the bottom must be dipped in water, to prevent the possibility of accidents.

How disposed. To give a portable form to compositions, they are enclosed in cases, cast in moulds, or attached to cotton yarn, rope, &c.

Cases. Cases are generally paper tubes, made by covering one side of a sheet of paper with paste, or gum-arabic, wrapping it around a *former*, and rolling it under a flat surface until all the layers adhere to each other (fig. 120). The quality of the paper, and the thickness of the sides of the case, should depend upon the pressure of the gases evolved in the burning.

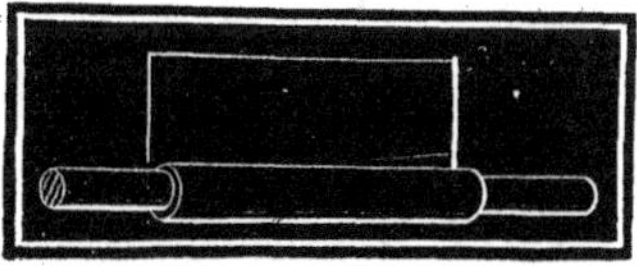

Fig. 120.

Filling. To fill a case, it is first cut to the proper length, and placed in a mould; the composition is then poured in, a ladleful at a time, and each ladleful is packed by striking a certain number of blows on a drift with a mallet of a given weight. The height of each ladleful of composition should be about equal to a single diameter of the bore of the case.

Drifts, &c. Small drifts, receiving heavy blows, should be made of steel, and tipped with bronze (fig. 121); large drifts may be made of wood or bronze, depending on the force of the blow. In driving highly inflammable compositions, as that of the rocket, care should be taken to settle the drift, so as to exclude the air before striking with the mallet, as the heat generated by the sudden condensation of air might be sufficient to ignite the composition.

Fig. 121.

Preliminary tests of all new materials should be made by burning one or more specimens of the composition, and the proportions of the ingredients corrected, if necessary.

Vent, &c. The length of the flame from a given composition depends on the size of the vent and the extent of the burning surface. The vent is made small by *choking* the end of the case with stout twine; and the burning surface is increased by driving the composition around a spindle, which, on being withdrawn, leaves a conical-shaped cavity. A vent may be also formed by driving in moist plaster of Paris or clay, and boring a hole in it with a gimlet. If the end of the case is to be closed up entirely the boring is omitted.

366. **For ammunition.** The preparations for the service of ammunition are *slow-match*, *quick-match*, *port-fires*, *friction-tubes*, and *fuzes*.

367. **Slow-match.** Slow-match is used to preserve fire. It may be made of hemp or cotton rope; if made

of hemp, the rope is saturated with acetate of lead, or the lye of wood-ashes; if made of cotton, it is only necessary that the strands be well twisted. Slow-match burns from four to five inches in an hour.

368. **Quick-match.** Quick-match is made of cotton yarn (candle-wick) saturated with a composition of mealed powder and gummed spirits; after saturation, the yarn is wound on a reel, sprinkled (dredged) with mealed powder, and left to dry.

It is used to communicate fire, and burns at the rate of one yard in thirteen seconds. The rate of burning may be much increased by enclosing it in a thin paper tube called a *leader*.

369. **Port-fires.** A port-fire is a paper case containing a composition, the flame of which is capable of quickly igniting primers, quick-match, &c.

The composition consists of—

NITRE.	SULPHUR.	MEALED POWDER.
13	4.5	2.5

A port-fire is about 22 inches long, and burns with an intense flame for ten minutes.

370. **Friction-tube.** The friction-tube is at present the principal preparation for firing cannon; its advantages are portability and certainty of fire. It also affords the means of firing a piece situated at a distance, and does not attract the notice of the enemy's marksmen at night.

It is composed of two brass tubes soldered at right

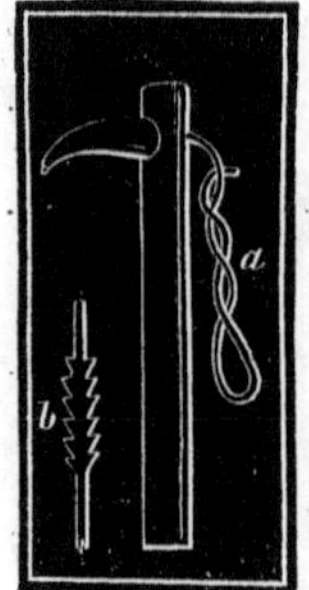

Fig. 122.

angles (fig. 122). The upper, or short tube contains a charge of friction powder, and the *roughed* extremity of a wire loop (*a*) (the extremity is shown by fig. *b*); the long tube is filled with rifle powder, and is inserted in the vent of the piece. When the extremity of the loop is violently pulled by means of a lanyard, through its hole in the long tube, sufficient heat is generated to ignite the friction powder which surrounds it, and this communicates with the grain-powder in the long tube. The charge of grained powder has sufficient force to pass through the longest vent, and penetrate several thicknesses of cartridge-cloth. The composition of friction powder is:—

CHLORATE OF POTASSA.	SULPHURET OF ANTIMONY.
1	2

formed into a paste with gum-water.

371. **Fuzes.** Fuzes are the means used to ignite the bursting-charge of a hollow projectile at any desired moment of its flight; they may be classified according to their mode of operation, as *percussion*, *concussion*, and *time-fuzes.*

Time-fuze. This fuze is composed of a case of paper, wood, or metal, enclosing a column of burning composition, which is set on fire by the discharge of the piece, and which, after burning a certain time, communicates with the bursting-charge.

Its successful operation depends on the certainty of

ignition, the uniformity of burning, and the facility with which its flame communicates with the bursting-charge.

Composition. The ingredients of all time-fuze compositions are the same as for gunpowder, but the proportions are varied to suit the required rate of burning. Pure mealed powder gives the quickest composition, and the others are derived from it by the addition of nitre and sulphur in certain quantities.

The rate of burning of a column of fuze composition depends on the purity and thorough incorporation of the materials, and on its density. These qualities are best secured by procuring the materials from the powder-mills ready mixed, and driving them with a press of peculiar construction.

Three kinds of time-fuzes are employed in the United States service, viz.: the *mortar-fuze*, the *Bormann-fuze*, and the *sea-coast fuze.*

Mortar-fuze. The case of the mortar-fuze is made of beech-wood, turned in a lathe to a conical shape, and bored out nearly to the bottom to receive the composition (fig. 123). The composition is driven with fifteen blows of the mallet. The bore is enlarged at the top to receive a priming of mealed powder moistened with alcohol. To protect the priming from injury by moisture, the top of the fuze is covered with a cap of water-proof paper, on which is marked the rate of burning of the composition. The exterior is divided into inches and tenths, to guide the gunner in regulating the time of burning. This operation is generally performed before the fuze is driven into the fuze-hole of the shell, by

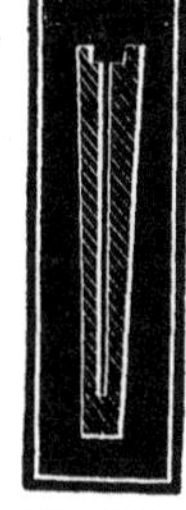
Fig. 123.

cutting it off with a saw, or boring into the composition with a gimlet.

If the fuze be driven, the column of composition may be shortened by taking a portion from the top with the fuze-auger.

372. **Bormann-fuze.** This fuze is the invention of an officer of the Belgian service. The case is made of an alloy of tin and lead, cast in iron moulds. Its shape is that of a thick, circular disk; and a screw thread is cut upon its edge, by which it is fastened into the fuze-hole of the projectile. (See figure 124.) The upper surface is marked with two recesses (*a a*), and a graduated arc. The former are made to receive the prongs of a screw-driver; and the latter overlies a circular groove, filled with mealed powder, tightly pressed in and covered with a metal cap. The only outlet to the groove containing the mealed powder is under the zero of the graduation; this outlet, or channel (*c*), is filled with rifle powder, and leads down to a circular recess (*b*), which is filled with musket powder, and covered with a perforated disk of tin. To enable this fuze to resist the shock of discharge, and at the same time to increase the effect of a small bursting-charge, the lower portion of the fuze-hole is closed with a perforated disk (*e*).

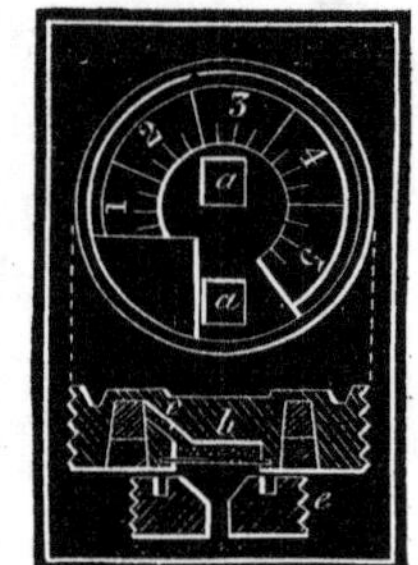

Fig. 124.

Before the projectile is inserted into the piece, a cut is made across the graduated portion, laying bare a small proportion of the mealed powder, which, being ignited by the flame of the charge, burns in both directions until the outlet is reached and the grain powder ignited. The graduations are seconds and quarter seconds, and the time of burning of the fuze depends on the length of the column of mealed powder included between the incision and outlet. If the metal covering be not cut, the projectile may be fired as a solid shot The Bormann-fuze is used for the field and siege services, and is found to be accurate and reliable, especially for spherical-case shot.*

373. **Sea-coast fuze.** The sea-coast fuze is principally distinguished from the mortar-fuze by having a metal cap, constructed to prevent the burning composition from being extinguished when the projectile strikes against water.

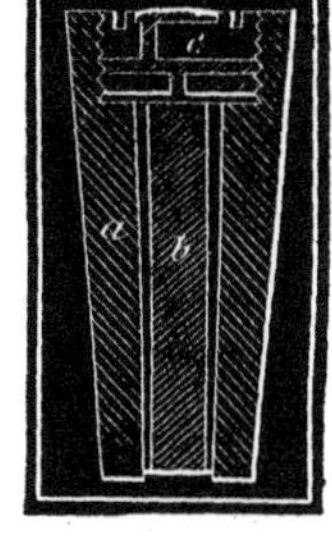

Fig. 125.

It is composed of a *brass plug* (*a*, fig. 125), which is firmly driven into the fuze-hole of the projectile; a *paper-fuze* (*b*), inserted into the plug, with the fingers, immediately before loading the piece; and a *water-cap* (*c*), screwed into the plug after the paper-fuze has been inserted.

The water-cap is perforated with a crooked channel, which is filled with mealed powder; the mealed powder communicates fire to the paper-fuze, and the angles of the channel break the force of the water.

The top of the cap has a recess which is filled with a priming of mealed powder, and is covered with a disk

* The time of burning of the Bormann-fuze, not being long enough for the general service of rifle projectiles, the paper time-fuze (*b*. fig. 125) is used instead of it for all of those projectiles which require the time-fuze. It is inserted into a zinc plug, which is screwed into the fuze-hole of the projectile.

of sheet lead to prevent accidental ignition; before loading, this disk is removed. The time of burning is regulated by the proportion of the ingredients of the composition; and this is indicated by the number of seconds marked on the paper case. In firing over land, the water-cap is omitted, and the brass plug may, for the sake of economy, be replaced by a wooden one. One advantage of this form of fuze is, that the bursting-charge may be put in, or taken out, after the fuze-plug has been driven.

374. **Concussion-fuze.** A concussion-fuze is made to operate by the shock of the discharge, or by the shock experienced in striking the object, and is applicable to spherical projectiles. One of the simplest, as well as one of the most effective, concussion-fuzes is that invented by Captain Splingard, of the Belgian service; and for the purpose of illustrating the principle of this class of fuzes, it may be proper to give an outline of its construction.

It is composed of two principal parts, the wooden plug, and the paper-fuze. The chief peculiarity lies in the arrangement of the paper-fuze. The case (*a*, fig. 126), is made of paper, rendered incombustible by a solution of sulphate of ammonia and alum, and filled with fuze composition (*b*) of variable quickness of burning. A long cavity is formed in the lower part of the composition, by driving it around a spindle, as in a rocket; this cavity is filled with moist plaster of Paris, and a long needle is inserted in it, nearly to the bottom of the plaster, forming a tube (*c*) enclosed in and supported

Fig. 126.

by the composition. The composition is ignited in the usual way, at the top, and, as it burns away, leaves a portion of the plaster tube unsupported; when the shell strikes its object, the shock breaks off the unsupported part of the tube, and the flame of the composition immediately communicates with the bursting-charge; if the tube do not break, the composition burns up, and the bursting-charge is ignited, as in an ordinary time-fuze. The upper portion of the composition burns away quickly, in order to leave the tube unsupported soon after the projectile leaves its piece.

375. **Percussion-fuse.** A percussion-fuse explodes by the striking of some particular point of a projectile against an object, as in the case of rifle-cannon projectiles.

One of the best and simplest forms of this kind of fuze is the ordinary percussion-cap placed on a cone affixed to the point of the projectile. The piece to which the cone is attached may be fixed or movable; in either case, the apparatus should be covered with a safety-cap to prevent the percussion-cap from taking fire by the discharge of the piece.

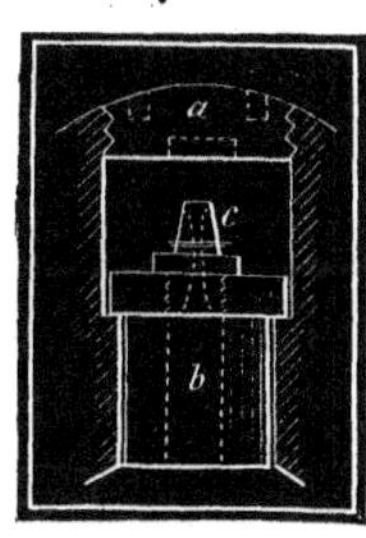

Fig. 127.

Fig. 127 represents a fuze of the percussion kind, in which *b* is a movable cone-piece, bearing a musket-cap (*c*); and *a* is the safety-cap which covers the fuze-hole. When the projectile is set in motion, the cone-piece, or "plunger," by its inertia, presses against the shoulders of the fuze-hole;* when its motion is ar-

* Late experience shows that the plunger should be enclosed in a tight metal case to prevent it from being fouled by the action of the powder; and to prevent premature explosions, the cone-piece should be confined by a screw or other device, to prevent it from moving until the projectile strikes its object.

rested, the inertia of the cone-piece causes the percussion-cap to impinge against the safety-cap, which produces explosion. The explosion of the projectile may be made to take place at any desired time, after the explosion of the cap, by interposing grain, or mealed powder, between the cap and bursting-charge.

FIREWORKS FOR SIGNALS.

The preparations employed for signals are *rockets* and *blue-lights.*

376. **Signal rockets.** The principal parts of a signal rocket are the *case* (a), the *composition* (b), the *pot* (c), the *decorations* (e), and the *stick* (f).

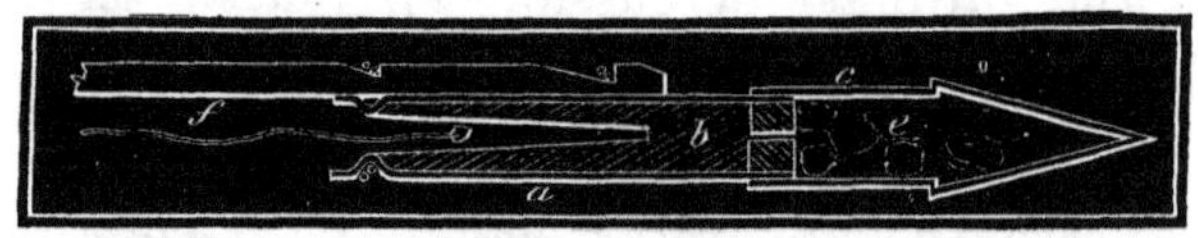

Fig. 128.

Case. The case is made by rolling stout paper covered on one side with paste, around a former, and at the same time applying a pressure until all the layers adhere to each other. The vent is formed by choking one end of the case, and wrapping it with twine. When the case is trimmed and dried, it is ready for driving the composition.

Composition. A variety of compositions are employed for signal rockets; the best can only be determined by trial, as it varies with the condition of the ingredients. The following proportions are used at the West Point laboratory:

NITRE.	SULPHUR.	CHARCOAL.
12	2	3

to increase the length and brilliancy of the trail, add steel, or cast-iron filings.

Driving. The case is placed in a copper mould which has a conical spindle attached to the centre of its base, to form the bore; the composition is driven with twenty-one blows of the mallet. The first and second drifts are made hollow to fit over the spindle, and the third is solid. The top of the case is closed by moist plaster of Paris, which is one diameter thick, and perforated with a hole for the passage of the flame from the burning composition to the pot. The rocket is primed by inserting a strand of quick-match into the bore, after which it is coiled up, and covered with a paper cap, until required for use.

Pot. The pot is formed of a paper cylinder slipped over, and pasted to the top of the case; it is surmounted with a paper cone, filled with tow. The object of the pot is to contain the decorations which are scattered through the air by the explosion which takes place when the rocket reaches the summit of its trajectory; the explosion is produced by a small charge of mealed powder.

Decorations. The decorations of rockets are *stars, serpents, marrons, gold rain, rain of fire, &c.*

Stars. Stars are formed by driving the composition, moistened with alcohol and gum-arabic in solution, in port-fire moulds. It is then cut into lengths about $\frac{3}{4}$ in., and dredged with mealed powder.

White.

NITRE.	SULPHUR.	MEALED POWDER.
7	3	2

Red.

CHLORATE OF POTASSA.	SULPHUR.	LAMPBLACK.	NITRATE OF STRONTIA.
7	4	1	12

Blue.

CHLORATE OF POTASSA.	SULPHUR.	AMMONIACAL SULPHATE OF COPPER.
3	1	1

Yellow.

CHLORATE OF POTASSA.	SULPHUR.	SULPHATE OF STRONTIA.	BICARBONATE OF SODA.
4	2	1	1

Serpents. The case of a serpent is similar to that of a rocket, but the interior diameter is only 0.4 inch. The composition is driven in, and the top is closed with moist plaster of Paris. It is primed by inserting a small piece of quick-match through the vent; it may be made to explode by driving mealed powder over the composition. The composition is—

NITRE.	SULPHUR.	MEALED POWDER.	CHARCOAL.
3	3	16	$\frac{1}{2}$

Marrons. Marrons are small paper shells, or cubes, filled with grained powder, and primed with a short piece of quick-match, which is inserted in a hole punctured in one of the corners. To increase the resistance of the shell, it is wrapped with twine, and dipped in *kit* composition.

Stick. The stick is a tapering piece of pine, about nine times the length of the case, and is tied to the side of the case to guide the rocket in its flight. The position of the centre of gravity depends on the diameter of the case; for a 2-in. rocket it should be $2\frac{1}{2}$ in. in rear of the vent; and it is verified by balancing on a knife-edge. The prescribed dimensions of the stick should be observed, for, if the stick be too heavy, the rocket will not rise to a proper height; if it be too light, it will not rise vertically.

377. **Blue-light.** A very brilliant bluish light may be made of the following ingredients, viz.:

NITRE.	SULPHUR.	REALGAR.	MEALED POWDER.
14	3.7	1	1

The brilliancy depends on the purity and thorough

24

incorporation of the ingredients. The composition may be driven in a paper case, and afterward cut off to suit the required time of burning. Both ends of the case are closed with paper caps, and primed with quick-match, in order that one or both ends may be lighted at pleasure. A light in which the composition is 1.5 inches diameter can be easily distinguished at the distance of 15 miles.

INCENDIARY FIREWORKS.

Incendiary preparations are *fire-stone*, *carcasses*, *incendiary-match*, and *hot shot*.

378. **Fire-stone.** Fire-stone is a composition that burns slowly, but intensely; it is placed in a shell, along with the bursting-charge, for the purpose of setting fire to ships, buildings, &c.

Composition. It is composed of—

NITRE.	SULPHUR.	ANTIMONY.	ROSIN.
10	4	1	3

Preparation. In a furnace of the second kind, or in a kettle in the open air, melt together one part of *mutton tallow* and one part of *turpentine;* the composition, having been properly pulverized and mixed, is added to the melted tallow and turpentine, in small quantities. Each portion of the composition should be well stirred with long wooden spatulas to prevent it from taking fire, and each portion should be melted before another is added.

How used. When fire-stone is to be used in shells, it is cast into cylindrical moulds, made by rolling rocket-paper around a former, and securing it with glue. A small hole is formed in the composition by placing a paper tube in the centre of each mould (*a*, fig. 129). When the melted composition has become hard, this hole is filled with a priming of fuze composition, driven as in the case of a fuze. The object of this priming is to insure the ignition of the fire-stone by the flame of the bursting charge.

Fig. 129.

There are two sizes of moulds, the largest for shells above the 8-in., and the other for the 8-in. and all below it.

379. **Carcass.** A carcass is a hollow cast-iron projectile filled with burning composition, the flame of which issues through four fuze-holes, to set fire to combustible objects.

Composition. The composition is the same as for portfires, mixed with a small quantity of finely-chopped *tow*, and as much *white turpentine* and *spirits of turpentine* as will give it a compressible consistency.

Preparation. The composition is compactly pressed into the carcass with a drift, so as to fill it entirely. Sticks of wood 0.5 in. diameter are then inserted into each fuze-hole, with the points touching at the centre, so that when withdrawn corresponding holes shall remain in the composition. In each hole, thus formed, three strands of quick-match are inserted, and held in place by dry port-fire composition, which is pressed around them. About three inches of the quick-match hangs out when the carcass is inserted in the piece; previously to that, it is coiled up in the fuze-hole, and closed with a patch of cloth dipped in melted kit.

A *common shell* may be loaded as a carcass by placing the bursting-charge on the bottom of the cavity, and covering it with carcass composition, driven in until the shell is nearly full, and then inserting four or five strands of quick-match, secured by driving more composition. This projectile, after burning as a carcass, explodes as a shell.

380. **Incendiary match.** Incendiary match is made by boiling slow-match in a saturated solution of nitre; drying it; cutting it into pieces, and plunging it into melted fire-stone. It is principally used in loaded shells.

381. **Hot shot.** For the purpose of setting fire to wooden vessels, buildings, &c., solid shot are heated in a furnace, before firing, to a red heat.* The time required to heat a 42-pdr. shot to a red heat is about half an hour. The precautions to be observed in loading hot shot are, that the cartridge be perfectly tight, so that the powder shall not scatter along the bore, and that a wad of pure clay, or hay, soaked in water, be interposed between the cartridge and the shot. When properly loaded, the shot may be allowed to cool without igniting the charge.

FIREWORKS FOR LIGHT.

The preparations for producing light are *fire-balls*, *light-balls*, *tarred links*, *pitched fascines*, and *torches*.

* In the British sea-coast service shells are used for incendiary purposes by filling them with molten iron drawn from a small cupola furnace. If the shell be broken on striking, the hot iron is scattered about; if it be not broken, the heat penetrates through the shell with sufficient intensity to set wood on fire.

382. **Fire-ball.** A fire-ball is an oval-shaped canvas sack, filled with combustible composition (fig. 130). It is intended to be thrown from a mortar to light up the works of an enemy, and is loaded with a shell to prevent it from being approached and extinguished.

Fig. 130.

Sack. The sack is made of sail-cloth, cut into three oval pieces or gores, and sewed together at their edges. Several thicknesses of cloth may be used, if necessary. One end of the sack is left open, and, after being sewed, it is turned to bring the seam on the inside.

Composition. The composition for a fire-ball consists of—

NITRE.	SULPHUR.	ANTIMONY.
8	2	1

After being pulverized, mixed, and sifted, the composition is moistened with one-thirtieth of its weight of water, and again passed through a coarse sieve. The ball is filled by pouring a layer of composition into the sack, and placing the shell (fuze down) upon it; after this, the composition is well rammed around and above the shell, and the sack is closed at the top.

Finishing. The bottom of the sack is protected from the force of the charge by an iron cup (*a*), called a *culot*, and the whole is covered and strengthened with a network of spun-yarn, or wire, and then overlaid with a composition of pitch, rosin, &c.

A fire-ball is primed by driving into the top of the composition a greased wooden pin about three inches deep, and filling the hole thus formed with fuze composition, driven as in a fuze; space is left at the top of each hole for two strands of quick-match, which are fastened by driving the composition upon them. The fuze-hole is covered with a patch saturated with kit composition, which is a mixture of rosin, beeswax, pitch, and tallow.

383. **Light-ball.** Light-balls are made in the same manner as fire-balls, except that, being used to light up our own works, the shell is omitted.

384. **Tarred links.** Tarred links are used for lighting up a rampart, defile, &c., or for incendiary purposes. They consist of coils of soft rope, placed on top of each other, and loosely tied together; the exterior diameter of the coil is 6 inches, and the interior 3 inches. They are immersed for about ten minutes in a composition of 20 parts of *pitch*, and one of *tallow*, and then shaped under water; when dry, they are plunged in a composition of equal parts of *pitch* and *rosin*, and rolled in tow or saw-dust. To prevent the composition from sticking to the hands, they should be previously covered with linseed oil.

How used. Two links are put into a rampart grate, separated by shavings. They burn one hour in calm weather, and half an hour in a high wind, and are not extinguished by rain. To light up a defile, the links are placed about 250 feet apart; to light up a march, the men who carry the grates should be placed to the leeward of the column, and about 300 feet apart.

385. **Pitched fascines.** Fagots of vine-twigs, or other

very combustible wood, about 20 in. long and 4 in. diameter, tied in three places with iron wire, may be treated in the same manner, and used for the same purposes as links. The incendiary properties of pitched fascines may be increased by dipping the ends in melted rock-fire; when used for this purpose, they are placed in piles intermingled with shavings, quick-match, bits of port-fires, &c., in order that the whole may take fire at once.

386. **Torches.** A torch is a ball of rope impregnated with an inflammable composition, and is fastened to the end of a stick, which is carried in the hand.

Preparation. Old rope, or slow-match, well beaten and untwisted, is boiled in a solution of equal parts of water and nitre; after it is dry, tie three or four pieces (each four feet long) around the end of a pine stick, about two inches diameter, and four feet long; cover the whole with a mixture of equal parts of sulphur and mealed powder, moistened with brandy, and fill the intervals between the cords with a paste of three parts of sulphur and one of quicklime. When it is dry, cover the whole with the following composition:

PITCH.	VENICE TURPENTINE.	TURPENTINE.
3	3	$\frac{1}{2}$

How used. Torches are lighted at the top, which is cracked with a mallet; they burn from one and a quarter to two hours. In lighting the march of a column, the men who carry torches should be about 100 feet apart.

OFFENSIVE AND DEFENSIVE FIREWORKS.

The principal preparations of this class, employed in modern warfare, are *bags of powder* and *light-barrels.*

387. **Bags of powder.** Bags or cases of powder may be used to blow down gates, stockades, or form breaches in thin walls. The *petard* was formerly employed for these purposes, but it is now generally thrown aside.

From trials made in England, it has been shown that a sand-bag (covered with tar, and sanded to prevent it from sticking) containing 50 lbs. of powder, has, sufficient force to blow down a gate formed of 4-inch oak scantling, and supported by posts 10 inches in diameter, and 8 feet apart; and a bag containing 60 lbs. of powder, and weighted with two or three bags of earth, has sufficient force to make a large hole in a 14-inch brick wall. The effect of the explosion may be much increased by making three sides of the bag of leather, and the fourth of canvas, which should rest against the object. A suitable means of exploding bags of powder is a *time-fuze*, or the ordinary *safety-fuze* for blasting rocks.

388. **Light-barrel.** A light-barrel is a common powder-barrel pierced with numerous holes, and filled with shavings that have been soaked in a composition of pitch and rosin; it serves to light up a breach, or the bottom of a ditch.

ORNAMENTAL FIREWORKS.

389. **Object, &c.** Ornamental fireworks are employed

to celebrate great events, as victories, treaties of peace, funerals, &c. They are divided into *fixed pieces*, *movable pieces*, *decorative pieces*, and preparations for *communicating fire* from one part of a piece to another. The different effects are produced by modifying the proportions of the ingredients of the burning composition, so as to quicken or retard combustion, or by introducing substances that give color and brilliancy to the flame.

The fixed pieces are *lances*, *petards*, *gerbes*, *flames*, &c.

390. **Lance.** Lances are small paper tubes from 0.2 to 0.4 in. diameter, filled with a composition which emits a brilliant light in burning (*a*, fig. 131). Instead of a single composition, each lance may contain two or more compositions, which, in turn, emit different-colored flames. The case should be as thin as possible, in order that the color of the flame of the composition may not be affected by that of the paper. Lances are generally employed to form figures; this is done by dipping one end in glue, and sticking them in holes arranged after a certain design, in a piece of wood-work.

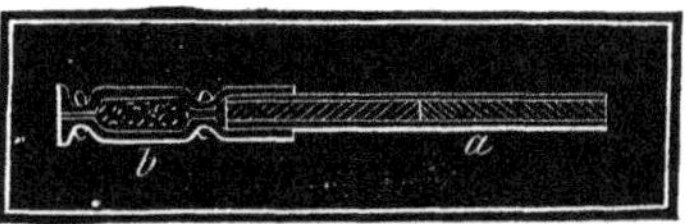

Fig. 131.

391. **Petard.** Petards are small paper cartridges filled with powder. One end is entirely choked, and the other is left partially open for the passage of a strand of quick-match, destined to set fire to the powder.

A petard is usually placed at the fixed end of a lance, that the flame may terminate with an explosion (*b*, fig. 131); they are also used to imitate the fire of musketry.

392. **Gerbe.** Gerbes are strong paper tubes or cases, filled with a burning composition. The ends are tamped

with moist plaster of Paris or clay; through one, a hole is bored, extending a short distance into the composition, that it may emit a long sheaf or *gerbe* of brilliant sparks.

The diameter of the case is about one inch, and the length depends upon the required time of burning. The number of blows to each ladleful of composition is ten.

Gerbes are secured to the frame of the piece with wire or strong twine, and pointed in the direction that the flame is to take.

Composition.

MEALED POWDER.	NITRE.	SULPHUR.	CAST-IRON FILINGS MIXED WITH SULPHUR.
32	16	10	26.4

393. **Flame.** Flames consist of lance or star composition, driven into paper cases or earthen vases. The diameter of the burning surface should be large, to give intensity to the flame. Lance composition is driven dry, and with slight pressure. Star composition should be moistened, and driven with greater pressure than the preceding.

MOVABLE PIECES.

The movable pieces are *sky-rockets*, *tourbillions*, *saxons*, *jets*, *Roman candles*, *paper shells*, &c.

394. **Sky-rocket.** Sky-rockets are the same as the signal-rockets before described, except that the com-

position is arranged to give out a more brilliant train of fire.

Composition.

MEALED POWDER.	NITRE.	SULPHUR.	CAST-IRON FILINGS.
122	80	40	40

395. **Tourbillion.** The tourbillion is a case filled with sky-rocket composition, and which moves with an upward spiral motion. The spiral motion is produced by six holes—two lateral holes (one on each side), for the rotary motion, and four on the under side, for the upward motion. It is steadied by two wings formed by attaching a piece of a hoop to the middle of the case, and at right angles to its length.

To give it a proper initial direction, a hole is made through the centre of the case to fit on a vertical spindle, which is fastened to an upright post.

396. **Saxon.** The saxon is the same as the tourbillion, except that it is only pierced with the central and two lateral holes, and has no wings. The central hole is placed on a horizontal spindle, and the piece has the appearance of a revolving sun.

397. **Jets.** Jets are rocket-cases filled with a burning composition; they are attached to the circumference of a wheel, or the end of a movable arm, to set it in motion. They also produce the effect of gerbes; and to increase the circle of fire, they are inclined to the radius at an angle of 20° or 30°.

Composition.

MEALED POWDER.	NITRE.	SULPHUR.	CAST-IRON FILINGS.
50	36.5	15	24.6

398. **Roman candles.** A Roman candle is a strong paper tube containing stars, which are successively thrown out by a small charge of powder placed under each star. A slow-burning composition is placed over each star to prevent all of them from taking fire at once.

Slow Composition.

MEALED POWDER.	CHARCOAL.
2	1

399. **Paper shell.** This piece is a paper shell filled with decorative pieces, and fired from a common mortar. It contains a small bursting-charge of powder, and has a fuze regulated to ignite it when the shell reaches the summit of its trajectory.

The shell is made by pasting several layers of thick paper over a sphere of wood, cutting the covering thus formed in halves, so as to remove the sphere, joining the halves again, and pasting paper over them until the thickness is sufficient to resist the charge of the mortar.

400. **Decorative pieces.** Decorative pieces are *stars*, *serpents*, *marrons*, &c., described under the head of rockets.

401. **For communicating fire.** Preparations for com-

municating fire from one piece to another are *quick-match*, *leaders*, *port-fires*, and *mortar-fuzes*.

The leader is a thin paper tube containing a strand of quick-match, and it is united to a piece by pasting pieces of paper over the joint. If the piece is to be fired at once, the leader may be omitted, and strands of quick-match tied together used in its place.

402. **General remark.** The foregoing pieces are generally mounted on pieces or frames of light wood, and are susceptible of being combined so as to produce a great variety of striking effects.

PART II.

CHAPTER VIII.

SCIENCE OF GUNNERY.

THE science of gunnery treats of the motion of projectiles, and of their effects. Ballistics is that branch of the science of gunnery which treats of the motion of projectiles.

403. **History of ballistics.** Ancient artillerists considered that the trajectory, or path described by a projectile after it left its piece, was composed of three distinct parts:—1st. The *violent*, which approached a straight line. 2d. The *middle*, or *mixed*, which was a circle. 3d. The last, or *natural*, which was also a right line.

Tartaglia, an Italian engineer, invented the quadrant for measuring elevations, which he divided into twelve parts, and by which he was able to compare the ranges of different cannon, fired under the same or different degrees of elevation. He demonstrated that no portion of the trajectory was a right line, and that the angle which gave the greatest range was 45°.

Galileo. About 1638, Galileo discovered the laws which govern the fall of bodies, and from these he demonstrated that the curve described by a projectile, thrown in a direction oblique to the horizon, is a parabola, the axis of which is vertical. He did not consider that the air offered any material resistance to the motion of artillery projectiles.

Newton. About 1723, Newton demonstrated that the curve described by a spherical projectile in the air, was far from being a parabola; that the two branches were dissimilar, and that the descending branch would become vertical if sufficiently prolonged. While he considered the resistance of the air proportioned to the square of the velocity, he did not conceal the fact that this was but an approximation to the true relation, which remained to be determined by experiment.

Robins. About 1765, Robins invented an instrument for determining the initial velocity of a projectile, called the ballistic pendulum, by which he was able to show that the range in vacuo was much greater than in air. He also discovered that the rotary motion which spherical projectiles generally assume around their centres of gravity, will cause them to deviate from their true direction.

Hutton. Hutton, who lived about the beginning of the present century, improved the ballistic pendulum, and applied it to determine the true law of the resistance of the air, as exemplified in projectiles of small calibre.

At Metz, in 1839 and '40, further experiments were made on the resistance of the air to projectiles of large size, moving with high velocities, and the law of variation was determined with great accuracy.

INITIAL VELOCITY.

404. **Fundamental questions.** The subject of ballis-

tics presents two fundamental questions: 1st. To determine the initial velocity of a projectile for a known piece and charge of powder. 2d. Knowing the initial velocity and angle of projection, to determine the range, time of flight, remaining velocity, and, in fact, all the circumstances of the projectile's motion.

405. **Definition of velocities.** The velocity of a projectile, at any point of its flight, is the space in *feet*, passed over in a *second* of time, with a continuous, uniform motion. *Initial* velocity is the velocity at the muzzle of the piece; *remaining* velocity is the velocity at any point of the flight; *terminal* velocity is the velocity with which it strikes its object; and *final* velocity of descent in air, is the uniform velocity with which a projectile moves, when the resistance of the air becomes equal to the accelerating force of gravity.

The initial velocity of a projectile may be determined by the principles of mechanics which govern the action of the powder, the resistance of the projectile, &c., or by direct experiment.

406. **By mechanical principles.** The instant that the charge of a fire-arm is converted into gas, it exerts an expansive effort which acts to drive the projectile out of the bore. If the gaseous mass be divided into elementary sections perpendicular to its length, it will be seen that, in their efforts to expand, each section has not only to overcome its own inertia, but the inertia of the piece and projectile, as well as the inertia of the sections which precede it. The tension of each section, therefore, increases from the extremities of the charge to some intermediate point where it is a maximum. The pressure on all sides of the section of maximum density

being equal, it will remain at rest, while all the others will move in opposite directions, constantly pressing against the projectile and piece, and accelerating their velocities.

As the projectile moves in the bore, the space, in which the gases expand is increased, while their density is diminished; it follows that the force which sets a projectile in motion in a fire-arm varies from several causes; 1st. It varies as the space behind the projectile increases, or as the velocity regarded as a function of the time; 2d. It varies throughout the column of gas for the same instant of time; and 3d. It varies from the increasing quantities of gas developed in the successive instants of the combustion of the powder.

Piobert has made the movement of a projectile in a fire-arm the subject of a very elaborate analytical investigation, based on the mechanical principles of the conservation of the motion of the centre of gravity, living forces, and Rumford's formula for the relation between the density and pressure of fired gunpowder.

Formula for initial velocity. By supposing the weight of the projectile to be nothing, compared to that of the piece and carriage combined, that the tension of the gases is proportional to the density, that the length of the bore is sufficient for the entire charge to be converted into gas, and that the projectile has no windage, the elaborate equations of Piobert may be reduced to

$$V = \lambda \sqrt{\frac{m}{p+\frac{1}{3}m} \cdot \log.\frac{M}{m}};$$

in which V is the initial velocity, λ a constant to be determined by experiment, m the weight of the powder, p

the weight of the projectile, and M the weight of powder (loose) which would fill the bore.

The above value of V should be diminished for the loss arising from windage; the loss of force from windage is directly proportional to the space between the bore and projectile, and inversely as the area of the bore. Hence we have

$$\Delta\frac{C^2-R^2}{C^2},$$

in which Δ is a constant to be determined by experiment, C is the radius of the bore, and R the radius of the projectile. For ordinary windage this may be replaced by

$$\Delta\frac{W}{C},$$

in which W is the windage, and the general expression for the initial velocity becomes

$$V=\lambda\sqrt{\frac{m}{p+\frac{1}{3}m}\log.\frac{M}{m}}-\Delta\frac{W}{C}.$$

There are three unknown quantities in this equation; V, λ, and Δ; V can be determined by direct experiment for two or more charges of powder, and projectiles, giving two equations containing the remaining unknown quantities λ and Δ. According to the experiments made at the Washington arsenal with the ballistic pendulum, the mean values of the co-efficients λ and Δ, for Dupont's powder in guns of various calibres (from 6-pdr. to 32-pdr.) are: $\lambda=3500$ feet, and $\Delta=3200$ feet.

M is equal to the gravimetric density of the powder (referred to pounds and inches) multiplied by the volume of the bore.

Example. What is the initial velocity of a 6-pdr. shot fired with a service-charge?

$m=1.25$ lbs.; $p=6.25$ lbs.; $C=1.83$ in; $W=.09$ in.; length of bore, 57.5 in.; weight of a cubic inch of powder, 0.0293 lbs.

$$V=3500\sqrt{\frac{1.25}{6.25+.42}\cdot\log\cdot\frac{17.82}{1.25}}-3200\frac{.09}{1.83}=1444.\text{ ft.}$$

The mean of 11 fires with the 6-pdr. gun pendulum at West Point, in November, 1860, was 1436.5 feet.

407. **Practical rule for initial velocity.** For the ordinary purposes of practice, where the weight of the powder and projectile alone vary, initial velocities may be considered *as directly proportional to the square root of the weight of powder divided by the square root of the weight of the projectile;* or

$$V:V_{,}::\frac{\sqrt{p}}{\sqrt{m}}:\frac{\sqrt{p'}}{\sqrt{m'}},\ V=V'\frac{\sqrt{m'}}{\sqrt{p'}}.\frac{\sqrt{p}}{\sqrt{m}}.$$

When V' is known for a given charge of powder p' and projectile m', the value of V can be obtained for any other charge of powder, p, and projectile, m, of the same calibre. This law however only holds true within certain limits, or when the powder is completely consumed before the projectile leaves the piece.*

408. **What affects initial velocity.** The principal

* *Table of Initial Velocities with service charges.*

KIND OF CANNON.	CHARGE OF POWDER.	KIND OF PROJECTILE.		
		SHOT.	SHELLS.	SPH'L CASE.
	Lbs.	Feet.	Feet.	Feet.
6-pdr. field,....................	1.25	1439		1357
12-pdr. "	2.50	1486		1486
12-pdr. " howitzer,...........	1.00		1054	953
24-pdr. siege-gun, ...	6.00	1680	1670	
	8.00	1870		
8-inch siege howitzer,..........	4.00		907	
32-pdr. sea-coast gun,..........	8.00	1640	1450	
15-inch columbiad...............	40.00		1000	

NOTE.—When the initial velocities of shot, shells, and spherical-case shot are given, the weight of the charge refers to the solid shot.

causes which influence initial velocity, are the character of the piece, powder, and projectile. Most of these have been considered under their appropriate heads, in treating of the construction of cannon; it will only be necessary, therefore, to recapitulate them here. They are the size and position of the vent, the windage, the length of the bore, the form of the chamber, the diameter and density of the projectile, the windage of the cartridge, and the form, size, density, and dryness of the grains of powder, and the barometric, thermometric, and hygrometric states of the atmosphere.

It has been found by late experiments that the initial velocity is unaffected by the angle of fire. Theoretically, varying the weight of the piece should exert an influence on the initial velocity; but, in consequence of the great disparity of the weight of the piece and projectile, this influence is inappreciable in practice.

409. **Determination of initial velocity by experiment.** A great variety of instruments have been invented to determine directly the initial velocity of a projectile, the most reliable of which are the *gun-pendulum*, the *ballistic pendulum*, and the *electro-ballistic machines*.

In the first, the velocity of the projectile is determined by suspending the piece as a pendulum, and measuring the recoil impressed on it by the discharge; the expression for the velocity is deduced from the fact, that the quantity of motion communicated to the pendulum is equal to that communicated to the projectile, charge of powder, and the air. The second apparatus is a pendulum, the bob of which is made strong and heavy, to receive the impact of the projectile; and the expression for the velocity of the projectile is deduced from the fact, that the quantity of motion of the projectile before impact is equal to that of the pendulum and projectile after impact. These machines have been carried to great perfection, both in this country and France, and very accurate and important results have been obtained by them; but they are very expensive, and cannot be easily adapted to the various wants of the service.

The employment of electricity to determine the velocity of projectiles, was first suggested by Wheatstone, in 1840. The application depends on the very great velocity of electricity, which, for short distances, may be considered instantaneous. The general method of applying this agent is by means of galvanic currents, or wires, supported on target frames, placed in the path of the projectile, and communicating with a delicate time-keeper. The successive ruptures of the wires mark on the time-keeper the instant that the projectile passes each wire, and knowing the distances of the wires apart, the mean velocities, or velocities at the middle points, can be obtained by the relation, velocity $=\frac{\text{space}}{\text{time}}$.*

The various plans in use differ only in the manner of recording and keeping the time of flight; one of the simplest and most common instruments employed is the pendulum. The ballistic machine of Captain Navéz, of the Belgian service, has been tried in this country, but has been found too delicate and complicated for general purposes.

* In consequence of the variable nature of the resistance of the air, this mean velocity does not strictly correspond to the middle point between the targets. The difference, however, is very slight, as is shown by Captain Navéz in the case of a 6-pdr. ball moving with an initial velocity of 600 meters, over a space of 50 meters; the difference between the mean velocity and the velocity which it should have at the middle point, is only 0.05 meter.

410. **The West Point ballistic machine.** Fig. 132 represents the front and end views of an electro-ballistic machine originally devised by the author for the use of the Military Academy, and since adopted by the ordnance department, for proving powder, &c.

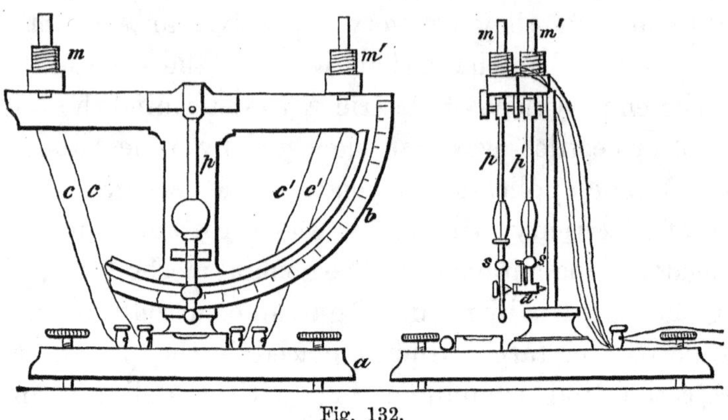

Fig. 132.

a is a bed-plate of metal, which supports a graduated arc (*b*). This arc is placed in a vertical position by means of *thumb-screws* and *spirit-levels* attached to it; and it is graduated into degrees and fifths, commencing at the lowest point of the arc, and ending at 90°.

pp' are two pendulums having a common axis of motion, passing through the centre, and perpendicular to the plane of the arc. The bob of the pendulum *p'* is fixed, but that of *p* can be moved up and down with a thumb-screw, so as to make the times of vibration equal.

m and *m'* are two electro-magnets attached to the horizontal limb of the arc, to hold up the pendulums when they are deflected through angles of 90°.

s and *s'* are pieces of soft iron attached to the prolongations of the suspension-rods, in such way as to be in contact with the lower poles of the magnets, when the pendulums are deflected.

d is an apparatus to record the point at which the pendulums pass each other, when they fall by the breaking of the currents which excite the magnets. It is attached to the prolongation of the suspension-rod *p'*; and consists essentially of a small pin enclosed in a brass tube; the end of the pin near the arc has a sharp point, and the other is terminated with a head, the surface of which is oblique to the plane of the arc. As the pendulums pass each other, a blunt steel point attached to the lower extremity of the suspension-rod *p*, strikes against the oblique surface of the head of the pin, which presses the point into a piece of paper clamped to the arc, leaving a small puncture to mark the point of passage. An improvement to the foregoing consists in attaching to the pendulum *p'* a delicate bent lever, which carries on its point a small quantity of printer's ink; the pendulum *p* presses upon this lever, causing the point to touch the arc and leave a small dot opposite to the point where the pendulums pass each other. The magnets are also so arranged that they can be transposed from one pendulum to the other, thereby affording the means of correcting errors arising from inequalities of magnetic power, by taking a mean of two observations. *c c* and *c' c'* represent the wires which conduct the two electric currents to the magnets *m* and *m'*.

Targets. The targets are two frames of wood placed so as to support the wires in a position to be cut by the projectile. For the purpose of obtaining the initial velocity, the first should be placed about 20 feet from the muzzle of the piece, that the flame may not break the current before the projectile reaches it; the position of the second depends on the velocity of the projectile. For cannon, it should be placed about 100 feet from the first target; and for small-arm and mortar projectiles, about 50 feet. The number of times that the wire should cross the targets depends on the size of the projectile and the accuracy of its flight.

Currents and batteries. The magnets should be made of the purest attainable wrought iron, in order that they shall retain no magnetic force after the exciting currents are broken; and for this purpose they are best made of bundles of wire. The batteries should be of nearly equal power and constancy, in order that, in case the magnets do retain a portion of their magnetism, the remaining portion may be as uniform as possible.

Grove's or Bunsen's batteries are the best that can be used for this purpose. The power of the battery is regulated by the distance of the targets and the size of the conducting wires. If a weak battery be used, the magnetic power may be increased by increasing the diameter of the wire, or by resting pieces of soft iron on the upper poles of the magnet.

In experimenting with cannon, the machine should be placed about 120 yards from the piece, to prevent any disturbance from the discharge; at this distance the record will have been made before the sound reaches the machine. For this distance, three cups, in which the zinc cylinders are 8 inches long and 3 inches diameter, and an insulated copper wire .06″ diameter, have been found to answer a good purpose.

The disjunctor. The disjunctor is a small auxiliary instrument for closing and breaking the currents at will.

It affords the means of verifying the accuracy of the pendulum machine by a succession of simultaneous ruptures of the wires; when the machine is in good condition, the position of the point of meeting seldom varies a *tenth* of a degree, an error which corresponds to only .000154 of a second of time.

When the currents are of equal strength, and the starting points are properly adjusted, the point of meeting will be found opposite to the zero of the graduated arc; if of unequal intensity, the point will be found near the zero point and on the side of the stronger magnet. As this position is nearly constant for the same currents, the error of the reading can be easily corrected. If the error be positive, subtract it; if negative, add it.

Arrangement, &c. Figure 133 shows the working arrangement of the several pieces: *a* represents the pendulum; *b* the disjunctor; *c c* and *c′ c′* the currents; *e e′* the batteries; and *d* the position of the gun. To operate them, the disjunctor is closed, the pendulums are deflected, the marking-pin revolved perpendicular to the arc, the piece is fired, and the position of the puncture in the paper, with reference to the graduated arc, noted.

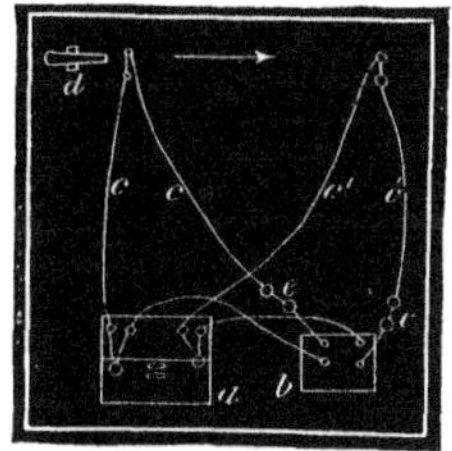

Fig. 133.

To determine the time. The velocity of the electric currents being considered instantaneous, and the loss of power of the magnets simultaneous with the rupture of the currents, it follows that each pendulum begins to move at the instant that the projectile cuts the wire, and that the interval of time corresponds to the difference of the arcs described by the pendulums up to the time of meeting.

Let m and m', fig. 134, represent the positions of the two magnets, and let the interval between the rupture be such that the centres of oscillation will pass each other at i. As the times of vibration are equal, the interval of time will correspond to the arc $i\ i'$, the arc $m'\ i$ being equal to $m\ i'$. A vertical line through the centre of motion bisects the arc $i\ i'$. The reading therefore corresponds to one-half of the required time, or time of passage of the projectile between the wires.

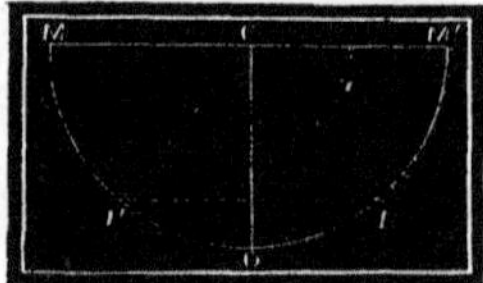

Fig. 134.

To determine a formula for the time that it takes for one of the pendulums to pass over a given arc, let l be the length of the equivalent simple pendulum, v the velocity of the centre of oscillation or point m', y the vertical distance passed over by this point, x the variable angle which the line of suspension makes with the horizontal, and t' the time necessary for the point m' to pass over an entire circumference, the radius of which is l, with a uniform velocity v, we have,

$$v=\sqrt{2gy}.$$

Substituting for y its value in terms of the constant angle of half-oscillation and the variable angle x, the above expression becomes,

$$v=\sqrt{2gl \cos. (90^\circ - x)}\,;$$

from which we see that the velocity of the pendulum increases from its highest to its lowest point, and *vice versa.*

The time t' is equal to the circumference of the circle, the radius of which is l, divided by the velocity, v;

again divide this by 360, we have the time of passing over each degree, or,

$$t=\frac{2\pi l}{360\sqrt{2gl}\cos.(90^\circ-x)}.$$

To determine l, it is necessary to change the cylindrical arms of suspension to knife-edges, in order to determine the time of vibration through a very small arc. The mean of 500 vibrations will be very near the exact time of a single vibration. Knowing the time of a single vibration, the length of the equivalent simple pendulum can be obtained by the relation $l=l't''^2$, in which t'' is this time, and l' is the length of the simple second's pendulum at the place of observation.

At West Point $l'=39.11448$ inches.
" " $g=32.17050$ feet.

In this way all the constants of the expression for t are known, and by assigning different values to x, a table can be formed, from which the times corresponding to different arcs can be obtained by simple inspection. The table in chapter XIII. is calculated for the West Point machine.

MOTION OF A PROJECTILE IN VACUO.

411. **Determination of equations of motion.** A *projectile* is a body thrown or impelled forward, generally in the air; and the *trajectory* is the line described by its centre of inertia. The movement of a projectile will be considered firstly in vacuo, and secondly in the air.

Let A (fig. 135) be the position of the muzzle of a fire-arm, and the line $A\ B$ its axis prolonged.

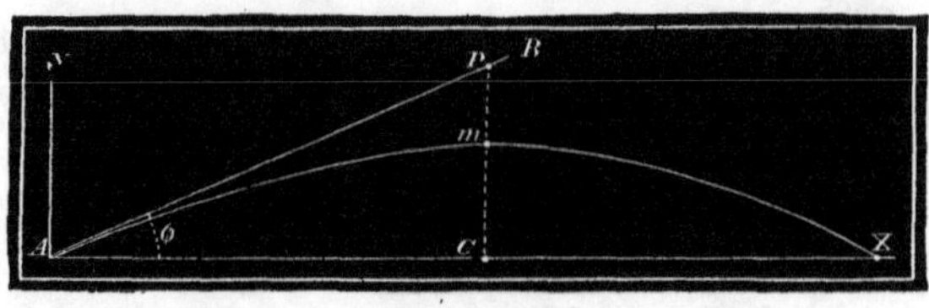

Fig. 135.

Let ϕ represent the angle which this line makes with the horizontal plane, or the *angle of projection.*

V the initial velocity.

v the velocity at the point m.

t the time of flight to the same point.

θ the inclination of the tangent at this point.

x, y, the co-ordinates of this point.

X the horizontal range.

Y the greatest height of ascent.

T the whole time of flight, or for the range X.

If the projectile were only acted upon by the force of the discharge, it would move in the straight line $A\ B$, and after a time, t, would reach the point P; but it is constantly drawn to the earth by the force of gravity, and instead of being found at the point P, it is found at the point m, situated at a distance below P equal to the distance which it would fall in the same time under the influence of gravity, or $\frac{1}{2}gt^2$, g being the velocity generated by gravity in a second of time.

The distance PC is equal to x tan. ϕ; the distance mC, or y, is equal to this distance diminished to $\frac{1}{2}gt^2$, or,

$$y = x \tan. \phi - \tfrac{1}{2}gt^2;$$

$x = t\,V \cos. \phi$, or $t = \frac{x}{V \cos. \phi}$. Substituting this for t in the preceding equation, it becomes,

$$y = x \tan. \phi - \frac{g}{2} \cdot \frac{x^2}{V^2 \cos.^2 \phi}.$$

From the laws which govern falling bodies, $V=\sqrt{2gH}$, or $V^2=2gH$; in which H is the height due to the velocity V. Substituting this value of V^2, the equation becomes,

$$y=x \tan. \phi-\frac{x^2}{4H \cos.^2 \phi}, \qquad (1)$$

which is the equation of a parabola.

From the same figure we obtain—

$$y= Vt \sin. \phi-\tfrac{1}{2}gt^2. \qquad (2)$$

$$x= Vt \cos. \phi. \qquad (3)$$

$$t=\frac{x}{V \cos. \phi}. \qquad (4)$$

2d. To determine the vertical ascent and horizontal range of the projectile, differentiate equation (1), and place the value of $\frac{dy}{dx}=0$; whence we obtain,

$$X=4 H \sin.\phi \cos.\phi=2H \sin. 2\phi. \qquad (5)$$

$\frac{1}{2}X$ being the abscissa of the highest point,

$$Y=H \sin.^2\phi. \qquad (6)$$

The first value of X shows, *that the range can be obtained with two angles of projection, provided they be complements of each other;* the second value shows, *that the greatest range corresponds to an angle of* 45°, *and that this range is equal to twice the height due to the velocity;* and, also, *that variations in the angle of fire produce less variations in range as the angle of fire approaches* 45°.

3d. If two projectiles be thrown under the same angle, with different initial velocities, V and V', the ranges being X and X', we have,

$$X=2H \sin. 2\phi=\frac{V^2}{g}\sin. 2\phi, \text{ and } X'=\frac{V'^2}{g}\sin. 2\phi;$$

and from these we have,

$$\frac{V}{V'}=\frac{\sqrt{X}}{\sqrt{X'}}. \qquad (7)$$

Therefore, under the same angle of fire, the ranges are proportional to the squares of the velocities; and reciprocally, the velocities are proportional to the square roots of the ranges.

4th. The velocity at any point is equal to $\frac{ds}{dt}$, or $v^2=\frac{dy^2+dx^2}{dt^2}$. Substituting the values of dy and dx, obtained by differentiating equations (2) and (3), we have

$$v^2=V^2-2\,Vgt\sin.\phi+g^2t^2.$$

Substitute for $-2\,Vgt\sin.\phi+g^2t^2$ its value $-2gy$, derived from equation (2), we have,

$$v^2=V^2-2gy.$$

Replace V^2 by $2gH$, and reducing, the expression becomes,

$$v=\sqrt{2g(H-y)}. \qquad (8)$$

This shows *that the velocity of a projectile, at any point, depends on its height above the muzzle of the piece; and that it is equal to that which is attained in falling through the height* $(H-y)$. It also shows *that the velocity is least when* y *is greatest, or at the summit of the trajectory; and that the velocities at the two points in which the trajectory cuts the horizontal plane are equal.*

5th. The total time of flight may be determined by substituting the value of $X=4H\sin.\phi\cos.\phi$, equation (5), in equation (4), which becomes

$$T=\frac{4H\sin.\phi}{V}=\frac{V\sin.\phi}{\frac{1}{2}g}. \qquad (9)$$

If $\phi=45°$, $\sin.\phi=\sqrt{\frac{1}{2}}$, and $V=\sqrt{gX}$. Calling $T_{,}$ the time of flight, we have,

$$T_{,}=\sqrt{\frac{X}{\frac{1}{2}g}}=\sqrt{\frac{X}{16.07}}=\tfrac{1}{4}\sqrt{X}.$$

Hence the time of flight for an angle of 45° *is equal to the square root of the quotient of the range divided by one-half of the force of gravity;* or, *it is approximately equal to one-fourth of the square root of the range expressed in feet.*

6th. The tangent of the angle made by a tangent line at any point of the trajectory is equal to $\frac{dy}{dx}$, which is obtained by differentiating equation (1); calling this angle θ, we have,

$$\tan.\theta=\tan.\phi-\frac{x}{2H\cos.^{2}\phi}. \qquad (10)$$

Substitute the value of $X=4\,H\sin.\phi\cos.\phi$, the angle of fall on horizontal ground is $\tan.\theta=-\tan.\phi$; that is to say, *the angle of fall is equal to the angle of projection, measured in an opposite direction.*

7th. The position of a point being given, to find the initial velocity necessary to attain it, let a and b be the horizontal and vertical co-ordinates of this point of the curve, and ε its *angle of elevation.* Substituting these quantities in equation (1), and recollecting that $\tan.\varepsilon=\frac{b}{a}$, we have,

$$H=\frac{a}{4\sin.(\phi-\varepsilon)\cdot}\,\frac{\cos.\varepsilon}{\cos.\phi},$$

$$\text{or,}\quad V=\sqrt{\frac{ag}{2\sin.(\phi-\varepsilon)\cdot\cos.\phi}\,\frac{\cos.\varepsilon}{}}. \qquad (11)$$

8th. The position of a point being given, to find the angle of fire necessary to attain it. Substituting a and b for x and y in equation (1), we have,

$$b = a \tan.\phi - \frac{a^2}{4H \cos.^2\phi},$$

from which to determine ϕ.

Making $\tan.\phi = \alpha$, we have, $\cos.^2\phi = \frac{1}{1+\alpha^2}$; which being substituted in the above equation gives—

$$\alpha = \tan.\phi = \frac{1}{a}\left(2H \pm \sqrt{4H^2 - 4Hb - a^2}\right). \qquad (12)$$

The two values of $\tan.\phi$ show *that the point may be attained by two angles of projection;* and the radical shows *the solution of the problem is possible when the quantity under it is positive;* or,

$$4H^2 > 4Hb + a^2.$$

412. **Practical application of formula.** The preceding formula will only be found to answer in practice for projectiles which experience slight resistance from the air, or for heavy projectiles moving with low velocities, as is commonly the case with those of mortars and howitzers.

The following table gives the difference between the observed and calculated times of flight of the French 8 and 10-inch mortar shells, weighing 64 and 119 lbs. respectively. The initial velocities being unknown, the times are calculated from the observed ranges.

The observed times are invariably greater than the calculated times, as might be expected from the resistance of the air, which retards the motion of projectiles.

Kind of projectiles.	Weight of powder.	Ranges at angles of		Times of flight.			
				45°		30°	
		45°	30°	Observed.	Calculated.	Observed.	Calculated.
	Kilog.	Meters.	Meters.	Seconds.	Seconds.	Seconds.	Seconds.
8-inch.	0.234	343	290	9.8	8.4	6.8	5.8
	0.351	629	561	12.9	11.3	10.0	8.1
	0.585	1146	1011	16.0	15.3	12.3	10.9
	0.994	1792	1690	20.8	19.2	16.9	14.1
10-inch.	0.468	457	383	11.0	9.7	7.5	6.8
	0.693	734	637	14.0	12.2	10.0	8.7
	1.054	1132	980	17.0	15.2	12.0	10.2
	1.405	1555	1355	20.0	17.8	14.0	12.6
	1.639	1757	1516	23.0	18.9	15.0	13.4

The next table shows the observed and calculated ranges, for 30° elevation, and the observed ranges for 45° elevation, for the above projectiles, the initial velocities being the same for each projectile.

Ranges of 10-inch Mortar Shells.				Ranges of 8-inch Mortar Shells.			
45° elevation.	30° elevation.			45° elevation.	30° elevation.		
Observed.	Observed.	Calculated.	Difference	Observed.	Observed.	Calculated.	Difference
Meters.	Meters.	Meters.	Meters.	Meters.	Meters.	Meters.	Meters.
457	383	396	+13	343	290	298	+ 8
734	637	637	0	629	561	545	—16
1132	980	982	+ 2	1146	1011	993	—13
1555	1355	1350	— 5	1792	1690	1552	—138
1757	1516	1522	+ 6				

It appears from the foregoing tables, that the ranges of mortars with different degrees of elevation, can be calculated up to about 1,400 yards from equation (5), or,

$$X = 2H \sin. 2\phi,$$

and the times from equation (4), or

$$T=\frac{X}{V\cos.\phi}.$$

RESISTANCE OF THE AIR.

413. **Importance of considering it.** A body moving in the air experiences a resistance which diminishes the velocity with which it is animated. That the retarding effect of the air, on projectiles moving with high velocities, is very great, is seen by comparing the actual ranges of projectiles with those computed under the supposition that they move in vacuo. Thus, it has been shown that certain cannon-balls do not range one-eighth as far in the air as they would if they did not meet with this resistance to their motion; and small-arm projectiles, which have but little mass, are still more affected by it.

414. **Law of resistance.** *Incompressible fluid.* The resistance experienced by a plane surface moving parallel to itself through an incompressible fluid, is equal to the pressure of a column of the fluid, the base of which is the moving surface, and its height that due to the velocity with which the surface is moved through the fluid, or, from the law of falling bodies, $h=\frac{v^2}{2g}$; in which h is the height, v the velocity, and g the force of gravity.

The resistance on a given area is therefore proportional to the square of the velocity, and the density of the fluid medium.

Let d, S, and v represent the density or weight of a

unit of volume of the fluid, the area pressed upon, and the velocity of the moving surface, respectively, and ρ the resistance in terms of the unit of weight, and we have,

$$\rho = kdS\frac{v^2}{2g}; \qquad (13)$$

in which k is a coefficient to be determined by experiment.

Compressible fluid. If the medium be formed of compressible gases, as the atmosphere, the density in front of the moving body will be greater than that behind it; and it will be readily seen that the body will meet with a resistance which increases more rapidly than the square of the velocity, in such a manner that the coefficient k, or the density of the medium, d, should be increased by a quantity which is a function of the velocity itself, or, what is the same thing, by adding another term to the resistance which shall be proportional to the cube of the velocity.

In examining the table of resistances, obtained by Hutton from firing a one-pound ball into a ballistic pendulum, at different distances, and with velocities varying from 300 to 1,900 feet, Piobert found, that if v^2 in the foregoing expression be replaced by the binomial term, $\left(v^2 + \frac{v^3}{r}\right)$, in which $\frac{1}{r} = \frac{1}{1427 \text{ ft.}}$, the expression would nearly satisfy the results of experiments.

Calling $A = \frac{kd}{2g}$, and πR^2 the area of the cross section of a projectile, the general expression for the resistance in air becomes,

$$\rho = A\pi R^2\left(1 + \frac{v}{r}\right)v^2. \qquad (14)$$

In this expression, A is the resistance, in pounds, on a square foot of the cross-section of a projectile moving with a velocity of one foot; r is a linear quantity depending on the velocity of the projectile. For all service spherical projectiles, A=.000514; and for all service velocities r=1,427 feet. The value of A for the rifle-musket bullet (page 312) has been determined at the Washington Arsenal, by the method laid down in the note on page 411, and found equal to 0.000358. *This shows that the resistance of the air is about one-third less on the ogeeval than on the spherical form of projectile.* This value has been found to answer well for calculating the ranges of rifle-cannon projectiles.

The coefficient A, being a function of the density of the air, its value depends on the *temperature*, *pressure*, and *hygrometric* condition; in the above value the weight of a cubic foot of air=.075 lb., at a temperature of 60° Fahr., and for a barometrical pressure of 29.5 inches.

If the surface of the projectile be rough or irregular, the value of this coefficient will be slightly too small.

Example.—What is the pressure of the air on a 42-pdr. shot moving with a velocity of 1,500 feet?

$$\rho = .000514 \times 3.1415 \times \overline{.29}^2 \times \overline{1500}^2 \left(1 + \frac{1500}{1427}\right) = 629.3 \text{ lbs.}$$

415. **Fall of a projectile in the air.** The motion of a body falling through the air, will be accelerated by its weight, and retarded by the buoyant effort of the air, and the resistance which the air offers to motion. As the resistance of the air increases more rapidly than the velocity, it follows that there is a point where the retarding and accelerating forces will be equal, and that beyond this, the body will move with a uniform veloc-

ity, equal to that which it had acquired down to this point.

The buoyant effort of the air is equal to the weight of the volume displaced, or $P\frac{d}{D}$; in which P is the weight and D the density of the projectile, and d the density of the air.

When the projectile meets with a resistance equal to its weight, we shall have,

$$P\left(1-\frac{d}{D}\right)=A\pi R^2v^2\left(1+\frac{v}{r}\right); \qquad (15)$$

in which the weight of the displaced air is transferred to the first member of the equation. As the density of the air is very slight compared to that of lead or iron, the materials of which projectiles are made, $\frac{d}{D}$ may be neglected. Making this change, and substituting for P, $\frac{4}{3}\pi R^3D$ (g having been divided out of the second member, should be omitted in the first), the expression for the *final velocity* reduces to

$$v^2\left(1+\frac{v}{r}\right)=\frac{4}{3}\frac{RD}{A}. \qquad (16)$$

The resistance on the entire projectile for a velocity of 1 foot, is $A\pi R^2$; dividing this by $\frac{P}{g}$, or the mass, we get the resistance on a unit of mass.

Calling this $\frac{1}{2c}$, we have,

$$\frac{1}{2c}=\frac{A\pi R^2}{\frac{P}{g}}, \text{ or } 2gc=\frac{P}{A\pi R^2}.$$

Substituting for P its value in the equation of vertical descent, we have,

$$2gc=v^2\left(1+\frac{v}{r}\right);$$

from which we see that v depends only on c; but

$$c=\frac{2}{3}\frac{RD}{gA} \qquad (17)$$

hence, *the final velocity of a projectile falling in the air is directly proportioned to the product of its diameter and density, and inversely proportional to the density of the air, which is a factor of A.*

The expression for the value of (c) shows, *that the retarding effect of the air is less on the larger and denser projectiles.* To adapt it to an oblong projectile of the pointed form, the value of D should be increased, (inasmuch as its weight is increased in proportion to its cross section,) while that of A should be diminished. It follows, therefore, *that for the same calibre, an oblong projectile will be less retarded by the air than one of spherical form, and consequently with an equal and perhaps less initial velocity its range will be greater.*

The value of (c) for service projectiles will be found ready calculated in the tables of fire, in Chap. XIII.

LOSS OF VELOCITY BY RESISTANCE OF THE AIR.

416. **Equations of motion.** For the purpose of determining the velocity which a projectile loses by the resistance of the air, in moving through a certain distance, x, the force of gravity may be disregarded; in which case the trajectory described will be a right line.

Let V be the initial velocity, and v the remaining velocity at the end of the distance x.

The expression for the resistance of the air is, as we have seen,

$$\rho = A\pi R^2\left(1+\frac{v}{r}\right)v^2.$$

But we know that the retarding force of the air is equal to the mass of the projectile against which it acts, multiplied by the first differential coefficient of the velocity, regarded as a function of the time, with its sign changed, and that $\frac{P}{g}$ is the mass of the projectile. We have, therefore,

$$\frac{\rho g}{P}=\frac{dv}{dt}=-\frac{g}{P}A\pi R^2\left(1+\frac{v}{r}\right)v^2.$$

Recollecting that $P=\frac{4}{3}\pi R^3 D$, and that $2c=\frac{4}{3}\frac{RD}{gA}$, the equation reduces to,

$$\frac{dv}{dt}=-\frac{v^2}{2c}\left(1+\frac{v}{r}\right)$$

Integrating this equation between the limits 0 and x, which correspond to V and v, we have,

$$t=2c\left(\frac{1}{v}-\frac{1}{V}\right)-\frac{2c}{r}\log.\frac{1+\frac{r}{v}}{1+\frac{r}{V}} \qquad (18)$$

To obtain a relation between the space and velocity, we have $v=\frac{dx}{dt}$, or $dt=\frac{dx}{v}$; substituting this in the equation for the intensity of the retarding force, and reducing, we have,

$$dx=-2c\frac{dv}{v\left(1+\frac{v}{r}\right)}.$$

Integrating between the same limits as in the preceding case, we have,

$$x=2c \log. \frac{1+\frac{r}{v}}{1+\frac{r}{V}} \text{ or } 1+\frac{r}{v}=\left(1+\frac{r}{V}\right)e^{\frac{x}{2c}}. \qquad (19)$$

Solving this equation with reference to v, we have,

$$v=\frac{r}{\left(1+\frac{r}{V}\right)e^{\frac{x}{2c}}-1} \qquad (20)$$

Substituting, in equation (18), x for its value given in equation (19), we have,

$$t=2c\left(\frac{1}{v}-\frac{1}{V}\right)-\frac{x}{r}. \qquad (21)$$

The logarithms in the above equations belong to the Napierian system, and are obtained by multiplying the corresponding common logarithm by 2.3026: $e=2.713$.

Practical remarks. Equation (19) gives the space passed over by a certain projectile when the velocities at the commencement and end of the flight, are known.

Equation (20) gives the remaining velocity when the initial velocity and the space passed over are known.

Equation (21) gives the time of flight when the velocities at the beginning and end and the space passed over, are known.

The distance at which the velocity V is reduced to v, and the duration of the trajectory, being proportional to c, are directly proportional to the product of the diameter and density of the projectile, and inversely proportional to the density of the air. This fact shows the great advantage, in point of range, to be derived *from using large projectiles over small ones, of solid projectiles over hollow ones, of leaden projectiles over iron ones, and of oblong projectiles over round ones.*

FORM OF PROJECTILE.

417. **Theory.** When a body moves through the air, the gaseous particles in front are crowded upon each other until they meet with a certain resistance, after which they move off laterally, and finally pass around and arrange themselves in rear of the moving body. It is evident that the difference of the densities, or pressures, front and rear, depends on the velocity with which the displaced particles rearrange themselves after displacement; and this, in turn, depends on the shape, and extent of the surfaces of the moving body. The best form for a projectile can only be determined by experiment, as theory and experiment do not agree in their results.

According to theory, if a plane of given area be moved through the air, it meets with a resistance which is proportional to the square of the sine of the angle which its direction makes with that of motion.

The experiments of Hutton with low velocities show that this is only true in cases of 0° and 90°; that from 90° up to 50° or 60°, the resistance is nearly proportional to the sine; beyond this, it decreases a little more rapidly than the sine, but not so rapidly as the square of the sine:

For an angle of 22° it is only $\frac{1}{2}$ the resistance proportional to the sine.
" " 14° " $\frac{1}{3}$ " " "
" " $9\frac{1}{2}$° " $\frac{1}{4}$ " " "
" " 4° " $\frac{1}{5}$ " " "
" " 2° " $\frac{1}{6}$ " " "

418. **Experiments of Hutton and Borda.** The following are the results of the experiments made by Hutton and Borda, on the resistances experienced by different forms of solids moving through the air with velocities varying from 3 to 25 feet per second.

HUTTON'S EXPERIMENTS. VELOCITY, 10 FEET.

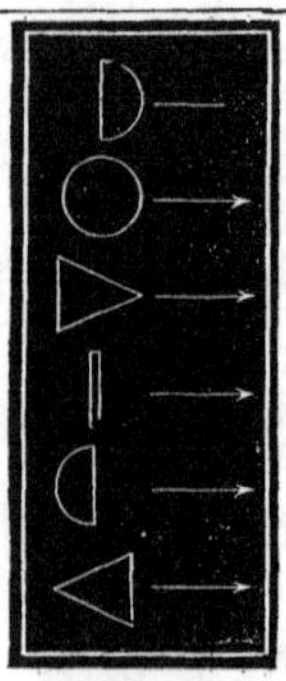

Kind of surface,	Experimental resistance.	Theoretical resistance.
No. 1 Hemisphere (convex surface in front),	119	144
No. 2 Sphere,	124	144
No. 3 Cone, elements inclined to the axis 25° 42′,	126	53
No. 4 Disk,	285	288
No. 5 Hemisphere (plane surface in front),	288	288
No. 6 Cone (base in front),	291	288

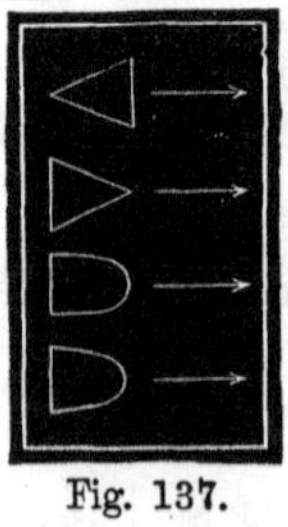

Fig. 136.

BORDA.

Kind of surface.	Experimental resistance.	Theoretical resistance.
No. 1, Prism, with triangular base,	100	100
No. 2, " "	52	25
No. 3, " semi-ellipse,	43	50
No. 4, " ogee,	39	41

Fig. 137.

419. **Conclusions.** The foregoing experiments show: 1st. That the results of theory do not agree with those of practice. 2d. That rounded and pointed solids suffer less resistance from the air than those which present flat surfaces of the same transverse area, but, at the same time, the sharpest points do not always meet with the least resistance. 3d. That where the front surfaces were the same, the resistance was least with those in which the posterior surfaces were the flattest. 4th. That the ogeeval form, or the form of the present rifle-musket bullet, experiences less resistance than any other tried.

These experiments, as before remarked, were made with low velocities, compared to those which ordinarily actuate projectiles, and the conclusions which have been drawn from them may not be strictly applicable in practice. Now that oblong projectiles are used in all kinds of fire-arms, it is important to determine that form which will be least affected by the resistance of the air. It is evident that that form will be the best which, on trial, is found to give the least value to A in equation (14), or, what is the same thing, to give the greatest value to c in equation (21).*

* The author proposes the following method of determining the value of c by the electro-ballistic machine.

Establish four targets in the line of fire, in such manner that the *first* shall be near the piece, the *second* shall be at a distance x from the first, the third at a distance $2x$ from the first, and the *fourth* at a distance of $4x$ from the first; let t, t', and t'' represent the intervals of time corresponding to the distances between the targets, respectively; let v be the velocity at the middle point between the first and third targets, or at the distance x, and let v' be the velocity at the middle point between the first and fourth targets, or at the distance $2x$.

Equation (21) becomes

$$t=\frac{2c}{v}-\frac{2c}{V}-\frac{x}{r}, \text{ or } \frac{2c}{V}=\frac{2c}{v}-\frac{x}{r}-t.$$

TRAJECTORY IN AIR.

420. **Difficulties of the problem.** In consequence of the variable nature of the resistance of the air, it has been found impossible to integrate the differential equations of the real trajectory, even under the supposition that this resistance varies in as simple a ratio as the square of the velocity. Several distinguished mathematicians have obtained expressions which approximate to the true results, but the expressions are generally too complicated to be of much practical value.

421. **Didion's method.** Captain Didion, professor of gunnery in the artillery school at Metz, however, furnishes an approximate solution to this difficult question, which may be used in practice. To do this, he considers the resistance of the air equal to

$$A\pi R^2\left(1+\frac{v}{r}\right)v^2;$$

and by assuming a mean value for the different inclinations of the elements of the trajectory to their horizontal projections, which makes $\frac{ds}{dx}$ constant, he is able to

Since $\frac{2c}{V}$ is the same for all the distances, we have

$$\frac{2c}{v}-\frac{x}{r}-t=\frac{2c}{v'}-\frac{2x}{r}-t', \text{ or } c=\frac{t-t'-\frac{x}{r}}{2\left(\frac{1}{v}-\frac{1}{v'}\right)}.$$

From the note on page 389, we are at liberty to place $v=\frac{2x}{t'}$ and $v'=\frac{4x}{t''}$; substituting these values in the preceding equation, reducing and changing the signs of both numerator and denominator of the second member, we have

$$c=\frac{2x\left(t'-t+\frac{x}{r}\right)}{t''-2t'}.$$

Which equation gives the value of c in terms of t, t', t'', and which can be determined by taking the mean of several shots, with the electro-ballistic machine, at the different distances, x, $2x$, and $4x$.

integrate the differential equations, and place them under the following forms:

$$y = x \tan.\phi - \frac{g}{2}\frac{x^2}{V^2 \cos.^2\phi} B;$$

$$\text{Tan.}\ \theta = \tan.\phi - g\frac{x}{V^2 \cos.^2\phi} I;$$

$$t = \frac{x}{V \cos.\phi} D; \quad v = \frac{V \cos.\phi}{\cos.\theta}\frac{1}{U}.$$

The same notation being preserved as in the equations in vacuo (page 397), it will be perceived that the equations in air differ from those in vacuo, by the multipliers B, I, D, and U, respectively.

The multiplier B relates to the fall of the projectile; I, to the inclination; D, to the duration; and U, to the velocity; they are each functions of $\frac{ax}{c}$ and $\frac{a V_{\prime}}{r}$; in which a is the constant relation of the arc to its projection, $V_{\prime} = V \cos.\phi$, and c and r are co-efficients of the formula for the resistance of the air. (See pages 403 and 406.) The general expression for a particular multiplier, B for instance, is $B\left(\frac{a\,x}{c}; \frac{a\,V_{\prime}}{r}\right)$.

The values B, I, D, and U, for such values of c and r as are likely to arise in service, have been computed, and arranged in tabular form; these tables, their construction, and use, are explained in chapter XIII.

So long as the inclination of the trajectory is slight, a differs but slightly from unity; for an angle of 15° it does not exceed 0.01; and as it only enters into the term which relates to the resistance of the air, the error does not exceed a pressure corresponding to 0.25 in. in the height of the barometer; it may, therefore, be regarded as unity, and $\frac{a x}{c}$ reduces to $\frac{x}{c}$. The same with regard to $\frac{a V_{\prime}}{r}$, or $\frac{a\, V \cos. \phi}{r}$; as a cos. ϕ, when $\phi = 10°$, differs only about 0.01 from unity; and this expression may be reduced to $\frac{V}{r}$. When the angle of projection does not exceed 3°, cos. ϕ differs only .001 from unity, and we can everywhere replace V cos. ϕ by V. Under this angle, $\frac{\cos. \phi}{\cos. \theta}$ differs but slightly from unity, and we have $v = \frac{V}{U}$, which is the same as if motion took place in a horizontal plane.

All cases of the movement of projectiles which occur in practice, may be divided into three distinct classes: 1st, When the angle of projection is slight, or does not exceed 3°, as in the ordinary fire of guns, howitzers, and small-arms; 2d, When the angle of projection is greater than this, but does not exceed 10° or 15°, as in ricochet fire, &c.; 3d, When the angle of projection exceeds 15°, as in the fire of mortars.

422. **1st Class.** *For small angles of projection, as in guns, howitzers, and small-arms.*

For slight variations of the angle of projection above or below the horizon, the form of the trajectory may be considered constant; and when the object is but slightly raised above, or depressed below the horizontal plane, it may be considered as in this plane.

In consequence of the windage, and the balloting of

the projectile which results from it, the projectile does not always leave the bore in the direction of the axis. The angle formed by the line of departure and the axis of the piece, is called the *angle of departure.* For guns in good condition, the vertical deviations do not exceed 5′, and for howitzers 10′; the side deviations never exceed 4′ 30″. To obtain exact results, therefore, it is necessary to correct the angle of projection for the angle of departure, when the latter is known.

Under the supposition that a, cos. ϕ, and $\frac{\cos. \phi}{\cos. \theta}$ are each equal to unity, the equations of the trajectory in air may be reduced to—

$$y = x \tan. \phi - \frac{g}{2} \frac{x^2}{V^2} B; \tag{22}$$

$$\text{Tan.}\, \theta = \tan. \phi - g \frac{x}{V^2} I; \tag{23}$$

$$t = \frac{x}{V} D; \tag{24}$$

$$v = \frac{V}{U}. \tag{25}$$

Knowing the weight and diameter of the projectile, c can be calculated by the formula $c = \frac{2RD}{3gA}$ if it be not found in the table which accompanies it. We know $\frac{x}{c}$ and $\frac{V}{r}$, and by means of the tables can determine the desired values of B, I, D, and U.

Of the three things, the initial velocity, V, the distance of the object, X, and the angle of projection, ϕ, two being known, to determine the third.

1st. *To determine the angle of projection,* ϕ.—Make $y=0$ in equation (22), and solve it with reference to tan. ϕ, we have,

$$\tan.\ \phi=\frac{g}{2}\frac{X}{V^2}B.$$

Example.—Find the angle of projection necessary to throw a 12-pdr. shot 1800 feet, with an initial velocity of 1500 feet. We have $V=1500$ feet; $\frac{x}{c}=\frac{1800}{3370}=0.5336$; $\frac{V}{r}=\frac{1500}{1427}=1.054$. From Table (1), $B=1.449$; $\tan.\ \phi=\frac{32.17}{2}.\frac{1800}{1500^2}1.449.=0.01864.$ $\phi=1°\ 05'$.

2d. *To determine the initial velocity, V, make $y=0$,* in equation (22), solve it with reference to V, and multiply both members by $\frac{1}{r}$, we have.

$$\frac{\frac{V}{r}}{\sqrt{B}}=\frac{1}{r}\sqrt{\frac{g}{2}\frac{X}{\tan.\phi}}=q.$$

Having the values of $\frac{X}{c}$ and q, seek in table (5) for the value of $\frac{X}{c}$, the value of $\frac{V}{r}$, which gives that of q; multiply $\frac{V}{r}$ by 1427 and we shall have V.

Example.—Find the initial velocity of a 12-pounder shot which, fired under an angle of 1° 05′, has a range of 1800 feet.

$$q=\frac{1}{1427}\sqrt{\frac{16.08\times1800}{0.01864}}=0.8732.$$

$$\frac{V}{r}=1.05.\quad V=1.05\times1427=1498.35\text{ feet.}$$

3d. *To determine the range, X.*—Make $y=0$ in equation (22), obtain the value of X, and divide both members of the equation by c, we have,

$$\frac{X}{c}B=\frac{\tan.\phi\ V^2}{c\frac{1}{2}g}=p.$$

Having the initial velocity, V, and angle of projection, ϕ, we can determine, $\frac{V}{r}$ and p; seek in table (4), for the value of $\frac{V}{r}$, that of $\frac{X}{c}$, which gives p; having $\frac{X}{c}$, multiply it by c, and we have X.

Example.—Find the range of a 12-pdr. ball, fired under an angle of 1° 05′, with an initial velocity of 1500 feet.

$$c=3370;\ \frac{V}{r}=1.0511;\ \text{tan.}\ \phi=0.01864.$$

$p=\frac{0.01864}{3370}\cdot\frac{\overline{1500}^2}{16.08}=0.774$; (from table 4), $\frac{X}{c}=.5340$; $X=.5340\times 3370=1800$ feet.

The slight discrepancies in the three preceding results, arise from the neglected decimals.

In firing spherical case-shot, it is important not only to know the time of flight, in order to regulate the fuze, but it is important to know that the projectile will have sufficient remaining velocity to render the impact of the contained projectiles effective.

4th. The *time of flight* can be obtained from equation (24), or, $t=\frac{x}{V}D$. Knowing $\frac{x}{c}$ and $\frac{V}{r}$, we can obtain the corresponding value of D from table (3).

Example.—Find the time of flight of a 12-pdr. spherical case-shot for a distance of 1500 yards, the initial velocity being 1500 feet.

$$\frac{X}{c}=\frac{4500}{3370}=1.335;\ \frac{V}{r}=\frac{1500}{1427}=1.051;\ D=1.859.$$

$$t=\frac{4500}{1500}1.859=5.58 \text{ seconds.}$$

5th. *The remaining velocity* can be obtained from

equation (25), or, $v=\frac{V}{U}$. Knowing $\frac{X}{c}$ and $\frac{V}{r}$, obtain from table (3) the corresponding value of U.

Example—Find the remaining velocity of a 12-pdr. spherical case-shot at the distance of 1500 yards, the initial velocity being 1500 feet.

$\frac{X}{c}=\frac{4500}{3370}=1.327$; $\frac{V}{r}=1.051$; $U=2.882$; $v=\frac{1500}{2.882}=520$ feet.

This velocity is more than sufficient for a musket-bullet to disable an animate object at the distance of 1500 yds.

423. **2d class.** *For angles of projection not exceeding 10° or 15°, as in the ricochet fire of guns, howitzers, and mortars.*

The formulas are:

$$y=x\ \tan.\phi-\frac{g}{2}\frac{x^2}{V^2\cos.^2\phi}B. \qquad (26)$$

$$\tan.\theta=\tan.\phi-g\frac{x}{V^2\cos.^2\phi}I. \qquad (27)$$

$$t=\frac{x}{V\cos.\phi}D. \qquad (28)$$

$$v=\frac{V\cos.\phi}{U\cos.\theta}. \qquad (29)$$

If the object be on a level with the piece, the solution of this class of problems is the same as those of class 1st, when the angle is very small; if not, it will be necessary to substitute for V, $V_{,}=V\cos.\ \phi$, and after having obtained $V_{,}$, divide it by the cos. ϕ, which gives V.

The object being situated at the distance a from the piece, and at the distance b above the horizontal plane passing through the centre of the muzzle, is seen under an *angle of elevation* ε, for which $\tan.\varepsilon=\frac{b}{a}$. One of the two things, the initial velocity or angle of projection being known, to determine the other.

1st. *To determine the initial velocity*, V. Substitute in equation (26) the co-ordinates a and b, and $V_{,}$; solve it with reference to $V_{,}$; substitute tan. ε for $\frac{b}{a}$, and divide both members by r, we have,

$$\frac{\frac{V_{,}}{r}}{\sqrt{B}}=\frac{1}{r}\sqrt{\frac{\frac{g}{2}a}{\tan.\phi-\tan.\varepsilon}}=q.$$

Having the value of q, seek in table (5) for the known value of $\frac{a}{c}$, the value of $\frac{V_{,}}{r}$ corresponding to it, and multiplying by $\frac{r}{\cos.\phi}$, we shall have V.

Example.—Find the initial velocity of an 8-inch siege-howitzer shell, which, being fired under an angle of 12°, will strike an object situated 1,000 feet from, and 20 feet above, the muzzle of the piece.

$$\text{Tan.}\phi=0.2125;\ \tan.\varepsilon=\frac{20}{1000}=0.0200;\ \tan.\phi-\tan.\varepsilon=0.1925;$$

$$\cos.\phi=0.9781;\ \frac{a}{c}=\frac{1000}{3570}=0.2801;\ q=\frac{1}{1427}\sqrt{\frac{16.08\,.\,1000}{0.1925}}=$$

$$0.2023;\ \frac{V_{,}}{r}=0.2150;\ V=\frac{0.2150\,.\,1427}{0.9781}=313\text{ feet.}$$

2d. *To determine the angle of projection.* The result will be sufficiently near the truth, if we substitute, in equation (26), V for $V_{,}$, or V cos. ϕ; and solving it with reference to tan. ϕ, we have,

$$\tan.\phi=\tan.\varepsilon+\frac{ga}{2V^2}B,$$

in which we substitute for B its value, corresponding to $\frac{a}{c}$ and $\frac{V}{r}$, obtained from table (1).

Example.—What angle of projection is necessary for an 8-inch siege-howitzer shell to strike an object situated 1000 feet from, and 20 feet above, the muzzle? The initial velocity being 313 feet,

$V=313$ feet; $\frac{a}{c}=\frac{1000}{3570}=0.2801$; $\frac{V}{r}=\frac{313}{1427}=0.2193$; $\tan.\varepsilon=\frac{20}{1000}=0.0200$; $\tan.\ \phi=0.0200+\frac{16.08\ .\ 1000}{\overline{313}^2}1{,}142=0.2084$; $\phi=11^\circ\ 28'$.

424. **3d class.** *Properties of trajectories under high angles of projection.*

As a projectile rises in the ascending branch of its trajectory, its *velocity* is diminished by the retarding effect of the air and the force of gravity: in consequence of the resistance of the air alone, the velocity continues to diminish to a point a little beyond the summit of the trajectory, where it is a minimum; and from this point it increases, as it descends, under the influence of the force of gravity, until it becomes uniform, which event depends on the diameter and weight of the projectile and the density of the air, or, in other words, upon the value of c.

The *inclination* of the trajectory decreases from the origin to the summit, where it is nothing; it increases in the descending branch from the summit to its termination, and if the ground did not interpose an obstacle, it would become vertical at an infinite distance. An element of the trajectory in the descending branch has a greater inclination than the corresponding element of the ascending branch.

Strictly speaking, the trajectory in air is an expotential curve with two asymptotes; the *first* is the axis of the piece, which is tangent to the trajectory when the initial velocity is infinite; the *second* is the vertical line toward which the trajectory approaches as the horizontal component of the velocity diminishes, and the effect of the force of gravity increases.

The *curvature* of the trajectory increases in the ascending branch, to a point a little beyond the summit. The point of greatest curvature is situated nearer the summit than the point of minimum velocity.

In the fire of mortar shells under great angles of projection, and at customary distances, the trajectory may be considered as an arc, in which the angle of fall is slightly greater than the angle of projection. In the ascending branch, the arc commences under an angle of ϕ, and terminates under an angle of 0; the ratio of the length of this arc to its projection, or a, is calculated for all arcs from 5° to 75°, and arranged, in groups of fives in the accompanying table.

The value of a is considered the same in the descending as in the ascending branch.

ARCS.	a	ARCS.	a	ARCS.	a
5°	1.00127	30°	1.05306	55°	1.27583
10	1.00516	35	1.07596	60	1.38017
15	1.01184	40	1.10730	65	1.53433
20	1.02165	45	1.14777	70	1.77772
25	1.03514	50	1.20189	75	2.20349

The multipliers, B, I, D, and the divisor, U, are calculated for the values $\frac{ax}{c}$ and $\frac{a V_{\prime}}{r}$, and they are employed in equations (26), (27), (28), (29), as in the preceding class of cases.

1st. *Find the initial velocity of a mortar shell, knowing the range and angle of projection.*

We know $\frac{ax}{c}$, and by solving equation (26) as before, we have,

$$\frac{\frac{aV_{\prime}}{r}}{\sqrt{B}}=\frac{a}{r}\sqrt{\frac{\frac{1}{2}gX}{\tan.\phi}}=q.$$

Having determined the value of q, seek in table (5) the value of $\frac{aV_{\prime}}{r}$ corresponding to it for $\frac{ax}{c}$; then multiply it by $\frac{r}{a\cos.\phi}$, and we have V.

Example.—What initial velocity is necessary to project a 10-inch shell 1,800 feet, under an angle of 45°?

For a 10-inch shell, $c=4677$; for 45°, $a=1.148$; $\frac{ax}{c}=\frac{1.148.1800}{4677}=0.4418$; $q=\frac{1.148}{1427}\sqrt{\frac{16.08\,.\,1800}{1.0000}}=0.1369$. By the aid of table (5) we find $\frac{aV_{\prime}}{r}=0.1490$; and from this we get $V=\frac{0.1490\,.\,1427}{1.148\,.\,0.7071}=262$ feet.

2d. *To determine the angle and velocity of fall, and the time of flight, knowing the initial velocity and range.* Let the projectile be the same as in the preceding case.

Example.—We have $\frac{aX}{c}=0.4418$; and $\frac{aV_{\prime}}{r}=0.1490$; from table (1) we have $I=1.291$; from table (2), $D=1.127$; and $U=1.272$. Substituting the proper values in equation (25) we have

$$\text{Tan. }\theta=1.0000-\frac{32.17\,.\,1800}{(262\,.\,0.7071)^2}.\,1.280=-1.159;\ \theta=-49^\circ\,12'.$$

The negative sign indicates that the angle of fall is measured in an opposite direction from the angle of projection. Making the proper substitutions in equations (28) and (29), we have

$$T=\frac{1800}{262\,.\,0.7071}1.127=10.95''.\ v=\frac{262\,.\,0.7071}{1,272.0.6534}=222\text{ feet.}$$

3d. *To determine the range, knowing the initial velocity and angle of projection.*

We have a, and $\frac{aV_{\prime}}{r}$; make $y=o$ in equation (26); solve it with reference to X, and multiply both members by $\frac{aX}{c}$, and we have,

$$\frac{aX}{c}B=\frac{aV^2}{gc}\sin.\ 2\phi=p.$$

Having found the value of $\frac{aX}{c}$, which for $\frac{aV_{\prime}}{r}$ gives p (table 4); multiply it by $\frac{c}{a}$, and we have X.

Example.—Find the range of a 10-inch mortar shell, the angle of projection of which is 45°, and the initial velocity is 262 feet.

Cos. $\phi=0.7071$; the sin. $2\phi=1.0000$; and $a=1.148$; $\frac{aV_{\prime}}{r}=$ $\frac{1.148\,.\,262\,.\,0.7071}{1427}=0.1490$; $p=\frac{aV^2}{gc}\sin.2\ \phi=\frac{1.148\,.\,\overline{262}^2}{32.17\,.\,4677}\ 1.0000=$ 0.5238 from table (4) $\frac{aX}{c}=0.4412$; $X=\frac{0.4412.4677}{1.148}=1798$ feet.

The slight discrepancies in these, as in the preceding results, arise from the neglected decimals.

425. **Comparison of true and calculated trajectories.** In consequence of considering the inclination of the trajectory as constant in the preceding equations, the resistance of the air is slightly underestimated in the more inclined portions of the trajectory, or at the beginning and end, and slightly overestimated in the less inclined portions, or about the summit. It follows that the calculated trajectory will at first rise above the true one, then pass below it, and again pass above it; the calculated ranges will therefore be found slightly in excess.

426. **Trajectory of oblong projectile.** From the law

of inertia, a rifle projectile moves through the air with its axis of rotation parallel to the axis of the bore. Hence, it follows, that an oblong projectile, fired under a low angle of projection, presents a greater surface toward the earth, and less parallel to it, than a round projectile of the same weight; consequently the vertical component of the resistance of the air is greater, and the horizontal component less, in the first case than in the second. The effect of this will be to give an oblong projectile a flatter trajectory and longer range than a round one.

DEVIATION OF PROJECTILES.

427. **Nature and causes.** The path described by the centre of inertia of a projectile, moving under the influence of gravity and the tangential resistance of the air, is called the *normal trajectory;* and it is this trajectory which has been the subject of the preceding discussions. In practice, various causes are constantly at work to deflect a projectile from its normal path, and it becomes necessary to study the nature of these causes, and their effects.

All deviating causes may be divided into two classes —those which act while the projectile is in the bore of the piece, and those which act after the projectile has left it. The first class includes all the causes which affect the initial velocity, and give rotation to the projectile; the second includes the action of the air.

428. **Causes which affect initial velocity.** The principal causes which affect initial velocity are variations in the weights of the powder and projectile, the manner

of loading, the temperature of the piece, and the balloting of the projectile along the bore. Experiments made by firing siege and field projectiles into the ballistic pendulum, show that, with care, the mean variation in the initial velocity, in a series of fires, does not exceed 20 feet.

A variation of 20 feet in initial velocity only produces a variation of $\frac{1}{2}$ a foot, in the vertical height of the trajectory of a 12-pdr. ball, at a distance of 1,000 yards.

429. **Rotation.** The principal cause of the deviation of a projectile is its rotation combined with the resistance of the air. It is proposed, in the first place, to show how rotation may be produced, and, in the second, to show how rotation, combined with the resistance of the air, produces deviation.

By balloting. If the projectile be spherical and homogeneous, rotation is produced by the bounding or balloting of the ball in the bore, arising from the windage. In this case the axis of rotation is horizontal, and passes through the centre of the ball; the direction of rotation depends on the side of the projectile which strikes the surface of the bore last; if it strike on the upper side, the front surface of the projectile will move upward; if on the lower side, this surface will move downward.

The velocity of rotation from this cause depends on the windage, or depth of the indentations in the bore, the charge being the same. It has been found to be, for ordinary windage, about 30 feet for a 24-pdr. shell fired with $2\frac{1}{4}$ lbs of powder.

By eccentricity. If, from the structure of the ball, or from some defect of manufacture, the centre of gravity do not coincide with the centre of figure, rotation generally takes place around the centre of gravity. This arises from the fact that the resultant of the charge acts at the centre of figure, while inertia, or resistance to motion, acts at the centre of gravity. The axis of rotation passes through the centre of gravity, and is perpendicular to a plane containing the resultant of the charge and the centres of figure and gravity. For the same charge, the velocity of rotation is proportional to the lever arm, or perpendicular, let fall from the centre of gravity to the resultant of the charge.

Knowing the position of the centre of gravity of the ball in the bore, it is easy to foretell the direction and velocity of rotation. In general terms, the front surface of the projectile moves toward the side of the bore on which the centre of gravity is situated, and the velocity of rotation is greatest when the line joining the centres of gravity and figure is perpendicular to the axis of the bore.

The position of the centre of gravity of a projectile is found by floating in a mercury bath; and by an instrument called the *eccentrometer*. The topmost point of the surface, when the projectile has settled to a state of rest in the bath, marks one point at which the line joining the centre of gravity and figure pierces the surface; the position of the centre of gravity along this line is determined by the eccentrometer, which is a peculiar kind of balance. w being the weight of the projectile, and x the distance of its centre of gravity from the fulcrum of the balance, and w' being the weight necessary to balance the projectile, and a its distance from the fulcrum, we have, from the equality of the moments—

$$aw' = wx, \text{ or } x = \frac{aw'}{w}.$$

The position of the projectile on the balance being known, by placing the marked point on the surface nearest the fulcrum, the position of the centre of gravity becomes known; for if b be the distance of the marked point from the fulcrum, and r the radius of the projectile, $x-b-r$ is the distance between the centres of gravity and figure.

430. **The effect of rotation.** The effect of rotation in producing deviation, may be studied under three heads: 1st. When the projectile is spherical and concentric. 2d. When it is spherical and eccentric; and 3d. When it is oblong.

Concentric projectiles. The simplest case is that of a homogeneous spherical projectile, rotating around a vertical axis passing through the centre of gravity.

Let $A\ B\ C\ D$ represent the great circle cut out of the sphere perpendicular to the axis of rotation, and suppose rotation to take place in the direction $A\ C\ B$, and the motion of translation in the direction $A\ B$; it is evident that each point of the circle moves in the direction $A\ B$, with a velocity which is equal to the velocity of translation, plus or minus the component of its velocity of rotation in the direction of the axis $A\ B$, which is equal to the projection of the arc over which the point moves in a unit of time, on the line $A\ B$. The points C and D have the greatest velocity in the direction of this line, $A\ B$, and the points A and B the least. All the points

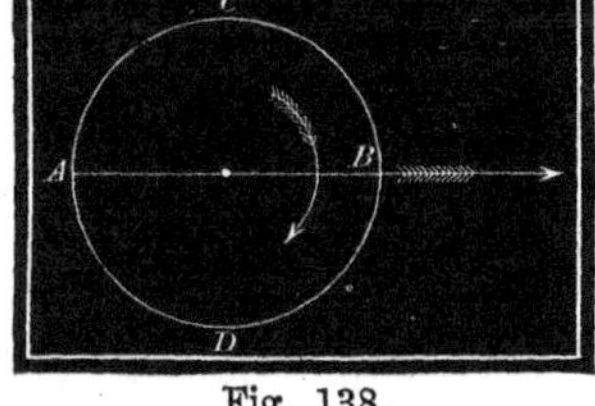

Fig. 138.

in the semi-circle *A C B* rotate in a forward direction, and the components of their velocities of rotation must be added to that of translation; while the points in the semicircle *B D A* move backward in rotation, and the components of their velocities must be subtracted from it. A body moving in the air draws with it a film of the particles which surround it, and these particles set in motion the adjacent particles, and so on from one layer to another; the number of particles set in motion and their reaction on the surface of the projectile, depend on the velocity of the moving surface; now it has been shown that the surface *A C B* moves with a greater velocity than the opposite side, the reaction, or pressure upon it, must be greater than upon the latter, and the projectile will be urged in the direction *C D*.

Eccentric projectiles. Let *A C B D* represent the great circle cut out of an eccentric projectile perpendicular to the axis of rotation, and containing the centre of figure O, and the centre of gravity O′. Suppose the motions of rotation and translation to take place as in the preceding case, it follows that the same cause will operate in this, as in the preceding case, to deviate the projectile in the direction *C D*; but there is another and more powerful cause operating to deviate the projectile in the same direction, and that is, the greater pressure on the side *A C B* arising from the greater surface offered to the air in consequence of the eccentricity.

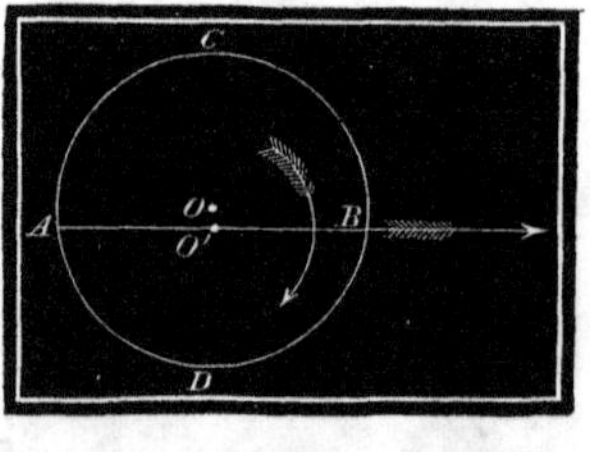

Fig. 139.

Prof. Magnus' apparatus. These phenomena may be

easily illustrated by the very simple and ingenious apparatus devised by Prof. Magnus, of Berlin. Let *C* (fig. 140) represent a light brass cylinder, delicately suspended in a ring, and made to revolve rapidly around its vertical axis, by means of a string, after the manner of a top; let this ring be suspended at the extremity of a wooden lever *B′*, which, in turn, is suspended by a delicate wire from the ceiling, so that it may rotate freely in a horizontal direction; let *P* be a counterpoise, and *R* the direction of a strong current of air blowing upon the cylinder from a fan-blower.

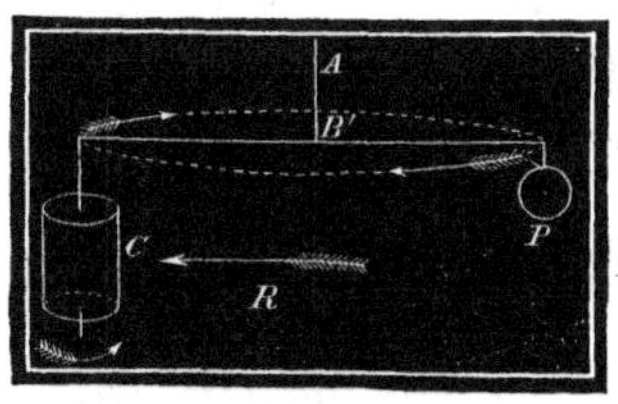

Fig. 140.

It is invariably found, that the axis of the cylinder will move in the opposite direction from the side which is moving toward the current of air from the blower (see direction of the arrows); but if there be no rotation of the cylinder, the axis will remain stationary.

Conclusions. If a projectile be spherical and concentric, rotation takes place from contact with the surface of the bore around a horizontal axis, and the effect will be to shorten or lengthen the range, as the motion of the front surface is downward or upward.

If the projectile be eccentric, the motion of the front surface is generally toward the side on which the centre of gravity is situated, and the deviation takes place in this direction.

The extent of the deviation for the same charge, depends on the position of the centre of gravity; the horizontal deviation being the greatest when the centres of gravity and figure are in a horizontal plane, and the line which joins them is at right angles to the axis of the piece; the vertical deviation will be the greatest when these centres are in a vertical plane, and the line which joins them is at right angles to the axis of the piece. If the axis of rotation coincide with the tangent to the trajectory throughout the flight, all points of the surface have the same velocity in the direction of the motion of translation, and there will be no deviation. This explains why it is that a rifle-projectile moves through the air more accurately than a projectile from a smooth-bored gun.

In the experiments of Major Wade with 32-pdr. field-shells, made purposely eccentric, the difference of the extreme lateral deviations, produced by placing the centre of gravity first on one side and then on the other, amounted to 100 yds., or one-fourth of the entire range.

The experiments of Captain Dahlgren with service 32-pdr. balls, show the following results when the centre of gravity is placed in different positions in the vertical plane through the axis of the bore.

POSITION OF CENTRE OF GRAVITY IN VERTICAL PLANE.			
90° up.	90° down.	Inward.	45° up. and in.
1415 yds.	1264 yds.	1329 yds.	1360 yds.

In accurate firing, therefore, it is important to know the true position of the centre of gravity: in ricochet firing over smooth water, the number of grazes may be increased or diminished by placing, in loading, the centre of gravity above or below the centre of figure.

The first person to call attention to the deviation

produced by rotation, was Robins, who illustrated it by bending a musket-barrel to the right, and firing through a succession of paper screens; the projectile was observed to deviate, first to the right, in the direction in which the muzzle was pointed, and then to the left, in the opposite direction from the side of the projectile which rotates toward the front.

431. **Deviation of oblong projectiles.** The cause of the deviation of an oblong rifle projectile is quite different from one of spherical form. An oblong projectile moving in the air is acted upon by two rotary forces, viz.; one which gives it its normal rotary motion around its axis of progression, and another the resistance of the air, which, in consequence of the deflection of the axis of progression from the tangent to the trajectory by the action of gravity, does not pass through the centre of inertia, but above or below it, depending on the shape of the projectile. From a law of mechanics, a body thus circumstanced, will not yield fully to either of the forces that thus act upon it, but its apex will move off with a slow uniform motion to the right or left of the vertical plane, depending on the relative direction of the two rotary forces. If the action of these forces be continued sufficiently long, it will be seen that the axis of the projectile before referred to, describes a cone around a line passing through the centre of inertia and parallel to the direction of the resistance of the air.

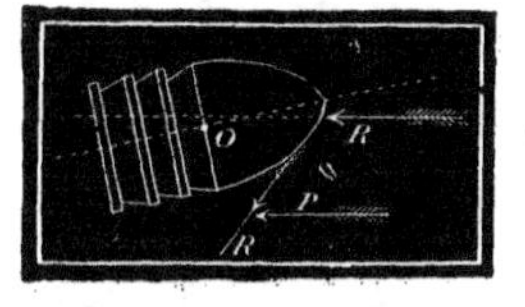
Fig. 141.

Owing to the short duration of the flight of an ordinary projectile, it is only necessary to consider the first part of this conical motion. If the projectile rotates in the direction of the hands of a watch to the eye of the marksman, and the resultant of the resistance of the air pass above the centre of inertia, as it does in the service bullet with a conoidal point, see fig. 141, then the point of the projectile will move to the right, which brings the left side of the projectile obliquely in contact with the current of the air. The effect of this position with reference to the air, will be to generate a component force that will urge the projectile to the right of the plane of fire, as a vessel sailing on the wind has a motion to the leeward.

If the bore be grooved with a left-handed twist, the deviation will be to the left of the plane of fire, as has been shown by actual experiment. This peculiar deviation was called by the French officers that first observed it, "*dérivation*" or "*drift.*" That it is not produced by the effect of the recoil on the shoulder of the marksman, as some assert, is shown by the fact that drift increases more rapidly than the distance.

The following table gives the drift at different distances, for the French rifle, model of 1842, with a twist of 4.37 feet, and a bullet with a single groove:

Distances in yards.	218	328	437	546	656	765	874	984	1093	1312	1421
Drift in feet and inches.	.5"	1'.1"	1'.9"	2'.0"	4'.9"	7'.6"	11'.6"	16'.1"	21'.0"	38'.4"	50'.6"

In consequence of the reduced calibre and twist, the drift of our present rifle-musket projectiles is less than the foregoing. The mean drift of 40 shots fired from two service rifle-muskets, at a distance of 1,150 yds., in a perfectly calm day, was about 18 feet; not a single shot deviated to the left of the point aimed at.*

Effect of wind. The deviating effect of wind depends on its force, and its direction with regard to the plane of fire; generally speaking, large and heavy projectiles, moving with high velocities, are deviated less than those of contrary character. It is difficult to calculate the effect of the wind in any particular case; in making allowance for it, therefore, the gunner should be guided by experience and judgment. For the same projectile, velocity, and wind, the deviation varies nearly as the square of the range.

432. **Summary of deviating causes.** The following summary may be considered as embracing nearly all the causes of deviation of cannon and small-arm projectiles.

1st. *From the construction of the piece.* These causes are, wrong position of the sight; bore not of the true size; crooked barrel; too hard on the trigger; windage; the recoil; and spring of the barrel.

2d. *From the charge of powder.* Improper weight; form of grain and variable quality of the powder; injury from dampness; more or less ramming; sticking along the bore from foulness and dampness.

3d. *From the projectile.* Not of the exact size, shape, or weight; disfiguration in loading, or on leaving the bore; eccentricity.

4th. *From the atmosphere, &c.* The effect of wind; variations in the temperature, moisture, and density of

* The subject of drift has been fully exposed in a learned analytical investigation by General Barnard, of the engineer corps, who shows that it is a particular case of the gyroscope. It has also been explained experimentally by Professor Magnus, of Berlin, a copy of whose apparatus may be found in the Museum of the United States Military Academy.

the air; position of the sun as regards the effect on the aim; difference of level between the object and piece; and rotation of the earth.

The latter source of deviation arises, 1st. From the fact that all points on the surface of the earth, not in the same parallel of latitude, move with different angular velocities; and 2d. That when a body is thrown from one point to another, it carries with it the angular velocity with which it started. Applying these facts, it is found that a projectile will deviate to the right of the object, whatever may be the direction of the line of fire, and at a distance from it, depending on the latitude of the place, and on the time of flight and the range of the projectile.

Poisson has shown that a 12-inch shell weighing 200 lbs., fired under an angle of 45°, with an initial velocity of 900 feet, will deviate from 15 to 20 feet to the right of the object—the range being about 4,400 yards.

CHAPTER IX.

LOADING, POINTING, AND DISCHARGING FIRE-ARMS.

433. **Loading.** In loading guns and howitzers, the powder is carefully put up in a cartridge-bag of woollen cloth, which is either attached to, or carried separate from the projectile, depending on the weight of the projectile. In ramming a charge, only a sufficient force should be used to send it home, as the space which the powder occupies affects the initial velocity. In loading mortars, the powder is poured from the cartridge-bag into the chamber, and levelled with the hand; the shell is then carefully lowered upon it with the hooks.

434. **Precautions.** After a piece has been discharged the bore should be well sponged, to extinguish any burning fragments of the cartridge that may remain; and to prevent the current of air from fanning any burning fragments that may collect in the vent, it should be kept firmly closed with a thumb-stall in the operation of sponging. Experience shows that the use of a wet sponge is dangerous, as it contributes to form, from the fragments of the cartridge-bag, a substance which retains fire.

435. **Use of projectiles not suited to the bore.** It may be sometimes necessary to fire projectiles that are either very much smaller or larger than the bore.

If it be desired to use a gun-shell, or solid shot, which

is much smaller than the bore, it is strapped to a stout sabot which fits the bore; if a mortar-shell, it is placed in the centre of the bore by means of wedges, and the surrounding space is filled up with earth.

Mortar-shells are fired from guns and howitzers, by digging a hole in the ground about 20 inches deep, and placing in it two pieces of stout plank inclined at an angle of 45°, for the support of the breech; the chase is supported on a movable wedge, which rests on skids firmly secured with platform stakes;* the charge of powder is then inserted in the bore, and the projectile is placed on the muzzle, and secured by passing strings over it, and tying their ends to a rope, which encircles the neck of the chase.

Pieces fired in this way should be elevated 40° or 45°; thus situated, the fuze of the 8-inch mortar-shell takes fire from very small charges; but the 10-inch fuze should be primed with strands of quick-match, which are allowed to hang over the sides of the shell.

POINTING.

To point or aim a fire-arm is, to give it such direction and elevation that the projectile shall strike the object. To do this properly, it is necessary to understand the relations which exist between the line of sight, line of fire, trajectory, &c.

436. **Definitions.** The *line of sight* is the right line containing the guiding points of the sights. The sights are two pieces, *A* and *B*, on the upper surface of the

* Pieces that have been disabled by breaking off a trunnion, may be fired in this manner.

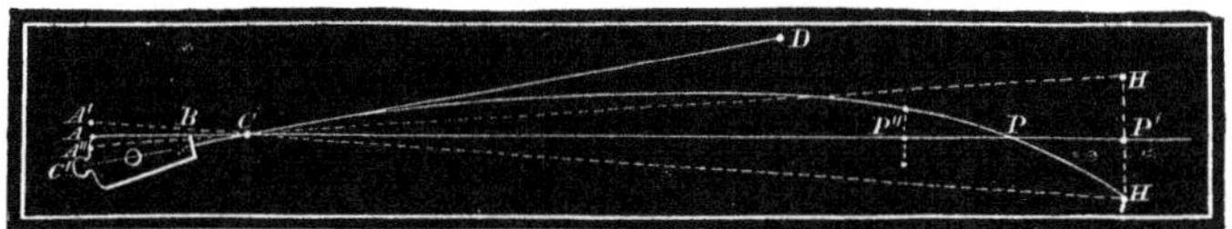

Fig. 142.

gun, the situation of which with regard to the axis of the bore is known. The *front* sight is situated near the muzzle, or on the right rimbase, and is generally fixed; the *rear* sight is placed near the breech, and is movable in a vertical, and sometimes in a horizontal direction. The *natural line of sight* is the line of sight nearest the axis of the piece; the others are called artificial lines of sight.

The *line of fire* is the axis of the bore prolonged in the direction of the muzzle, or $C'D$.

The *angle of fire* is the angle included between the line of fire and horizon; on account of the balloting of the projectile, the angle of fire is not always equal to the angle of departure, or projection. See section 268.

The *angle of sight* is the angle included between the line of sight and line of fire; angles of sight are divided into natural and artificial angles of sight, corresponding to the natural and artificial lines of sight which enclose them.

The *plane of fire* is the vertical plane containing the line of fire.

The *plane of sight* is the vertical plane containing the line of sight.

The *point-blank* is the point at which the line of sight intersects the trajectory, or P. Strictly speaking, the line of sight intersects the trajectory at two points, C

and P; but, in practice, the point P is only considered. The distance, $B\ P$, is called the point-blank distance.

The *natural point-blank* corresponds to the natural line of sight; all other point-blanks are called *artificial point-blanks.* In speaking of the point-blank of a piece, the natural line of sight is supposed to be horizontal.

In the British service, the point-blank distance is the distance at which the projectile strikes the level ground on which the carriage stands, the axis of the piece being horizontal. It is evident that this definition of point-blank distance conveys a better idea of the power of the piece than the former, which makes it depend on the form of the piece, as well as on the charge.

As the angle of sight $A\ C\ C'$ is increased, the point-blank distance is increased; as it is diminished, the intersections of the line of sight and trajectory approach each other until they unite, when the line of sight and trajectory are tangent to each other; beyond this, the point-blank is imaginary.

As the angle of fire increases, the force of gravity acts more in opposition to the force of projection, and the point-blank distance is diminished, until at 90° it becomes zero. Under an angle of depression, the force of gravity acts more nearly in the direction of gravity, and the point-blank distance is increased, becoming infinite when the angle of depression is equal to 90° minus the angle of sight.

In ordinary firing, it is not considered that the trajectory changes its position with reference to the lines of sight and fire, for angles of elevation and depression, less than 15°. In aiming at an object, therefore, the *angle of elevation* of which is less than 15°, aim as though it were in the same horizontal plane with the piece.

For the same piece, the point-blank distance increases with the charge of powder; for the same initial velocity, a large projectile has a greater point-blank distance than a small one; a solid shot than a hollow one; an oblong projectile than a round one; or, in other words, it varies with the value of c, before referred to.

Range is the distance at which a projectile first strikes the ground on which the carriage is situated; *extreme range* is the distance to the point at which the projectile is brought to a state of rest.

437. **Pointing guns and howitzers.** In pointing guns and howitzers under ordinary angles of elevation, the piece is first directed toward the object, and then elevated to suit the distance. The accuracy of the aim depends—1st. On the fact that the object is situated in the plane of sight; 2d. That the projectile moves in the plane of fire, and that the planes of sight and fire coincide, or are parallel and near to each other; and 3d. On the accuracy of the elevation.

The first of these conditions depends on the eye of the gunner, and the accuracy and delicacy of the sights; the errors under this head are of but little practical importance.

When the trunnions of the piece are horizontal, and the sights are properly placed on the surface of the piece, the planes of sight and fire will coincide; but when the axis of the trunnions is inclined, and the natural line of sight is oblique to the axis of the bore, the planes are neither parallel nor coincident, and the aim will be incorrect. If the natural line of sight be made parallel to the line of fire, by making the height of the front sight equal to the dispart of the piece, the planes of sight and fire will be parallel, and at a distance from each other equal to the radius of the breech multiplied by the sine of the angle which the axletree makes with the horizon. To show this, let the circle *A C B D* represent the section of the breech of the piece taken at right angles to the axis, and *C* the projection of the natural line of sight upon this plane; let *A′ B′* be the inclined position of the axletree, or trunnions, *C′* marks the revolved position of the natural line of sight, and *C′ D′* the trace of the plane of sight, which is parallel to *C D*, the trace of the plane of fire. As the lines of sight and fire are parallel in their revolved position, the planes of sight and fire must also be parallel. The angle $COC' = BOB'$, therefore $CC' = OC' \sin. BOB'$. It is easily seen that with this arrangement of the front sight, the error of pointing can never exceed the radius of the breech. By an inspection of the figure, it will also be seen, that in the revolved position of the line of sight, the elevation is diminished by a small quantity, which is equal to the versed sine of the arc CC'.

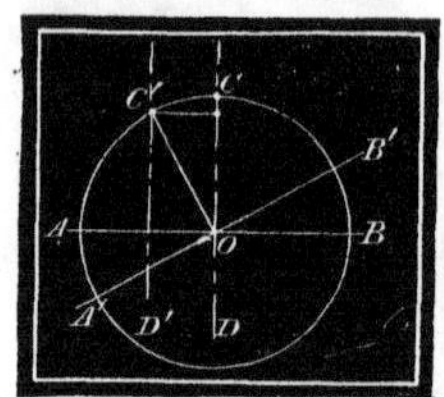

Fig. 143.

By referring to the construction of the pendulum hausse, on page 255, we see that if its centre of motion coincide with the point *C′*, and the scale coincide with the line *C′ D′*, the error of aiming with an artificial line of sight is practically no greater than with the natural line of sight.

If the natural line of sight be not parallel to the axis of the piece, the planes of sight and fire intersect at a short distance from the muzzle; hence, it follows, that as the object is situated in the plane of sight, the projectile will deviate from the object to the side on which the lower wheel is situated, and at a distance from it, which is proportional to the distance of the object from the piece; to correct for this source of error, the line of sight should be pointed to the side of the higher wheel, and at a distance from the object, which is proportional to the distance of the object from the piece.

Siege and sea-coast cannon are generally fired from fixed platforms, which renders the axis of the trunnions horizontal; they are, therefore, not furnished with pendulum sights.

In case the axis of the trunnions is not horizontal, and the piece has not a pendulum hausse, the highest points of metal at the breech and muzzle may be determined by the gunner's level (see page 254), and marked with chalk; the centre line of the tangent scale, or breech-sight, is placed on the mark at the breech, the slider is placed at the proper elevation, and the aim is taken along the notch of the slider and the mark on the muzzle. This method, however, does not give a perfectly accurate aim.

In the absence of a breech-sight, the piece can be pointed with the natural line of sight so as to strike objects not situated at point-blank distance; if the object be within point-blank range, as at P'' (fig. 142), the natural line of sight should be depressed below the object as much as the trajectory is above it; if it be beyond point-blank, as at P', the natural line of sight should be directed to a point H, which is as much above the object, as the point H', of the trajectory, is below it.

Owing to the shape and size of the reinforce of sea-coast cannon, the natural line of sight is formed by affixing a front sight to the muzzle, or to a projection cast on the piece between the trunnions. Although the latter arrangement does not give quite so long a distance between the sights as is desirable, it permits the use of a shorter breech-sight, and the front sight does not interfere with the roof of the embrasure, when the piece is fired under high elevation.

438. **Pointing mortars and small-arms.** In pointing small-arms and mortars, the piece is first given the elevation, and then the direction necessary to attain the object.

Pointing mortars. Mortars are generally fired from behind epaulements, which screen the object from the eye of the gunner.

The elevation is first given by a gunner's quadrant, applied as described on page 256; and the direction is given by moving the mortar-bed with a handspike, so as to bring the line of sight into the plane of sight, which, by construction, passes through the object and the centre of the platform. The plane of sight may be determined in several ways; the method prescribed is to plant two stakes, one on the crest of the epaulement, and the other a little in advance of the first, so that the two shall be in a line with the object, and the gunner standing in the middle of the rear-edge of the platform; a cord is attached to the second stake, and held so as to touch the first stake; a third stake is driven in a line with the cord, in rear of the platform, and a plummet is attached to this cord so as to fall a little in rear of the mortar. It is evident that the cord and plummet determine the required plane of sight into which the line of sight of the mortar must be brought.

The usual angle of fire of mortars is 45°, which corresponds nearly with the maximum range. The advantages of the angle of greatest range are: 1st. Economy of powder; 2d. Diminished recoil, and strain on the piece, bed, and platform; 3d. More uniform ranges.

When the distance is not great, and the object is to penetrate the roofs of magazines, buildings, &c., the force of fall may be increased by firing under an angle of 60°. The ranges obtained under an angle of 60° are about *one-tenth* less than those obtained with an angle of 45°.

If the object be to produce effect by the bursting of the projectile, the penetration should be diminished by firing under an angle of 30°.

When the object is not on a level with the piece, the angle of greatest range is considered in practice to be $45°+\frac{1}{2}\theta$, or $45-\frac{1}{2}\theta$, θ being the angle of elevation or depression of the object. Thus to attain a magazine, for instance, situated on a hill, for which $\theta=15°$, the angle of greatest range is $52\frac{1}{2}°$ instead of 45°.

The angle of fire being fixed at 45° for objects on the same level with the piece, the range is varied by varying the charge of powder. The practical rule is founded on the knowledge of the amount of powder necessary to diminish or increase the range 10 yards. For the French 8 and 10 inch siege-mortars, this amount is about 60 grains for the former, and 125 grains for the latter.

A practical rule for finding the time of flight by which the length of the fuze is regulated, is to take the square root of the range in feet, and divide it by four; the quotient is the approximate time in seconds.

Stone-mortars are pointed in the same manner as common mortars: the angle of fire for stones is from 60° to 75°, in order that they may have great force in falling; the angle for grenades is about 33°, in order that their bursting effect may not be destroyed by their penetration into the earth.

439. **Night-firing.** Cannon are pointed at night by means of certain marks, or measurements, on the carriage and platform, which are accurately determined during the day.

In the case of guns and howitzers, the elevation may be determined by marking the elevating screw where it enters the nut, or by measuring the distance between the head of the screw and stock. In the case of mortars, the position of the quoin may be determined by marking, or by nailing a cleat on the bolster.

The direction of a carriage or mortar-bed is determined by nailing strips of boards along the platform, as guides to the trail and wheels; to prevent the strips from being injured by the recoil, they should be nailed at a certain distance from the carriage, or bed, and the space filled up with a stick of proper width, which should be removed before firing. The chassis of a sea-coast carriage can be secured in a particular direction by firmly chocking the traverse wheels.

440. **Pointing small-arms.** The rear-sights of small-arms are graduated with elevation marks for certain distances, generally every hundred yards; in aiming with these, as with all other arms, it is first necessary to know the distance of the object. This being known, and the slider being placed opposite the mark corresponding to this distance, the bottom of the rear-sight notch, and the top of the front sight, are brought into a line joining the object and the eye of the marksman. The term *coarse-sight* is used when a considerable portion of the front-sight is seen above the bottom of the rear-sight notch; and the term *fine-sight*, when but a small portion of it is seen. The graduation marks being determined for a fine-sight, the effect of a coarse-sight is to increase the true range of the projectile.

441. **Graduation of rear-sights.** If the form of the trajectory be known, the rear-sight of a fire-arm can be graduated by calculation; the more accurate and reliable method, however, is by trial. Suppose it be required to mark the graduation for 100 yards; the slider is placed as near the position of the required mark as the judgment of the experimenter may indicate; and, with this elevation, the piece is carefully aimed, and fired, say ten times, at a target placed on level ground, at a distance of 100 yards. If the assumed position of the slider be correct, the centre of impact of the ten shot-holes will coincide with the point aimed at; if it be incorrect, or the centre of impact be found below the point aimed at, then the position of the slider is too low on the scale. Let P be the point aimed at, and P' the centre of impact of the cluster of shot-holes; we have, from close similarity of the triangles, $A'F : FP ::$

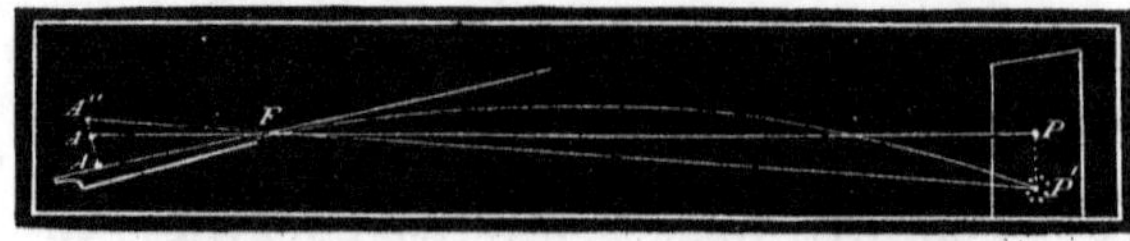

Fig. 144.

$A'A'' : PP'$, from which we can determine $A'A''$, the quantity that must be added to AA', to give the correct position of the graduation mark for 100 yards. If the centre of impact had been above P, the trial mark would have been too high. Lay off the distance AA'' above A'', on the scale, and we obtain an approximate graduation for 200 yards, which should be corrected in the same way as the preceding, and so on. The distance PP' is found by taking the algebraic sum of the distances of all the shots from the point P, and dividing it by the number of shots. It will be readily seen that an approximate form of the trajectory may be obtained by drawing a series of lines through the different graduation marks of the rear-sight, and the top of the front-sight, and laying off from the front-sight, on each line, the corresponding range. The points thus determined are situated in the required trajectory.

442. **Distance of object.** Various instruments have been devised to determine the distances of objects, based on the measurement of the visual angles subtended by a foot or cavalry soldier, of mean height, at different distances; but these instruments are considered of little practical value, especially in the excitement of action. Every officer and soldier should be taught to estimate distances by the eye, and in so doing much assistance is derived from knowing what parts of a soldier's dress, or equipments, are visible at certain distances. These data vary with the power of the eye, and each soldier should

be required, by comparison and reflection, to create a standard for his own.

In firing cannon, the point at which the projectile strikes the ground or bursts, can generally be observed, and from it, the error of aim can be corrected in a few fires; this, however, does not hold true for small-arm projectiles, which are seldom seen to strike the ground, unless the soil be dusty.

In the defence of sea-coast batteries, the distances of objects may be determined by their proximity to known objects, as fixed buoys, or by their bearing with reference to prominent landmarks. Plane-tables may be also used to determine the distances of objects.

The degree of accuracy with which the distance of an object should be known, depends somewhat on the size of the object and the inclination of the trajectory to the line of sight; if the object be large and the trajectory vary but slightly from the line of sight, it is not necessary to know the exact distance, provided the aim be accurately taken.

TABLES OF FIRE.

443. **Purpose.** The nature and purpose of a table of fire should be explained in connection with the subject of pointing cannon. A properly constructed table of fire, for a particular piece, contains the range and time of flight for each elevation, charge of powder, and kind of projectile. Its purpose is to assist the artillerist in attaining his object without waste of time and ammunition, and also when the effect of shot cannot be seen on account of the dust and smoke of the battle-

field. The first few shots generally produce a great effect on the enemy, and it is very important that they should be directed with some knowledge of their results, which, in the field, can only be attained by experience, or from the data afforded by a table of fire.

The following is the form of a table of fire for guns and howitzers:

KIND OF ORDNANCE.	POWDER.	PROJECTILE.	ELEVATION.	RANGE.	TIME.
	Lbs.	Lbs.	° ′	Yards.	Sec'ds.
Armstrong gun, 4-inch bore.	$3\frac{5}{8}$	29, (solid.)	5° 0′	2099	7.5
			7° 0′	2894	9.1
			10° 0′	3700	11.6
			12° 0′	4196	14.2
			15° 0′	4776	17.1
			20° 0′	6070	21.4
			25° 0′	6580	25.
			30° 0′	7555	31.
			35° 0′	9000	

The ranges in the foregoing table were determined at West Point, in 1860, and are the mean of five shots for each angle of elevation. The ranges obtained with the best American muzzle-loading rifle-cannon compare favorably with these.

Tables of fire, for the different service cannon, may be found in the Ordnance and Artillery Manuals, and the XIII. chapter of this work.

RAPIDITY OF FIRE.

444. **Depends on size of piece, &c.** The rapidity with which cannon can be loaded and discharged depends on the size of the piece, the construction of the carriage, and the care required in aiming.

Field-cannon. Field-cannon can be discharged with careful aim, about twice per minute; in case of emergency, when closely pressed by the enemy, canister-shot may be discharged four times per minute. The 12-pdr. boat-howitzer of the navy, with experienced gunners, can be discharged at the rate of sixteen times per minute.

Siege-cannon. Siege-guns are generally discharged about twelve times per hour; if necessary, they can be discharged as rapidly as twenty times per hour. Iron cannon can be fired more rapidly than bronze, as the latter metal is softened by the heat, and the piece is liable to bend. Siege-mortars can be conveniently fired twelve times per hour, and more rapidly than this if the object be large, as a city. Siege-howitzers can be fired about eight times in an hour.

Sea-coast cannon. The fire of a sea-coast cannon depends much on the ease with which its carriage can be manœuvred. The heaviest, or 15-in. gun, mounted on the new iron carriage, can be loaded and fired in 1′ 10″; the time required in aiming depends on the angle through which the chassis is to be traversed, and piece elevated, or depressed; it can be traversed through an angle of 90° in 2′ 20″.

Small-arms. Muzzle-loading small-arms can be discharged two or three times in a minute, and breech-loading arms about ten times; the revolver can be discharged much more rapidly for six shots.

This quality of a military fire-arm should be carefully guarded, as it is found that soldiers are prone to discharge their pieces in the excitement of battle without taking proper aim, and consequently to waste their ammunition.

CHAPTER X.

DIFFERENT KINDS OF FIRES.

445. **Classification.** Artillery fires are distinguished by the manner in which the projectile strikes the object—as *direct*, *ricochet*, *rolling*, and *plunging* fires; by the nature of the projectile, as *solid shot*, *shell*, *shrapnel*, *grape*, and *canister* fires; and by the angle of elevation, as *horizontal fire*, or the fire of guns and howitzers under low angles of elevation, and *vertical* fires, or the fire of mortars, under high angles of elevation.

446. **Direct fire.** A fire is said to be direct when the projectile hits its object before striking any intermediate object, as the surface of the ground, or water. This species of fire is employed where great penetration is required, as the force of the projectile is not diminished by previous impact; it is necessarily employed for spherical-case shot, and for rifle-cannon projectiles, which, from their form, are liable to be deflected, by previously striking a resisting substance; it is also used for all field-cannon projectiles, when the nature of the ground does not insure a regular rebound.

To point a piece in direct fire, bring the line of sight to bear upon the object, and then elevate the piece according to the distance.

447. **Ricochet fire.** When a projectile strikes the ground, or water, under a small angle of fall, it penetrates obliquely to a certain distance, and is then re-

flected at an angle greater than the angle of fall; the reason for this is, that the projectile, in forming the furrow, loses a portion of its velocity, making the distance from *A* (fig. 145), the point at which it enters the ground, to *C*, or the vertical drawn through the deepest point, greater than the distance from *C* to *D*, the point where it leaves the ground.

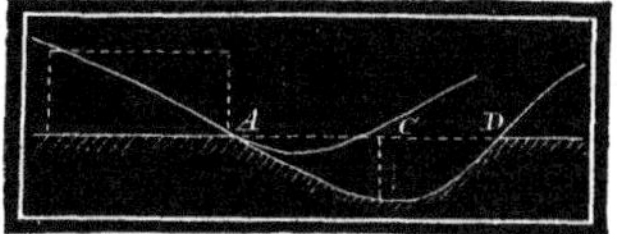

Fig. 145.

As this recurs every time the projectile strikes the ground, it follows that the trajectory is made up of a series of rebounds, or *ricochets*, each one shorter and more curved than the preceding one.

The number, shape, and extent of the ricochets, depend on the nature of the surface struck, the initial velocity, shape, size, and density of the projectile, and on the angle of fall.

A spherical projectile ricochets well on smooth water, when the angle of fall is less than 8°, but if the surface of the water be rough, very little dependence can be placed on the extent of the ricochet. Captain Dahlgren cites a case as coming under his observation where the distance between the first and second rebound was increased from 400 to 800 yards by a strong wind; at the same time, the height of the highest point of the curve was increased from a very small distance above the water, to more than 50 feet, which would have rendered it ineffective against the hull of a ship. From the same causes the lateral deviations in ricochet fire will be very considerable, amounting, in some cases, to between 100 and 200 yards in the entire range.

In general, those projectiles which present a uniform

surface, and have the least penetrating power, are most suitable for ricochet firing; hence, large shells fired with small charges are more suitable than solid shot, and round projectiles more suitable than those of an oblong form. The distance at which the larger size shells will ricochet on water is about 3,000 yards, the axis of the piece being horizontal and near the water.

Where used, &c. Ricochet fire is employed in siege operations to attain the face of a work in flank, or in

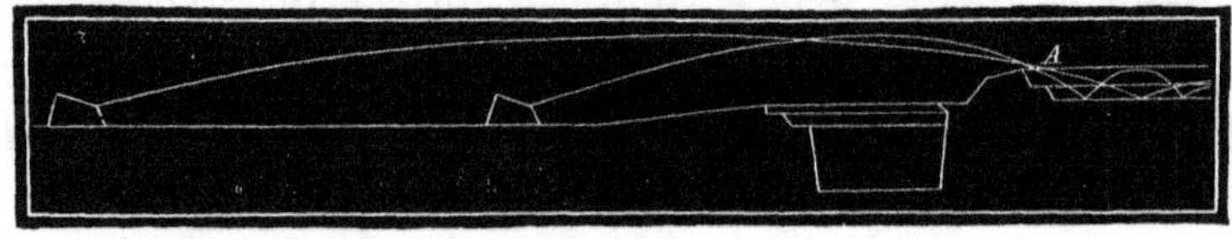

Fig. 146.

reverse (see fig. 146), and on the field, or on water, when the object is large and its distance is not accurately known.

The character of ricochet fire is determined by the angle of fall, or the angle included between the tangent of the trajectory and horizon at the point of fall. There are two kinds of ricochet fire—the *flattened*, in which the angle of fall is between 2° and 4°; and the *curvated*, in which the angle of fall is between 6° and 15°.

The principal pieces employed in ricochet fire in siege operations are the 8-inch howitzer, and the 8 and 10-inch common mortars; the first two may be used when the angle of fall is less than 10°, and the 10-inch mortar when the angle of fall is less than 15°—the proper elevation being given to the mortar by raising the rear portion of the bed. With these pieces, the limit of ricochet is about 600 yards. Solid shot should not be used in ricochet fire for any distance less than 200 yards, as it would then be necessary to diminish its velocity so much as to destroy its percussive effect. In ricochet firing against troops in the open field, the angle of fall should not exceed 3°.

448. **Practical rules for ricochet fire.** In enfilading the face of a work, the form of the trajectory and point of fall should be such that the projectile will strike the surface of the terreplein the greatest number of times; the object being to destroy the men, carriages, and traverses situated upon it. To do this, the projectile should be made to graze the crest of the adjacent parapet, and strike the terreplein as near the foot of the interior slope as possible; the distance of the crest, and its height above the terreplein and battery, should therefore be known.

The formulas in chapter VIII. furnish accurate means for calculating the various elements of ricochet fire, but they are too complicated for use in the field; it is therefore proposed to deduce simple and practical rules for this purpose.

1st. *To find the angle of arrival.* The angle of arrival is the angle which the tangent to the trajectory at the crest of the parapet makes with the horizon. Let A be the crest, and B the point of fall (fig. 147); the distance A B being short, the portion of the trajectory included between these two points may be considered a right line, and the angle of fall

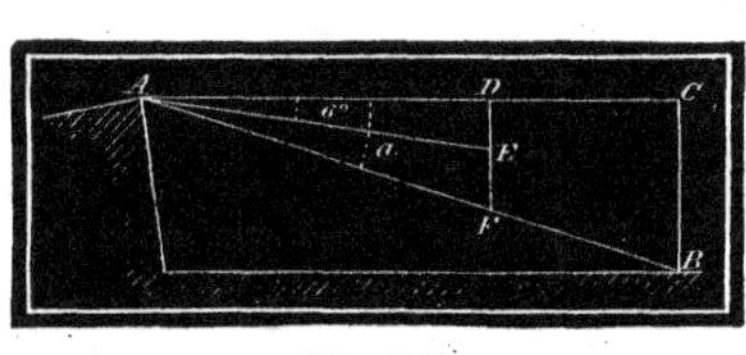

Fig. 147.

and arrival will be equal. Calling a the angle of fall, and erecting the perpendicular $B\ C$, we have,

$$\tan. a = \frac{BC}{AC},$$

or, *the tangent of the angle of arrival is equal to the vertical distance of the point of fall below the crest, divided by the horizontal distance.*

Within the limits of ricochet fire, the angles may be supposed proportional to their tangents; calling the tangent of 6° (which is 0.1051) 0.1, we have the following proportion:

$$a : 6^\circ :: \frac{BC}{AC} : 0.1,$$

or,

$$a = 60^\circ \frac{BC}{AC},$$

or, *the angle of arrival is equal to* 60° *multiplied by the ratio of the horizontal and vertical distances of the point of fall from the crest of the parapet.*

This rule gives the angle of arrival without the aid of a table of natural tangents.

2d. *To find the angle of fire.* The distance of the parapet is always known, and the angle of elevation of the crest can be determined by sighting along the long branch of a gunner's quadrant, and observing the position of the plummet on the arc.

In consequence of the nearness of the object, and the large size and low initial velocity of the projectile, the resistance of the air in this species of ricochet fire may be neglected, which makes the trajectory a parabola. In this case the angle of fall is equal to the angle of fire, when the object is situated in the same horizontal plane with the piece; if it be not in the same horizontal plane, let $B\ A\ M$ (fig. 148), which is the angle of elevation

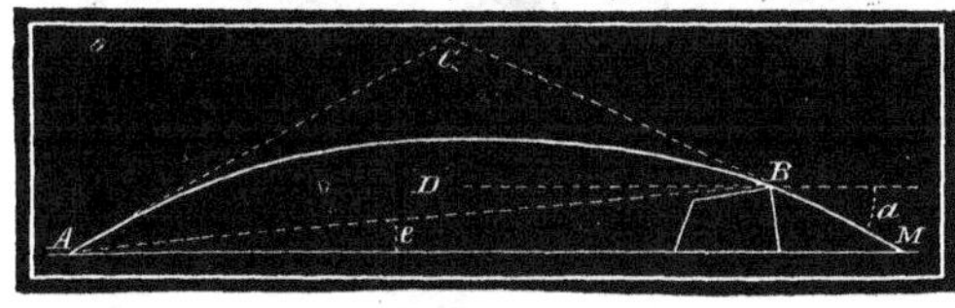

Fig. 148.

of the crest, be represented by e. As the angle e is very small, we are at liberty to suppose $C\ A\ B = A\ B\ C$. Through the point B draw the horizontal line $B\ D$, the angle $C\ B\ D$ is equal to the angle of arrival a; the lines $B\ D$ and $A\ M$ being parallel, the angle $A\ B\ D = e$; therefore $C\ A\ B = C\ B\ A = a + e$, but the angle $C\ A\ M = C\ A\ B + e = a + 2e$: or, *the angle of fire is equal to the angle of arrival increased by twice the angle of elevation of the crest of the parapet.*

From the erroneous suppositions made in the course of the preceding demonstrations, it will be seen that the rules deduced should give too great an angle of fire. In practice, this angle should be somewhat greater than the true angle, in consequence of the deviations, which render the projectile liable to strike against the parapet, and, of course, destroy its effect.

3d. *To find the charge.* When a projectile moves in vacuo, we have seen that the distance which it falls below the line of fire, in the time t, is $\frac{1}{2}\ gt^2$; and for a given distance, t is inversely proportional to the initial velocity V; hence the distances which the same projectile, fired with different velocities, would fall below the line of fire, in the distance $A\ C$ (fig. 149), will be *inversely proportional to the squares of the initial velocities.*

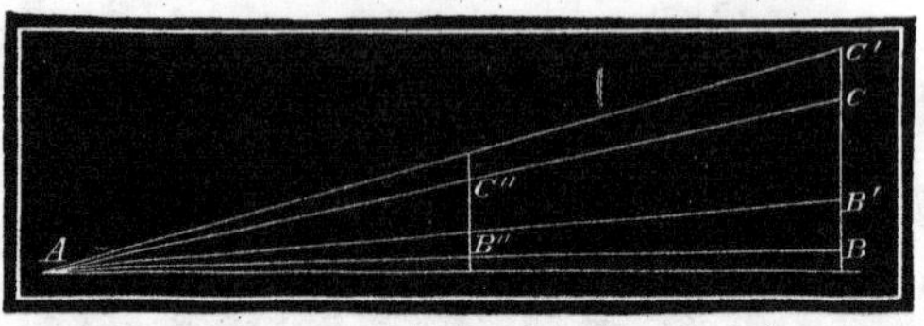

Fig. 149.

If we suppose the lines of fire of two projectiles be $A\ C$ and $A\ C'$, and the initial velocities, V and V', to be such that they will fall the distances $B\ C$ and $B'\ C'$, and that the angles subtended by these lines be proportional to the lines themselves, we shall have

$$B\ A\ C : B'\ A\ C' :: V'^2 : V^2.$$

It has been seen that the initial velocities of small charges are nearly proportional to the square roots of the weight of the charges. Calling the corresponding charges C and C', we have

$$B\ A\ C\ : B'\ A\ C' :: C' : C,$$

or, *for the same distance of the object, the charges should be inversely proportional to the difference between the angle of fire and angle of elevation of the object.*

Take the case in which the objects are situated at different distances, as B and B'' (fig. 149), but have the same angle of elevation e; and suppose we wish to strike them with the same angle of fire; what should be the relation between the charges?

Substitute in the expression $\frac{1}{2}gt^2$, the value of t, which is $\frac{D}{V}$, in which D is the distance, and V the initial velocity of the projectile, we have $\frac{1}{2}g\frac{D^2}{V^2}$, which shows that the distance which a projectile falls below the line of fire is directly proportional to the square of the distance measured on the line of fire, and inversely proportional to the square of the velocity. But the distances $B'' C''$ and $B C$ are proportional to $A C''$ and $A C$, or D'' and D, and, recollecting that the squares of the initial velocities are proportional to the charges C and C'', we have

$$D'' : D :: \frac{D''^2}{C''} : \frac{D^2}{C},$$

or,

$$D'' : D :: C'' : C,$$

or, *for the same difference between the angle of fire and the angle of elevation of the object, the charges are proportional to the distances.*

In arriving at the foregoing rules, we have committed three errors: 1st. Supposing the sides of the triangles proportional to the angles. 2d. Considering the resistance of the air nothing; and, 3d. That the initial velocities are proportional to the square roots of the charges. The errors resulting from these suppositions are not only small in themselves, but the 2d and 3d are of a nature to counteract each other.

By means of the foregoing relations suitable charges can be calculated for every case of practice, when we know the charge corresponding to a given distance, and to a given difference between the angle of fire and the angle of elevation of the object. Represent by C' the charge corresponding to a distance, D', and to a difference, E', between the angle of fire and the angle of elevation of the object; we have the charge, C, corresponding to the distance, D, and the difference, E, between the two angles, by means of the formula

$$C = \frac{D}{E} \times \frac{C' E'}{D'}.$$

The factor, $\frac{C'E'}{D'}$ is a constant number for each calibre. This number may be considered as the charge corresponding to the distance of 1 yard, and to a difference of 1° between the angle of fire and of elevation of the object.

For the French 8-inch siege howitzer, the value of this factor has been found by careful experiment to be 0.31 oz. Making an allowance for difference of weight of projectile and unit of distance, it becomes 0.28 oz. for the American 8-inch siege howitzer.

Example.—Find the angle of arrival, angle of fire, and charge of powder, necessary to hit, with an 8-inch howitzer shell, a point on a terreplein, 12 yards behind a traverse which is 2.5 yards high and 350 yards from the battery—the angle of elevation of the crest being 1°, and the command 6 yards.

For the angle of arrival we have

$$a = 60^\circ \frac{BC}{AB} = \frac{60^\circ \times 2.5}{12} = 12^\circ\ 30'.$$

For the angle of fire we have

$$\phi = a + 2e = 12^\circ\ 30' + 2^\circ = 14^\circ\ 30'.$$

For the charge we have

$$C = \frac{D}{E} 0.28 \text{ oz.} = \frac{350.}{13.5} 0.28 = 7.25 \text{ oz.}$$

449. **Rolling fire.** Rolling fire is a particular case of ricochet fire, produced by placing the axis of the piece parallel, or nearly so, with the ground. It is generally used in field service. When the ground is favorable for ricochet, the projectile, in rolling fire, has a very long range, and never passes at a greater distance above the ground than the muzzle of the piece; it is therefore more effective than direct fire, as may be seen by inspecting fig. 150.

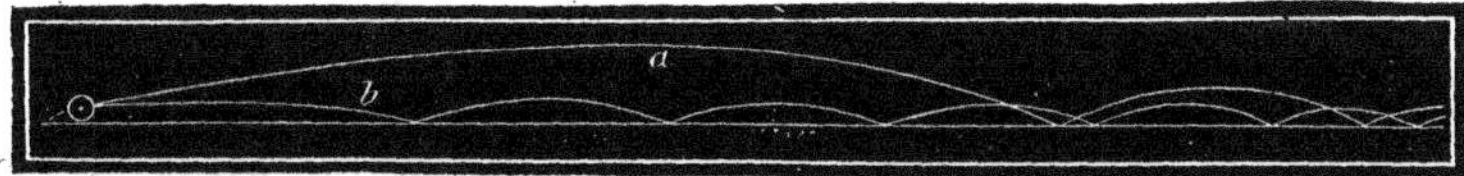

Fig. 150.

To point a piece in rolling fire, *direct it at the object, and depress the natural line of sight so as to pierce the surface of the ground about* 80 *yards in front of the muzzle;* if the piece be sighted for the pendulum hausse, *aim directly at the object with the lowest line of sight, or with the slider fixed at the zero point of the scale.*

450. **Plunging fire.** A fire is said to be plunging when the object is situated below the piece. This fire is particularly effective against the decks of vessels.

451. **Effect of fire in general.** Before proceeding to describe the fires of different kinds of projectiles, it may be proper to explain what is meant by accuracy of fire, and to determine a suitable measure for it. It has been seen that there are causes constantly at work to deviate nearly every projectile from its true path. As the effect of these deviating forces cannot be accurately foretold, there is only a probability that the projectile will strike the object against which the piece is pointed. The degree of probability is called accuracy of fire.

For all projectiles of the same nature, the chance of hitting an object increases with the velocity and weight of the projectile, whereby the effects of the deviating forces are diminished; it also increases as the size of the object is equal to, or greater than, the mean deviations, and as the trajectory more nearly coincides with the line of sight. If the size of the object be greater than the extreme deviation, and the trajectory coincide with the line of sight, the projectile will be certain to hit the object at all distances.

452. **Measure of deviation.** For the same trajectory, therefore, the mean deviation of a projectile at a given distance may be taken as an indirect measure of its accuracy at this distance.

To obtain this mean deviation, let the piece be pointed at the centre of a target, stationed at the required distance, and fired a certain number of times—say ten—and let the positions of the shot-holes, measured in vertical and horizontal directions, be arranged in the following tabular form:

No. of shot.	Distances from centre of target, in feet.				Distances from centre of impact, in feet.			
	Vertical.		Horizontal.		Vertical.		Horizontal.	
	Above.	Below.	Right.	Left.	Above.	Below.	Right.	Left.
1	3		4		4.33		2.66	
2		6		2		4.66		3.33
3		1	2		.33		.66	
	3	7	6	2	4.66	4.66	3.33	3.33
	4 ÷ 3 = 1.33		4 ÷ 3 = 1.33		9.32 ÷ 3 = 3.11		6.66 ÷ 3 = 2.22	

The algebraic sum of the distances in each direction, divided by the number of shots, gives the position of the centre of impact in this direction. In the above table the position of the centre of impact is found to be 1.33 ft. below, and 1.33 ft. to the right, of the centre of the target. To obtain the mean deviation, it is necessary to refer each shot-hole to the centre of impact as a new origin of co-ordinates; and this is done by subtracting the tabular distance from the distance of the centre of impact, if both be on the same side of the centre of the target, and adding them, if on different sides. The sum of all the distances thus obtained in one direction, divided by the number of shots, gives the mean deviation in that direction; which in the present case is 3.11 ft. vertically, and 2.22 horizontally.

The foregoing affords a measure for the accuracy of fire of the piece and projectile, but it does not afford a measure for marksmanship, the object of which is to direct a projectile so as to strike a given point or surface. In target-practice with sporting rifles, the *string*, or sum of the distances of a certain number of shots, from the point aimed at, is taken as the measure of accuracy. In military arms, marksmanship is measured by the greatest number of projectiles out of a certain number, placed in a target of given size, or placed within a given space surrounding the centre of the target.

453. **Targets.** Targets for heavy cannon are made of cotton cloth (or light boards) stretched over two upright poles firmly secured in the ground. The size varies with the distance: for 1,000 yards and upward, it should be about 20 feet high and 40 feet long. Targets for the field service are made of the same materials, about 8 feet high, and from 30 to 40 feet long. Targets for small arms, if permanent, are made of cast-iron; if portable, of a wrought-iron frame covered with cotton cloth. For distances less than 200 yards, they should be 6 feet high and 22 inches broad; beyond this distance, the breadth of a target may be increased by placing two or more of these targets side by side.

454. **Deviations.** The vertical deviation of a projectile is generally greater than its corresponding hori-

zontal deviation, and this difference increases with the range. As objects against which military projectiles are directed, present a greater extent of surface in a horizontal than in a vertical direction, it becomes necessary to exercise great care in the selection of the proper angle of fire. If the ground or water in front of the object be favorable to ricochet, the difficulty will be diminished by aiming so that the projectile will strike the object after one or more rebounds.

455. **Solid-shot firing.** Solid shot are generally used for percussion and penetration, and, when heated to a red heat, for the purpose of setting fire to wooden vessels or buildings. From their great strength, they can be fired with a large charge of powder, which gives them great initial velocity, and having great density, which diminishes the effect of the resistance of the air, they have great range and accuracy. In firing hot shot, the charge should be reduced, to prevent too great penetration, which would exclude the air and render combustion impossible.

The extreme range of field artillery is about 3,000 yards; it is not very effective, however, beyond 1,700 yards for the 6-pdr., and 2,100 yards for the 12-pdr. At 600 yards the horizontal deviation of the 12-pdr. is about 3 feet, and at 1,200 yards it is about 12 feet. For the 6-pdr. the deviations are somewhat greater at both distances.

The service of solid shot demands less skill than that of shells and spherical case-shot, and they are often effective when the latter are rendered non-effective by untimely explosion.

456. **Shell-firing.** The diameter and velocity of two

projectiles being the same, the retarding effect of the air is inversely proportional to their weight (see page 406); hence a shell has less accuracy and range than a solid shot of the same size, in the proportion of 3 to 2—these numbers representing the weights of a solid shot and shell, respectively.

Field shells. As shells act both by percussion and explosion, they are particularly effective against animate objects, earthworks, buildings, block-houses and shipping, posts and villages occupied by troops, and against troops sheltered by accidents of the ground; but against good masonry they have but little effect, as they break on striking. Against troops, especially cavalry, they possess a certain moral effect which solid shot do not possess. They are used to form breaches in intrenchments, in which case they act as small mines. The 32-pdr. shell is the most effective field projectile for this purpose; and, when fired with a large charge, has a penetration of from 5 to 8 feet in fresh earth.

The extreme range of field shells is from 2,500 to 3,000 yards. The 24 and 32-pdr. shells burst into about eighteen effective fragments, some of which are thrown to a distance of 600 yards. All field shells have considerable lateral deviation; it is stated that the 24-pdr. shell is sometimes deviated as much as 30 yards in 1,200.

Mountain shells. The extreme range of the mountain howitzer is about 1,200 yards, after three or four rebounds. The 12-pdr. shell employed in this service bursts into twelve or fifteen fragments, some of which are thrown to a distance of 300 yards.

Siege shells. The great weight of an 8-inch shell, and the large quantity of powder which it contains, render it a very formidable projectile against the traverses and epaulements of siege works.

Sea-coast shells. In sea-coast defence, the 8, 10, and 15-inch shells are very destructive to vessels built of timber. They range from 3 to 3½ miles; but the angle which the trajectory makes with the line of sight at this distance (about 40°) renders their fire very uncertain against individual objects of the size of a ship; but it is presumed that they would have the effect to prevent a blockading fleet from lying at anchor within their range, as it is well known that a single 10-inch shell, striking on the deck of a vessel, has sufficient force to penetrate to the bottom and sink her. The 8-inch shell bursts into 28 or 30 fragments; and from the experiments made at Brest, some years ago, it was inferred that three of four of these shells, properly timed and directed, were capable of disabling a ship of war.

Mortar shells are employed to break through the roofs of magazines, &c., and to blow them up; to destroy the surface of the terrepleins, ditches, &c., by forming deep hollows, which are produced by explosion, and to interrupt the communications from one part of a work to another. The great depth to which mortar shells penetrate in earth, almost entirely destroys the effect of their fragments; some remain buried in the ground, and the others are thrown out at too high an angle to be dangerous. One of the principal objects of traverses, on a terreplein, is to confine the bursting effects of shells within narrow limits. Mortar shells penetrate from half a yard to one yard in earth; and the amount of earth thrown up by explosion is about one cubic yard for each pound of the bursting-charge. Ordinarily, the diameter of the crater at the top is two or three times the depth. The 13-inch shell will often break in falling on a pavement. Roofs of good masonry, little more than a yard thick, are sufficient to resist the penetration of mortar shells.

The effect of mortar-firing is generally in favor of the besiegers, as the works of the besieged present a larger and more favorable surface for the action of shells. About one-fifth of the shells fall inside of a demi-lune at a distance of 650 yards, and about one-third at a distance of 450 yards. The line of fire should be taken in the direction of the greatest extent of the part to be shelled. The fire of mortars at sea is very uncertain, unless the object be very large.

Stone mortars. The charge of a stone mortar should be small, to prevent the stones and grenades from being too much scattered. A charge of stones is generally scattered over a space varying from 30 to 50 yards broad, and from 60 to 100 yards long. The dispersion of grenades is somewhat less than this; the larger portion, however, are found within a radius of 12 or 15 yards. Each grenade furnishes from 12 to 15 fragments in a radius of 10 or 20 yards; some of the fragments are projected to a distance of 300 yards.

457. **Shrapnel firing.** When a shrapnel or case-shot bursts in its flight, the fragments of the case and the contained projectiles are influenced by two forces, viz., the force of propulsion, which moves each piece in the direction of the trajectory, and the force of rupture, which moves it in the direction of a normal to the sur-

face of the case. The path described by each fragment and projectile depends on the angle which the normal makes with the trajectory, and on the relative velocities

Fig. 151.

generated by the two forces; and, when taken together, these paths form a species of cone, called the cone of dispersion, the apex of which coincides with the point of rupture, and the axis is the trajectory, prolonged. Fig. 151.

The velocity of a projectile diminishes from the time it leaves its piece, while the velocity generated by the rupturing force remains constant. It follows, therefore, that the dispersion of a spherical case-shot increases with the distance, while the force of impact is diminished.

The distance at which a spherical case-shot ceases to be effective depends on the relation between the remaining velocity and the velocity generated by the force of rupture. The improvements which have lately been introduced into this species of projectile, have for their objects, to increase the remaining velocity at any point by increasing the propelling charge, and to diminish the force of rupture, and at the same time increase the number of contained projectiles by diminishing the bursting-charge. By filling the interstices of the bullets with sulphur or rosin, the propelling charge of a spherical case-shot can be made the same as that of a solid shot. (See chapter II.)

It is considered that a spherical case-shot is effective when a large portion of the projectiles have sufficient force to penetrate one inch of soft pine. The present 12-pdr. spherical case-shot, fired with a charge of $2\frac{1}{2}$ pounds of powder, has a remaining velocity of about 500 feet at a distance of 1,500 yards, which renders it effective at this distance.

The principal difficulty experienced in firing a spherical case-shot is, to burst it at the proper distance in front of the object. This arises from the difficulty of estimating the correct distance of the object, the rapid flight of the projectile, and the difficulty of observing the effect of a shot in order that correction may be made for the succeeding one, if necessary. To overcome these difficulties requires skill and judgment on the part of the gunner, and great accuracy and delicacy in the operation of the fuze.

The proper position of the point of rupture varies from 50 to 130 yards in front of, and from 15 to 20 feet above, the object.

The mean number of destructive pieces from a 12-pdr. spherical case-shot, which may strike a target 9 feet high and 54 feet long, situated at a distance of 800 yards, is 30.

The effect of elongated case-shot from rifle-cannon is said to extend upward of 2,000 yards. This arises from the fact that an oblong projectile preserves its velocity for a much longer distance than a round one.

The weight of a spherical case-shot is about the same as a solid shot of the same size, and being fired with the same charge of powder, it can be used for attaining long ranges, in the absence of solid shot. For this purpose the fuze should not be cut.

Spherical case-shot should not be used for a less distance than 500 yards; although in cases of emergency the fuze may be cut so short that the projectile will burst at the muzzle of the piece, in which case it will act like grape or canister shot.

458. **Grape and canister firing.** In grape and canister firing, the apex of the cone of dispersion is situated in the muzzle of the piece, and the destructive effect is confined to short distances. The shape of this cone is the same as in spherical case-shot; its intersection by a vertical plane is circular, while that of a horizontal plane, as the ground, is an oval, with its greatest diameter in the plane of fire. The greatest number of projectiles are found around the axis of the cone, while the extreme deviations amount to nearly one-tenth of the range.

The most suitable distance for field canister-shot is from 350 to 500 yards; if the ground be hard and the surface be uniform, the effect may extend as far as 800 yards. In cases of great emergency a double charge of canister, fired with a single cartridge, may be used for distances between 150 and 200 yards.

Under favorable circumstances, one-third of the whole number of contained projectiles will strike the size of a half-battalion-front of infantry, and one-half, the front of a squadron of cavalry.

Grape and canister shot are employed in siege and sea-coast operations; in the latter, they are effective against boats, and the rigging, &c., of vessels. Grape-shot, being larger than canister-shot, are effective at greater distances.

Canister-shot for the mountain service are not effective beyond 250 and 300 yards.

459. **Small-arm firing.** Beyond 200 yards, the fire of the smooth-bored musket becomes very uncertain against individual objects, as the lateral deviations often exceed four feet; but by aiming high it may be made effective against troops in mass at 400 yards. The fire of the rifle-musket is effective at 1,000 yards; the angle of fall, however, is so great (about 5°) that great care must be exercised in determining the exact distance of the object. If the ground be favorable, the projectile will ricochet at 1,000 yards, which increases the *dangerous space*, and therefore the chances of hitting the object. The limit of any fire is determined by the distinctness of vision;—the limit of distinct vision for a foot-soldier is about 1,100 yards; that for a mounted soldier is about 1,300 yards.

The effect of small-arm firing depends much on the skill and self-possession of the soldier in action; for, without these qualities, the most powerful and accurate arms will be of little avail. The number of cartridges expended for each person disabled in previous European wars has been variously stated to be from 3,000 to 10,000. In the late Mexican war, where an unusually large proportion of the American troops were armed with rifles, this number has been estimated to be from 300 to 400.

Where a soldier discharges his piece from the back of a horse, as in the cavalry service, the effect of fire is much less than in the dragoon and mounted-rifle services, where he rides from point to point, but discharges his piece on foot.

At short distances, and against troops in mass, two or three round bullets may be employed with effect;

the bullets should be so small that they will readily drop to their place without the aid of the ramrod. Buckshot have very little effect beyond 100 yards.

The following are the mean deviations of the rifle-musket fired from a shoulder and rest.

DISTANCE.	VERTICAL.	HORIZONTAL.
Yards.	Inches.	Inches.
100	1.9	1.5
600	22.2	14.6
1000	55.9	25.5

CHAPTER XI.

EFFECTS OF PROJECTILES.

460. **General considerations.** A knowledge of the destructive effects of projectiles on iron, wood, earth, and masonry, the materials of which covering masses are made, is of very great importance in a military point of view. In general, these effects, and particularly that of penetration, depend on the nature of the projectile, its initial velocity, and the distance of the object.

461. **Cast-iron.** Experiment shows that cast-iron, even in large masses, is easily broken by projectiles; it should not, therefore, be used for gun-carriages, nor the revetment of fortifications.

462. **Wrought-iron.** The following deductions have been made from trials with armor plates, extending over several years.

1st. The best material to resist projectiles is soft, tough wrought-iron; and to attain these qualities it should be pure — free from sulphur, phosphorus and carbon. Steely iron, commonly known as homogeneous iron, puddled steel, etc., when in large masses, is easily cracked by shot, and is not, therefore, suitable for armor plates. Thin plates afford a greater comparative resistance. Soft steel may be used for armor plates; but when cost is taken into consideration, it is doubtful if it possesses any advantages over wrought-iron.

2d. Rolled iron does not offer quite so much resistance as hammered iron, yet if the size of the plate admit of it, it is to be preferred on the score of economy. Plates should be as large as possible to reduce the number of joints, which are lines of weakness.

3d. A solid plate offers, for the same thickness, a greater resistance to a projectile than a laminated one, or one made up of several thinner plates; but when the surface is rounded in shape and of small extent, as in the Monitor turrets, the latter may be used to great advantage, as great thickness of metal may thereby be easily obtained.

4th. The resistance of a plate to perforation is very much increased by a suitable backing. Cast-iron, granite, and brick in masses, while they enable a plate to offer a very great resistance, are soon broken up by the blows of heavy projectiles, and their fragments thrown off with great force. Oak and teak are the most suitable timbers for backing plates, and are used as such on vessels. A yielding backing is found to occasion less strain on the fastenings than a very hard one.

5th. Where projectiles are made of the same material and are similar in shape, their penetration into unbacked plates is nearly in proportion to their *living force*, or *their weight multiplied by the square of the velocity of impact.* The resistance which an unbacked plate offers to penetration is nearly in proportion to the *square of its thickness*, provided this thickness be confined within ordinary limits. In the case of oblique plates the penetration diminishes nearly with the *sine of the angle of incidence.*

6th. The most suitable material for shells to be used against iron plates is tempered steel. These projectiles should be made of cylindrical shape, with thick sides and bottom, to direct the explosive effect of the charge forward after penetration is effected (see figs. 163, 164). The most suitable material for solid shot is hard and tough cast-iron. Captain Pallisers' chilled shot are made of this material, and so are the shot made by the Army and Navy Ord. Departments of this country. Round shells made of cast-iron will be broken in passing through an inch plate, and an ordinary cast-iron shot will be broken in passing through a two inch plate. Late experience shows that the pointed, or ogeeval, is better than the flat form of head for penetration of iron plates.

7th. It follows from the preceding that the most suitable covering or shield for cannon, is a conical-shaped turret made of wrought-iron plates, as large as it is practicable to make them, backed with oak or teak. To protect the gunners from the fragments of projectiles which may penetrate completely through this covering, there should be an "inner skin" of thick boiler plate placed behind the wood.

The Chief of the Naval Ordnance Bureau, in a late annual report uses the following language, viz.:—"As "a general summary of the results obtained in practice "against iron targets, it may be stated that the XI-inch "gun is capable of piercing 4½ inches of the best iron "plating, backed by 20 inches of solid oak. The new "X-inch solid shot gun will penetrate any iron-clad of "which we have, at present, any information. And in "the target experiments, the XV-inch gun has broken "up and shattered plates exceeding in thickness any-"thing hitherto proposed as defensive armor."

He also states, "that cast-iron projectiles made of the

"best charcoal iron, poured and worked in a peculiar "manner so as to obtain hard and solid masses, have "been found by recent experiment able to penetrate at "close range any given thickness of iron armor which "can be worn on the sides of ships of war. In fact, the "penetration is quite as great and uniform as that ob- "tained with *steel* shot of equal weights propelled by "similar charges, the only difference being that the iron "breaks after passing through, while the steel is only "compressed or flattened,—a result rather in favor of "the iron shot if entrance is made between decks, where "men are exposed to its fragments."

They may also be cast in strings on Gen. Rodman's plan, see page 89.

463. **Effect on wood.** The effect of a projectile fired against wood varies with the nature of the wood and the direction of the penetration. If the projectile strike perpendicular to the fibres, and the fibres be tough and elastic, as in the case of oak, a portion of them are crushed, and others are bent under the pressure of the projectile, but regain their form as soon as it has passed by them. It is found that a hole, formed in oak by a ball 4 inches in diameter, closes up again, so as to leave an opening scarcely large enough to measure the depth of penetration. The size of the hole and the shattering effect increases rapidly for the larger calibres. A 9-inch projectile has been found to leave a hole that does not close up, and to tear away large fragments from the back portion of an oak target representing the side of a ship of war, the effect of which, on a vessel, would have been to injure the crew stationed around, or, if the hole had been situated at

or below the water line, to have endangered the vessel. If penetration take place in the direction of the fibres, the piece is almost always split, even by the smallest shot, and splinters are thrown to a considerable distance.

In consequence of the softness of white pine, nearly all the fibres struck are broken, and the orifice is nearly the size of the projectile; for the same reason, the effects of the projectile do not extend much beyond the orifice; pine is therefore to be preferred to oak for structures that are not intended to resist cannon projectiles, as block-houses, etc.

464. **Effect on earth.** Earth possesses advantages over all other materials as a covering against projectiles; it is cheap and easily obtained, it offers considerable resistance to penetration, and to a certain extent regains its position after displacement. It is found by experience that a projectile has very little effect on an earthen parapet unless it passes completely through it, and that injury done by day can be promptly repaired at night. Wherever masonry is liable to be breached, it should be masked by earth works with natural slopes.

General Gillmore states, as the result of his experience before Charleston, that the powers of resistance of pure, compact, quartz sand to the penetration of projectiles, very much exceeds that of ordinary earth, or mixture of several earths.

The size of the openings formed by the passage of a projectile into earth is about one-third larger than the projectile, increasing however towards the outer orifice.

Rifle-projectiles are easily deflected from their course in earth. They are sometimes found lying in a position at right angles to their course, and sometimes with the base to the front; hence their penetration is variable. Unless a shell be very large in proportion to the mass of earth penetrated, its explosion will produce but little displacement,—generally a small opening is formed around an exploded shell by the action of the gas in pressing back the earth. Experience at Fort Wagner showed that it took one pound of metal to permanently displace 3.27 lbs. of the sand of which the Fort was made. Time fuzes, being liable to be extinguished by the pressure of the earth, are inferior to percussion fuzes, which produce explosion when the projectile has made about three-fourths of its proper penetration.

The penetration in earth of oblong, compared to round projectiles, when fired with service charges, and at a distance of about 400 yds., is at least *one-fourth greater*. This difference, however, is less at short and greater at long distances.

The penetration of the smallest, or 3-in. cannon projectile, at a distance of 400 yds., in a newly made parapet of loam mixed with gravel, is about 6 feet. The 100-pdr. projectile, under similar circumstances, penetrates about 16 feet. A penetration as great as 31½ ft. has been obtained at the Washington Navy Yard by firing a 12-in. rifle-projectile into a natural clay bank, at a short distance.

The greatest penetration of a 15-in. solid shot, fired with 50 lbs. of powder, in well rammed sand, at a distance of 400 yds., is 20 ft.

465. **Penetration.** The resistance which a projectile encounters in penetration, arises from the cohesion and inertia of the particles, and the friction of the particles against the surface of the projectile.

To obtain an expression for the penetration, we will suppose that the resistance is proportional to the area of the cross-section of the projectile, and independent of the velocity. Let E be the penetration expressed in calibres, D the density of the projectile, v its velocity at the commencement of penetration, r the radius of the projectile, and R the constant resistance experienced by a unit of surface; the quantity of work done in overcoming the resistance is,

$$\pi r^2 . E . 2r . R;$$

and the living force of the projectile is

$$\frac{4}{3}\pi r^3 \frac{D}{g} v^2.$$

But the living force is equal to twice the quantity of work; hence we have,

$$4\pi r^3 RE = \frac{4}{3}\pi r^3 \frac{D}{g} v^2;$$

by making $K = \frac{1}{3Rg}$ we obtain,

$$E = Kv^2 D.$$

Take another projectile, having a velocity v', a density d, and a penetration e, and we have the expression, $e = Kv_{\prime}^2 d$. Dividing the preceding expression by this, member by member, and we have,

$$E = e\frac{v^2 D}{v_{\prime}^2 d};$$

that is to say, *the penetrations of similar projectiles into a given substance, are proportional to the squares of the velocities of impact, and to the diameters* and densities of the projectiles.*

* The diameters of all service projectiles are given in the Ordnance Manual.

Knowing, therefore, the penetration e, for a given velocity and projectile, we can obtain the penetration E, for another projectile and velocity. Let e represent the penetration for a shot moving with a velocity of 1,650 feet, the expression becomes,

$$E = e\frac{v^2}{1650^2}\frac{D}{d}.$$

This formula has been found by experience to give, with sufficient accuracy, the penetration of projectiles in hard substances, as wood, cast-iron, and masonry, for all velocities up to 1,000 feet per second. The following penetrations, or values of e, have been found for round shot moving with a velocity of 1,650 feet, viz.:

Cast-iron (depending on its nature),	$\frac{1}{5}$ to $\frac{1}{6}$
Lead, ...	$3\frac{1}{4}$ to $3\frac{1}{2}$
Calcareous rocks (particular kind), .	2
Masonry of good quality, ...	4
" rubble, ...	5 to $5\frac{1}{2}$
" brick, ...	8

Substituting the above values of e in the value of E, we obtain the penetrations, expressed in terms of the diameter of the projectile, for any velocity not exceeding 1,000 feet.

Solid shot are broken, when fired against very hard substances, with charges exceeding the following, viz.:*

Against cast-iron, ...	$\frac{1}{128}$
" lead, ...	$\frac{1}{8}$
" calcareous rock (oolitic), .	$\frac{1}{4}$
" masonry, ...	$\frac{1}{3}$

* Some shot resist these charges.

The velocity v, is the velocity which the projectile possesses at the commencement of penetration; if the piece be situated at a distance, it is necessary to determine the remaining velocity, by making allowance for the resistance of the air. Equation (20), page 408.

Wood. The formula $E=e\frac{v^2}{1650^2}\frac{D}{d}$, may be used to calculate penetrations in wood, for velocities which do not exceed 1,000 feet, making use of the following values of e for penetrations perpendicular to the fibre:

For oak of ordinary quality, . .	$e=12\frac{1}{2}$
" elm, ...	16
" pine, ...	23

For velocities exceeding 1,000 feet, the formula just employed gives results which are too large; from this it is inferred that penetration really increases less rapidly than the square of the velocity.

Earth. In the experiments made at Metz, in 1834, on various kinds of earths, it was found necessary to modify this expression for penetration. Calling p the weight of the powder, and m the weight of the projectile, the expression becomes, for all charges between $\frac{1}{2}$ and $\frac{1}{24}$,

$$E=e\frac{\log.\left(1+480\times\frac{p}{m}\right)}{\log.\ (1+480\times\frac{1}{3})}\frac{D}{d}=e\frac{\log.\left(1+480\times\frac{p}{m}\right)}{2.20683}\frac{D}{d}.$$

The following values of e, for a charge of $\frac{1}{3}$, were found for different earths:

For sand mixed with gravel, .	$e=10\frac{3}{4}$
For earth, settled, . . .	$17\frac{1}{2}$

Potter's clay, saturated with water, .	36
Light earth newly dug over, . .	$32\frac{1}{2}$

The penetration being given in terms of the weight of the powder and projectile, the piece should be sufficiently long to obtain the full force of the charge, or from 17 to 20 calibres; or, in other words, the expression is only suited to field and siege guns.

In general, sand, sandy earth mixed with gravel, small stones, chalk, or tufa, resist shot better than the productive earths, or clay, or earth that retains moisture.

Water. To obtain penetration in water, replace $480\frac{p}{m}$ by $4800\frac{p}{m}$, and make e equal to 275 calibres.

In some late experiments, it was found that the Whitworth projectile had sufficient force, at short distances. to pass through 33 feet of water and then penetrate 12 or 14 inches of oak beams or scantling. The penetration of a rifle-projectile in water, depends much on the direction of its axis with respect to penetration, for instance, penetration rapidly diminishes at long distances, as the axis of the projectile strikes the surface of the water under a diminished angle.

466. **Effect on masonry.** The effect of a projectile against masonry, is to form a truncated conical hole, terminated by another of a cylindrical form. (See fig. 153.) The material in front of and around the projectile is broken and shattered, and the end of the cylindrical hole even reduced to powder. Pieces of the masonry are sometimes thrown

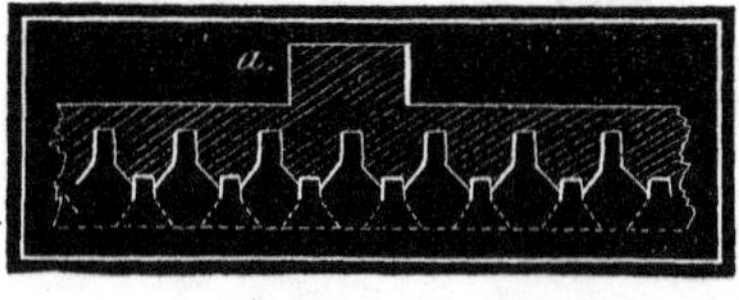

Fig. 153.

50 or 60 yards from the wall. The elasticity developed by the shock, reacts upon the projectile, sometimes throwing it back 150 yards, so as to be dangerous to persons in a breaching battery. The exterior opening varies from 4 to 5 times the diameter of the projectile, and the depth, as we have seen, varies with the size and density of the projectile, and its velocity.

With charges of $\frac{1}{2}$, $\frac{1}{3}$, $\frac{1}{4}$, and $\frac{1}{6}$, a projectile ceases to rebound from a wall of masonry when the angles, formed by the line of fire and the surface of the wall, exceed 20°, 24°, 33°, 43°, respectively. With these angles, the angle of reflection is much greater than the angle of incidence, and the velocity after impact is very slight.

When a projectile strikes against a surface of oak, as the side of a ship, it will not stick if the angle of incidence be less than 15°, and if it do not penetrate to a depth nearly equal to its diameter.

Solid cast-iron shot break against granite, but not against freestone or brick. Shells are broken into small fragments against each of these materials.

467. **Breaching.** Escalade being ordinarily very difficult, particularly when the besieged are aware of the intention of the besiegers, the latter are generally compelled to destroy a portion of the face of the work to obtain an entrance. Such an opening is called a *breach;* and to effect it with artillery, particularly in a well-constructed work, where no part of the scarp-wall is visible from the adjacent ground, within effective range of siege-cannon, *breaching-batteries* are established either on the crest of the *covered way*, or on the *glacis*.

When the walls of fortified places were very high and not supported by terraces or ramparts, stone projectiles were used. From the want of sufficient hardness in these projectiles, the besiegers were forced to commence battering at the top of the wall where the least resistance was offered, and gradually to lower the shot until the breach reached the wrecks already formed at the base of the wall. When the style of fortification was changed, this operation became very laborious, the ascent was very steep, and the breach was often impracticable. This method was abandoned and mining substituted. Iron projectiles superseded stone, and then a more rapid mode of effecting a practicable breach was suggested and confirmed by experience.

Vauban recommended increasing the size of the hole first formed, by continually firing at its sides until the wall should fall; but the ball was found to glance into it, and injure but slightly the untouched portion of the revetment. The best mode, however, as found by experiment, is to cut the wall up into detached parts, by making one horizontal and several vertical fissures, and *battering* each part down separately. (Fig. 154.)

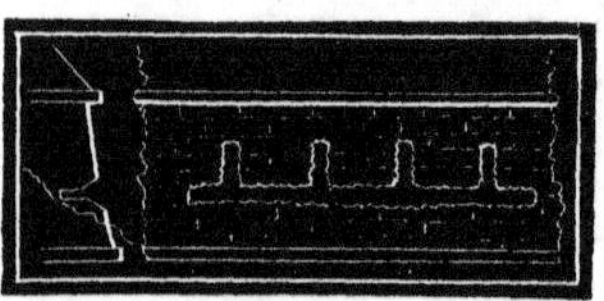

Fig. 154.

The easiest mode of making the cut is to direct the shots upon the same line, and form a series of holes (fig. 154), a little greater than a diameter apart, and then to fire a second series of shots, directed at the intervals between the first, and so on, until an opening is made completely through the wall.

The first cut is made horizontally, and finished, which will be known by the earth falling through it; the vertical cuts are then made, there being one at each end of the intended breach. These cuts are commenced at the horizontal cut, and raised until the wall, isolated from its supports, sinks, overturns, and breaks into pieces, which become covered by falling earth. If the earth be sustained by its tenacity, loaded shells are fired into it, which, acting like small mines, cause it to fall, and make the breach *practicable*, or of easy ascent.

If the portion of the wall between the vertical cuts should not be overthrown by the pressure of the earth behind, it must be detached by a few volleys of solid shot, fired at its centre. This will speedily bring it down in a mass. The moment the wall is down, and the parapet destroyed, the breach will be as perfect, and the slope as easy of ascent, as it can be made by the fire of the batteries. It is important to determine the height of the horizontal cut above the bottom of the ditch, for, if this height be not properly chosen, the breach may be difficult, if not impracticable. If too high, the ramp composed of the *debris* will be intercepted by a portion of the wall; if too low, the opening will be masked by the debris, and the formation of the cut impeded. The most suitable height is nearly equal to the thickness of the wall where the cut is established. The thickness, where not known, can be deduced from the dimensions necessary to be given to the wall, to resist the pressure of the earth of the rampart and parapet.

The time necessary to make a breach, depends on the size of the breach to be made, the material of the scarp, the number of guns, &c. For a breach of 20 to 30 yards in length, at forty yards from the battery, 1,500 shot of large calibre will be required; but when the battery is at a greater distance, a greater number of projectiles will be necessary, on account of the diminished accuracy and penetration. Thus, at 500 or 600 yards 9,000 to 10,000 may be needed.

Rules. The following general rules should be observed in firing to effect a breach:—

1. Ascertain as accurately as possible the widths of the ditch and covered way, the height of the scarp-wall, the thickness of the parapet, the height of the counterscarp, and crest of the covered way. By the aid of a profile that can be constructed from this data, determine the height of the horizontal cut to be made in the scarp, so that the slope of the ramp shall be 45°. This height should never be below a fourth of that of the scarp, and, to avoid interference from the wrecks, it should be nearly equal to the presumed thickness of the wall at the cut. If the ditch be a wet one, commence cutting at the water's edge.

2. From the number of pieces with which the battery is to be armed, and the length of the breach, determine the field of fire of each piece, and the length of cut that it is to make.

3. Ascertain the angle of elevation, or depression, for each piece, to strike the cut, and mark it unalterably on the elevating screw.

4. Direct each piece on the right or left of the part to be cut, and space the shot from right to left, or from left to right, at $1\frac{1}{4}$ to $1\frac{1}{2}$ yards for the 24-pdr., and 1 yard for the 18-pdr. Mark on the platform the direction of the stock and wheels at each shot. Returning then from left to right, or from right to left, fire at the middle of the intervals left by the first shot, and mark the directions as before. Continue this firing

regularly at the most prominent points, and make the cut progress equally throughout.

5. Fire at the horizontal cut until the earth falls throughout the cut.

6. Determine the number of vertical cuts to be made, at the rate, at most, of one to a piece, without spacing more than 10 yards apart, in order that no part shall be sustained by more than one counterfort. Fire as in the case of the horizontal cut, commencing at the upper line.

7. See that the extreme vertical cuts progress as rapidly as the interior ones, and direct the adjoining guns upon them if necessary.

8. If the wall do not fall after the cuts are made, fire a few volleys at the middle of the spaces thus outlined.

9. After the fall of the wall, break down the counterforts, and, if time or resources permit, replace the guns by 8-inch howitzers, and fire upon the earth with loaded shells, or fire shells from the guns.

468. **Breaching with rifle-cannon.** The foregoing section has reference particularly to breaching masonry with smooth-bored guns. The principles and, to a certain extent, the rules laid down, are applicable to rifled guns, the only difference being that, from their superior penetration and accuracy, the latter are effective at much longer distances. This has been fully shown by trials in England and actual experience in this country.

The subjects of experiment in England were two Martello towers, 30 feet high and 48 feet diameter; the walls were from 7¼ to 10 feet thick of solid brick masonry of good quality. The distance was 1,032 yds.,—

more than twenty times the usual breaching distance. The pieces were Armstrong guns, throwing projectiles weighing 40, 80 and 100 lbs., and the 32-pdr. and 68-pdr. smooth bored guns.

The 80-pdr. shot passed completely through the masonry (7¼ feet), and the 40-pdr. percussion shells lodged in the brick work at a depth of 5 feet. After firing 170 projectiles, a small portion of which were loaded shells, the entire land side of the tower was thrown down, and the interior space was filled with the debris of the vaulted roof, forming a pile, which alone saved the opposite side from destruction.

The average depth of penetration of the smooth bored guns was 1.91 feet, and was produced by 163 projectiles. It was found that the penetration of the 80-pdr. solid shot with 10 lbs. of powder in this brick masonry was 7½ feet, while that of the 68-pdr., with 16 lbs. of powder, was only 1⅔ feet.

Fort Pulaski, a casemate work made of brick masonry of good quality, and having walls 7¾ feet thick, was breached by batteries situated at distances varying from 1650 to 1740 yds. The pieces employed were three 10-inch and one 8-inch columbiads, five 30-pdr. Parrott and one 24-pdr., two 32-pdr. and two 42-pdr. sea-coast guns rifled on the James plan. There were several 10 and 13-inch sea-coast mortars, and 8 and 10-inch columbiads placed further off, or at distances varying from 2400 to 3400 yds.; but these pieces, and especially the mortars, produced but little effect, owing to the small size of the Fort and the inaccuracy of their fire. Less than one-tenth of the mortar shells fell inside of the Fort, and no shell penetrated through the casemate roofs.

Subsequently, Fort Sumter, a work similar to the foregoing in its construction, was breached at distances varying from 3428 to 4290 yds. by nine 100-pdrs., five 200-pdrs., and one 300-pdr. Parrott rifle-guns. Against Fort Pulaski 110,643 lbs. of metal were fired at the breach, 58 per cent. of which was from the rifle-guns. Against Fort Sumter there were fired 552,683 lbs. of metal, about one-half of which reached the Fort to form the breech. The most destructive projectile against masonry is the elongated percussion shell.

469. **Effect of bullets.** The penetration of the new breach-loading rifle-musket bullet, in a target made of pine boards, one inch thick, is as follows: At 100 yds., 13 inches; at 500 yds., 9 inches. If bullets are hardened by the addition of a little tin or antimony to the lead, their penetration is very much increased.

From experiments made in Denmark, the following relations were found between the penetration of a bullet in pine and its effects on the body of a living horse, viz.:

1st, When the force of the bullet is sufficient to penetrate .31 inch into pine, it is only sufficient to produce a slight contusion of the skin; 2d, When the force of penetration is equal to 0.63 inch, the wound begins to be dangerous, but does not disable; 3d, When the force of penetration is equal to 1.2 inch, the wound is very dangerous.

A plate of wrought-iron three-sixteenths of an inch thick, is sufficient to resist a rifle-musket bullet at distances varying from 20 to 200 yds. That a rope mantlet may give full protection against rifle-musket bullets, it should be composed of five layers (three vertical and two horizontal), of four and a half inch rope.

CHAPTER XII.

EMPLOYMENT OF FIELD-ARTILLERY.*

470. **Greatest range.** The extreme range of field-artillery has been stated to be about 3,000 yards;† a somewhat greater range than this can be obtained by sinking the trail of the carriage into the ground, thereby increasing the elevation of the piece; but in consequence of the great strain thus thrown upon the carriage, and the great inaccuracy of the fire, it should be seldom resorted to, unless it be to produce a moral effect on an army in retreat, or passing a defile. If employed against an enemy acting on the offensive, it would have the effect, from its extreme inaccuracy, to give him increased confidence.

In general terms, firing at long range should only be employed when the nature of the ground, or the shortness of the time, does not permit a nearer approach to the object; and it should always cease when the object of the fire is attained.

Effective range. The greatest effective range of field-artillery varies from 1,400 to 1,800 yards. Batteries of position belonging to an army acting on the defensive, should open fire at a distance of 1,300 or 1,400 yards. The object of this fire is not so much to arrest, as to retard, the movement of the enemy, and compel him to establish batteries to cover his approach. The distances

* Vide Decker's Instruction Pratique, &c.

† The ranges in this chapter refer to smooth-bored rather than rifled guns. The principles, however, involved in it, are equally applicable to both.

should be carefully estimated, and the firing should take place slowly, in order that the effect of each shot may be observed, and the aim corrected, if necessary.

Rapid and continuous firing should commence at a distance of 800 or 1,000 yards; the attacking party should, at the same time, establish his batteries to cover the deployment of his columns, and to enable him to make the necessary preparations for attack.

At a distance of 600 or 700 yards, or point-blank distance, the fire becomes very destructive; generally not more than six or eight shots can be fired before one of the parties will either advance or retire. As the distance closes, canister-shot should replace round-shot, which generally ends in producing disorder.

Against infantry. Formerly artillery could take up a position about 300 or 400 yards in front of infantry without serious loss; but the introduction of the rifle-musket has produced a very great change in the relative powers of these two arms. The experiments made at the musketry-school at Hythe, show conclusively that artillery cannot long maintain a position within half a mile of properly instructed skirmishers, as the fire of rifle-musketry at this distance is as effective as that of canister at 250 or 300 yards.

Should the surface of the ground be broken, or of such nature as to afford shelter to skirmishers, the preponderance will be still more in their favor. And should the artillery not succeed in silencing the fire of the skirmishers by well served case-shot, it will be obliged to retire beyond the reach of the rifles, and trust to the effect of round and spherical case-shot upon the enemy's masses.

Against cavalry. Cavalry, in charging upon an enemy situated at a distance of 1,000 yards, pass over the intervening space in about seven minutes. Each piece may fire nine rounds of solid shot, or spherical case-shot, in the first 400 yards, two solid and three canister shot in the next 400 yards, and two rounds of canister-shot while passing over the remaining 200 yards, making a total of eleven round and five canister shot. Neither spherical case-shot nor shells should be fired against cavalry in rapid motion; and care should be taken not to cease firing solid shot too soon in order to commence firing canister.

471. **Employment of different kinds of fire.** The following circumstances should be known, to enable the artillerist to select the most suitable fire for a particular occasion: 1st. The distance of the enemy. 2d. The conformation and quality of the intervening ground. 3d. The formation of the enemy, as far as can be seen or judged of.

Direct fire. Direct fire should be employed wherever the surface of the ground is uneven and the quality of the soil varied, or wherever a portion fired over is smooth and the remainder broken, or the soil soft and light. There are other special cases where direct fire should be employed:

1st. When the enemy is so situated as to conceal the depth of his formation; otherwise the ground in rear of his front line may be such that the ricochet will not take effect;

2d. When the enemy is about to pass a defile, and the head of the column only is seen; or when the depth of the column can be seen, by being commanded or overlooked; in this case, the projectile which would miss the head might strike the middle or the rear of the column;

3d. It should be employed in all sustained cannonades, because the effect of its shots can be more easily distinguished than that produced by the shots of a rolling fire. The aim should be corrected by observing the point of fall of the projectile; and, for this purpose, it is desirable to take the mean of three shots. If a rolling fire be employed under these circumstances, the character of the ground and formation of the enemy may be such, that the cannonade may be carried on for hours without knowing what effect is produced.

To produce good results with direct fire, it is absolutely necessary to ascertain the exact distance of the enemy, which can only be done by a practised eye. This circumstance will be appreciated when we consider that, if a shot only strike the ground fifteen yards in front of a target six feet high, it will pass completely over it.

When the object is not on the same level with the piece, the character of the fire will be determined by the nature of the intervening ground.

If the surface be uniform, and have an inclination to the horizon not exceeding 15°, above or below, no change need be made in the kind of fire, or elevation of the piece, from what they would be on horizontal ground.

If the enemy be posted on a mountain, or in a valley, the direct fire can only be used. As it is often difficult to estimate the distance, the pieces should be aimed with great precision, and the point of fall should be carefully noted; the firing should be deliberate, and it should be recollected that a different height of sight is necessary than when the object is on level ground.

472. **Ricochet fire.** Ricochet fire should never be used for a less distance than 1,000 yards, even when the ground is favorable; for, in order that this fire may produce its greatest effect, it is necessary that the projectile should make two or three rebounds in front of the enemy, which it rarely does at a less distance than 1,000 or 1,100 yards. If the ground, for 300 or 400 yards in front of the pieces, be soft and uneven, or if it be soft and uneven for 100 or 300 yards in front of the enemy, rolling fire, which is a species of ricochet fire, cannot be employed with effect.

Large and deep objects, as a mass of troops, a park, or a column of artillery on the march, are the most suitable objects for ricochet fire, as these objects present several lines, one behind the other.

473. **Canister fire.** The fire of canister does not always produce the effect anticipated for it, for the following reasons, viz.:

1st. The object is thought to be nearer than it really is, and the firing sometimes commences too soon.

2d. The danger is often thought to be more imminent than it really is, and, consequently, proper care is not observed in aiming.

3d. The character of the ground is not properly appreciated; and too much confidence is reposed in the effect of the projectiles thrown over unfavorable ground.

474. **Field-howitzers.** The extreme range of shells fired from field-howitzers has been stated to be from 2,500 to 3,000 yards. The deviation of shells at ex-

treme distances is so great that they should only be employed against large objects, as cities, camps, &c.

The greatest effective range of howitzer-shells is about 1,500 yards; shells should only be employed at this distance in the offence, and then, rather as an exception to the general rule.

The gun should always be employed when capable of producing the same effect as the howitzer.

Shells act by percussion, by explosion, and by moral effect; and they should be employed in preference to shot under the following circumstances, viz.:

1st. When the enemy is stationary and under cover.

2d. When the ground is much broken, or cannot be seen.

3d. When troops are posted in woods.

4th. From one mountain to another.

5th. When the enemy is posted on higher or lower ground.

6th. When on a road leading through a valley.

7th. For incendiary purposes.

8th. In pursuit.

9th. Whenever it is necessary to produce a moral rather than a physical effect.

EMPLOYMENT OF SIEGE-CANNON.

In siege operations, the same fires are employed as in the field, but under different circumstances. The position of the object is generally fixed and known, and there is sufficient time to consider the best means of attaining it.

475. **Long ranges.** The greatest range of the 24-pdr.

siege-gun, mounted on its appropriate carriage, is about 3,500 yards; but the defence should not, without good reason, make use of a greater distance than 950 yards, or point-blank distance, for it is his duty to economize his ammunition, if it cannot be replaced.

It will be proper to fire at a reconnoitring party at a distance of 1,000 or 1,100 yards, to prevent a nearer approach, and against strong attacking columns, provided they offer sufficient surface to render the chances of hitting probable.

In the attack. Firing at long ranges, on the part of the besiegers, should be strictly forbidden, as it would disclose to the enemy the proposed front of attack, without any compensating advantage.

In the siege service, it is more important to avoid useless firing than in the field, for every shot that does not contribute to the progress of the attack, by weakening the defence, is a shot lost.

476. **Enfilading and counter fires.** An *enfilading* fire is directed along a particular portion of a work, and a *counter* fire is directed toward it.

In the defence. Solid shot are used in enfilading and counter fires under the following circumstances:

1st. To destroy the head of a sap, or the parapet of a trench.

2d. When the enemy passes from the first to the second parallel, and before he has completed the batteries intended to dismount the artillery of the garrison.

3d. To batter vigorously the lateral works of attack as soon as they are finished.

4th. To protect and support sorties. The guns placed on the parapet of the place keep up a warm fire of solid shot against the batteries of attack, and the heads of saps, until they are masked by the troops making the sortie.

5th. To prevent the enemy from following too closely upon the heels of the party, which, having made the sortie, are returning, successful or otherwise.

6th. From the guns placed on the flanks of the bastions when the besiegers attempt to pass the ditch; in this case the fire is plunging.

7th. To drive the besiegers from any outwork that they may have taken.

8th. In a cannonade, the object of which is to dismount the besiegers' guns.

In the attack. The object of enfilading fire in the attack of a place, is to rake the terrepleins of the faces, curtains, &c., and to render them untenable; for this purpose the batteries should be established on the prolongation of, and at right angles, or nearly so, with the direction of the part to be enfiladed. As the portion of the works to be attained is not commanded by the besiegers' cannon, enfilading fire, under these circumstances, becomes ricochet fire, the nature and treatment of which have already been described.

Enfilading and counter batteries are generally established at 300 or 600 yards from the place, or at the first and second parallels. As the object of a counter battery is to silence the fire of the place by dismounting the guns, its pieces should be directed against the embrasures.

This demands great care in aiming, and great accuracy of fire; the heaviest smooth-bored or rifled guns should therefore be employed for this purpose.

477. **Firing in breach.** When the besiegers have approached to a suitable distance to commence the breach, the opposing artillery will have been silenced; but they will be subjected to flank and rear fires, against which they will protect themselves by traverses. Counter-batteries will also be established with the breaching-batteries, the object of which will be to silence the artillery bearing on the breaching-batteries, and the passage of the ditch. The method of forming a breach has already been described.

478. **Fire of case-shot.** Case-shot should be employed in the defence of a work under the following circumstances, viz.:

1st. In sorties, where field-artillery can be employed.

2d. At all points liable to sudden attacks, as on avenues leading toward gates, or on bridges. Pieces situated on the flanks are particularly suited to this fire.

3d. Against the gorge of an outwork which the enemy may make a bold attempt to seize. For this purpose, pieces on the curtains, or shoulder angles, should be employed, taking care, at the same time, not to fire over works occupied by the defence.

4th. This fire may be safely employed in the defence of dry ditches, reveted with masonry.

5th. Against the batteries of the first parallel during their erection, and after their position has been disclosed by means of fire-balls.

6th. Against the head of a sap at night.

7th. Against the workmen engaged on the construction of the second parallel.

8th. Against the workmen engaged on the third parallel, against the works leading to the covered way, and against the crowning of the covered way.

9th. Against craters formed by the explosion of mines, to prevent the enemy from crowning them.

10th. Against the passage of the ditch.

11th. Against the breach.

12th. All cannon on the flanks which remain mounted, fire rapidly grape or canister shot at the moment of assault.

In the attack. The besiegers are much more restricted in the use of case-shot than the besieged. It should be principally employed under the following circumstances, viz.:

1st. By cannon placed on the flanks of attack whenever the besieged make a sortie, and come within suitable range.

2d. At night, against the embrasures which have been cannonaded during the day with solid shot, to prevent them from being repaired.

3d. Against the flanks, during the night.

4th. Against the breach during the day or night, as soon as completed, to prevent the enemy from erecting means for defending it.

5th. Against the besieged, if he attempt to pass out through the breach, after the assault has been repelled.

479. **Fire of the siege-howitzer.** The siege-howitzer should be employed in the defence,—

1st. Against an attacking column, when the ground in front of the place affords a shelter against the fire of guns.

2d. Against the works of the besiegers. Howitzers

are placed on the salients to blow up, with shells, the works situated on the prolongations of the capitals.

3d. Against the batteries in process of construction on the three parallels.

4th. Against the heads of saps; this fire should be executed with small charges.

5th. The counter approaches are armed with howitzers.

6th. Against troops opposing sorties, and especially against cavalry.

7th. Against the enemy's depôts, when their position is known, and when they are within effective range.

8th. Against the enemy's convoys, when they can be reached, and they offer sufficient surface.

In the attack. Howitzers are employed by besiegers—

1st. In a bombardment, by day and night.

2d. During all periods of the siege, when occasion requires.

3d. In the half-parallels established between the second and third; against the covered-ways and places of arms. The fire is executed with small charges.

4th. For ricochet fire, in preference to cannon.

480. **Use of fire-balls.** Fire-balls are used by the defence—

1st. Against columns of attack.

2d. Against the opening of parallels, so soon as it is ascertained that preparations are made for this purpose.

3d. Against points in the space occupied by the besiegers, where a remarkable noise may be heard, and there is reason to suspect that it proceeds from preparations for attack.

4th. Particularly when it is thought that the besiegers are about to move forward from one parallel to another.

5th. To discover the movements of the enemy after he has repulsed a sortie, and to prevent him, by the fire of the guns of the place, from following too closely in pursuit.

In attack. As it is for the interest of the besiegers to conduct their operations as silently and unobserved as possible, they will seldom have occasion to use fire-balls.

481. **Fire of mortars.** Mortars generally perform a more important part in siege operations than howitzers; there are times, even, when they play a very decided part; too much care, therefore, cannot be employed to render them effective.

In the defence. Mortars are employed in the defence—

1st. Concurrently with howitzers, when the shape of the ground in front shelters the enemy from the fire of the guns.

2d. Against batteries and heads of saps.

3d. Against places sheltered from the fire of flanking guns. Mortars, and particularly light mortars, can be suitably placed at all points, and without interfering with the establishment of gun and howitzer batteries.

4th. Against the works of the besiegers generally, and especially against the opening of parallels, and the passage from one parallel to another.

5th. When the besiegers' fire has silenced the fire of the guns, the fire of the mortars continues in full activity, not only in the body of the place, but in the demilunes and lateral works.

6th. In covered batteries, during the entire siege, but

particularly during or after the construction of the third parallel.

7th. Light mortars should be employed in the counter approaches.

8th. Against the workmen who are engaged in running the sap up the glacis, for the purpose of crowning the covered way.

9th. To prevent the construction of counter and breaching batteries.

10th. To prevent the besiegers from establishing themselves in the craters formed by the mines.

11th. To drive the besiegers from any exterior work which they have taken.

12th. To prevent the passage of the ditch, or render it difficult.

13th. To prevent the besiegers from effecting a lodgment in the breach, by firing from the interior retrenchment.

In the attack. It is very difficult to specify all the circumstances which should govern the besiegers in carrying on a bombardment, since they depend on a variety of causes; the following, however, may be enumerated:

1st. In a regular attack, mortars are the first to open fire, which should be kept up night and day whenever a result can be obtained.

2d. Heavy, and sometimes medium-sized, mortars, can be employed to retard the enemy's works on the front of attack, the armament of his batteries, the transportation of his cannon, and to shower shells upon the places where his troops assemble, and to burn his principal buildings, etc. Light mortars are rarely used for these purposes, in consequence of the distance of the object and the lightness of the shells, which have little force of percussion.

3d. Mortars are employed to throw shells over the entire surface of the ramparts of the front of attack; and, for this purpose, the fire should be taken in the direction of their length.

4th. They are also employed against the lateral works as soon as the enemy seeks to establish his guns there for the purpose of retarding the works of attack.

5th. The curved or mortar fire of the second parallel is as efficient as that of the first parallel, at all periods of the siege. Light mortars here begin to be usefully employed.

6th. Light mortars are also used with great advantage in the half-parallels. From this period of the siege, the covered-way and places of arms are showered with shells.

7th. From the period of the third parallel, the enemy's flanks are plied with mortar shells, to support the fire of the counter batteries.

8th. As soon as the covered-way is crowned, and subsequently, when a lodgment in the breach shall have been effected, Coehorn mortars are employed against the enemy, who has withdrawn to the interior retrenchment of the bastion.

482. **Mortar case-shot, &c.** Stones and case-shot from mortars, should be thrown by the defence as soon as the besiegers pass to the construction of the third parallel, and the batteries pertaining to it. This should be continued during the crowning of the covered-way, and during the assault.

The besiegers, on the contrary, employ these projectiles in all the batteries of the third parallel, and, by this means, seek to drive the enemy from the covered way and places of arms, thus preparing the way for the assault.

SEA-COAST DEFENSES.

488. **Nature of defenses.** The means employed for the defense of harbors, are: 1st. Artillery mounted on fortifications, floating batteries, monitors and ships of war. 2d. Channel obstructions, such as rafts, heavy chains, and sunken vessels. 3d. Torpedoes and torpedo vessels, both of which act under water. These means may be used singly, but the most effective defense is made by combining them, according to circumstances.

The introduction of steam into vessels of war, and their protection by armor plates have necessitated considerable change in the construction of sea-coast forts and their armament. The best temporary fortifications are made of earth, with large masses in the way parapets, traverses, etc., for covering the guns. It is now proposed to secure a still better covering for the guns by mounting them in iron turrets, similar in shape and construction to the Monitor turrets, but to be manœuvred by man instead of steam power. Late experiments show that the plan of reveting the masonry of casemated works with wrought-iron plates will not answer, as the iron affords but little protection to the masonry. Monitors as a means of harbor defense have the advantage of being able to select their position of attack and of following an attacking vessel if it should succeed in passing the guns of a fort. The range and accuracy of guns mounted on floating defences are not so great as those of guns mounted on fortifications, and they are liable to be destroyed by rams.

Obstructions. The object of obstructions is to detain attacking vessels under the fire of the guns of the forts and monitors. They may consist of rafts of strong timbers, or heavy chains sustained by buoys, or rows of piles, with suitable arrangements to let in or out friendly vessels. Sunken vessels should only be employed when there is very little commerce to be interfered with, or when it is not practicable to employ other means. Ropes and netting may be placed in a channel to foul the propellors of hostile vessels.

Torpedoes. Torpedoes and torpedo vessels were used in the late war in a great variety of forms and with some success. Torpedoes may be made of boxes or barrels covered with pitch, but the largest are made of boiler plate riveted together. They are exploded by contact, by electricity, by clock-work or by time-fuzes.

484. **Armament.** The armament of sea-coast batteries depends on their importance and on the depth and width of the channel to be defended. Deep channels which permit the entrance of large vessels and wide channels requiring long ranges should be defended with guns of heavy calibre. The salients and flanks which generally have an enfilading fire on a channel, should be armed with the heaviest rifle-cannon. The curtains and faces which bear directly upon it may be armed with heavy smooth-bored cannon.

In addition to the cannon enumerated on page 192 as properly belonging to sea-coast armament, each fort should be provided with a certain number of field-pieces, principally howitzers, to prevent a landing, or to act in close engagements against the rigging and small boats of vessels. Forts should be provided with furnaces for heating shot which have been found to be effective in protracted engagements with wooden vessels.

485. **Fires.** Direct, ricochet, and plunging fires are principally employed in sea-coast defence.

Direct fire should be used when the surface of the water is rough, and the accuracy of the rebound cannot be depended upon. The accuracy of sea-coast fire is generally greater than that of the field or siege service, for the reasons, that, the distance of the object, though moving, can be readily and accurately determined by its relation to known objects, the effect of shot can be more easily observed on water than on land, and the size of the object is large, and its appearance, generally, well defined. In aiming at a vessel with direct fire, the piece should be pointed at the water-line; for, if the projectile strike the water, it will either penetrate the hull below the water-line, or rebound and strike above it.

The range of effective direct fire does not much exceed one mile and a quarter; the extreme range of sea-coast mortars is about two and a half miles; that of the columbiads, about three and a quarter miles, and the heavy rifle-guns about five miles.

The accuracy of ricochet-fire depends on the surface of the water; under favorable circumstances, the larger sea-coast shells have a range of about 3,000 yards in rolling fire; their penetrating force, however, is very much diminished toward the extremity of this range.

The fire of mortars, from ship-board, is very uncertain, if the surface of the water be much disturbed.

CHAPTER XIII.

TABLES OF MULTIPLIERS.

B, I, D, V, &c.

487. **Explanation.** It would exceed the limits of this work to enter into a discussion of the formulas from which the values of the multipliers used in the equations of motion in air (page 412) are calculated; it will be sufficient to explain how these tables are used in practice.

The pupil will find this subject, as well as all others relating to Ballistics, ably and fully treated in Didion's *Traité de Balistique.*

488. **Table 1.** *Multiplier B.* The decimals are carried out to three places, which is sufficient for ordinary purposes. The values of $\frac{x}{c}$ are given in the first horizontal line, the value of $\frac{V_{\prime}}{r}$ in the first vertical column, and the values of the corresponding multipliers are set opposite to them.

To find the multiplier B for two intermediate values of $\frac{x}{c}$ and $\frac{V_{\prime}}{r}$, not given in the tables, we seek, in the absence of the proper numbers, the corresponding values of the nearest tabular numbers. We add to these, parts proportional to the differences, as though each part were to be considered separately.

Example.—Find the value of B for $\frac{x}{c}=0.5755$, and $\frac{V_{\prime}}{r}=1.1219$, *i. e.* B (0.5755; 1.1219). Starting with 0.55 in the first horizontal column, and 1.10 in the first vertical column, we find $B=1.479$; the difference between this and the next number of the horizontal line is 0.054; the difference between the same and the next number of the vertical column is 0.013. The difference between 0.5755 and 0.55 is 0.0255, and between 1.1219 and 1.10 is 0.0219. The value of $B\ (0.5755;\ 1.1219)=1.479+\frac{0.0255}{0.05}0.054+\frac{0.0219}{0.05}0.013=1.479+$ $0.027+0.006=1.512$.

Or, for greater convenience, the foregoing may be placed in the following form, the differences being written as whole numbers:

$$
\begin{array}{lcr}
B\ (0.5755;\ 1.1219) & = & 1.512 \\
B\ (0.55;\ 1.10) & = & 1.479 \\
\frac{255}{500}54\ \ .\ \ . & = & 27 \\
\frac{219}{500}13\ \ .\ \ . & = & 6
\end{array}
$$

Multiplier, I. The values of I are given in the same table as those of B; except that it is necessary to commence in the lower horizontal line, and subtract from them the product of $\frac{V_{\prime}}{r}\left(1+\frac{V_{\prime}}{r}\right)$, by the corresponding number of the line called "correction."

Example.—To find the value of I (0.5755; 1.1219), take $\frac{x}{c}=$ 0.545, which is less than the proposed number by 0.305, and which differs by 0.035 from the next number in the table; $\frac{V_{\prime}}{r}=1.10$ is the nearest number to 1.1219 in the first vertical column; for these two numbers we have $I=1.771$. This number differs from the adjoining horizontal and vertical numbers in the table by 0.066 and 0.022, respectively. The value sought is 1.830, as is thus shown:

$I\,(0.5755;\ 1.1219) = 1.830$
$I\,(0.545;\ 1.10) = \overline{1.771}$
$\frac{305}{350}66 = 58$
$\frac{219}{500}22 = 10$
$-1.1219\,.\,2.1219.4 = -9$

Table 3. *Values of U and D.* This table is calculated for differences of 0.10 in case of $\frac{x}{c}$, in the upper line, and for differences of .05 in case of $\frac{V_{\prime}}{r}$. For U, the values of $\frac{x}{c}$ are found in the upper horizontal line, and for D, in the lower line.

Example.—Find the values of U (0.5755; 1.1219) and D (0.5755; 1.1219).

$U\,(0.5755;\ 1.1219) = 1.707$	$D\,(0.5755;\ 1.1219) = 1.336$
$U\,(0.50;\ 1.10) = \overline{1.597}$	$D\,(0.393;\ 110) = \overline{1.221}$
$\frac{755}{1000}138 = .104$	$\frac{1825}{1920}119 = .113$
$\frac{219}{500}14 = .006$	$\frac{219}{500}5 = .002$

We have $U = 1.707$, and $D = 1.336$.

Table 4. *Values of* $\frac{x}{c}B$ *for the calculation of Ranges.* This table gives the value of $\frac{x}{c}B$ for values of $\frac{x}{c}$ and $\frac{V_{\prime}}{r}$, for differences of 0.05 and 0.05; the unknown quantity to be determined is $\frac{x}{c}$ when $\frac{V_{\prime}}{r}$ and $\frac{x}{c}B = p$, are given.

Arrange the calculations as in the preceding cases. Only one of the proportional parts is unknown, and this is determined by the condition, that if it be added to the other proportional part, and to the number in the table, the sum is equal to the required number.

Examples.—Having $\frac{V_{,}}{r}=1.1219$ and $\frac{x}{c}B$, or $p=0.8729$, find $\frac{x}{c}$.

Starting with $\frac{V_{,}}{r}=1.10$, and following the horizontal line, we come upon 0.8135, the nearest approach to the proposed number, 0.8729. Find the corresponding value of $\frac{x}{c}$, which is 0.55; the unknown value of $\frac{x}{c}$ surpasses 0.55 by a certain quantity which we shall call Δ; following the previous arrangement of the calculation, and observing that the differences of 0.8135 with the adjacent horizontal and vertical tabular numbers are 0.1065 and 0.0071, respectively, and representing by p the result, we have—

$$
\begin{aligned}
&p\,(0.55+\Delta;\ 1.1219)=\overline{0.8720}\\
&p\,(0.55\quad;\ 1.10\quad)=0.8135\\
&\frac{\Delta.}{0.05}1065 \qquad = .0559\\
&\frac{0.0219}{0.0500}71 \qquad = .0035\\
&\text{We have } \Delta=\frac{559}{1065}0.05 \qquad =0.0263\\
&\frac{x}{c}=0.55+0.0263 \qquad =0.5763
\end{aligned}
$$

The proportional part 559 is equal to 8729—(8135+35).

Table 5. *Values of* $\dfrac{\frac{V_{,}}{r}}{\sqrt{B}}$ *for initial velocities.*

This table gives the quotient arising from dividing $\frac{V_{,}}{r}$ by $\sqrt{B}$ for values of $\frac{x}{c}$ and $\frac{V_{,}}{r}$; the quantity to be determined is $\frac{V_{,}}{r}$. The method is the same as in the preceding table; if the value of the quotient q diminishes as $\frac{x}{c}$ increases, the sign of the difference should be changed.

Example.—Having $\frac{x}{c}=0.5755$, and $q=\dfrac{\frac{V_{,}}{r}}{\sqrt{B}}=0.9110$, find $\frac{V_{,}}{r}$.

The vertical column nearest to $\frac{x}{c}=0.5755$ is that which corresponds to 0.55 ; the number in this column nearest to 0.9110 is 0.9045, which corresponds to 1.10, and the difference between this and the required number is 0.0065 ; the differences with the neighboring numbers to the right and above, are — 0.0162 and 0.0370, respectively. We therefore have,

$$
\begin{aligned}
q\,(0.5755;\ 1.10+\Delta) &= \underline{0.9110} \\
q\,(0.55;\ 1.10) &= 0.9045 \\
\frac{255}{500}162 &= -.0082 \\
\frac{\Delta}{0.05}370 &= .0147 \\
\text{or } \Delta=\frac{147}{370}\,0.05 &= 0.0199 \\
\text{and } \frac{V_{\prime}}{r}=1.10+0.0199 &= 1.1199
\end{aligned}
$$

The proportional part 147 is equal to 9110—(9045—82), giving $\Delta=0.0199$, which, added to 1.10 gives $\frac{V_{\prime}}{r}=1.1199$.

TABLE 1.—Values of B and I.

For B	$\frac{x}{c}$	0.00	0.05	0.10	0.15	0.20	0.25	0.30	0.35	0.40	0.45	0.50
For $\frac{aV_1}{r}$	0.00	1.000	1.017	1.034	1.052	1.070	1.089	1.108	1.128	1.148	1.169	1.190
	0.05	1.000	1.018	1.036	1.055	1.074	1.093	1.114	1.134	1.156	1.177	1.200
	0.10	1.000	1.019	1.038	1.057	1.077	1.098	1.119	1.141	1.163	1.186	1.210
	0.15	1.000	1.020	1.039	1.060	1.081	1.103	1.125	1.148	1.171	1.195	1.220
	0.20	1.000	1.020	1.041	1.063	1.085	1.107	1.130	1.154	1.179	1.205	1.231
	0.25	1.000	1.021	1.043	1.065	1.088	1.112	1.136	1.161	1.187	1.214	1.241
	0.30	1.000	1.022	1.045	1.068	1.092	1.117	1.142	1.168	1.195	1.223	1.252
	0.35	1.000	1.023	1.046	1.071	1.096	1.121	1.148	1.175	1.203	1.232	1.262
	0.40	1.000	1.024	1.048	1.073	1.099	1.126	1.153	1.182	1.211	1.241	1.273
	0.45	1.000	1.025	1.050	1.076	1.103	1.131	1.159	1.189	1.219	1.251	1.283
	0.50	1.000	1.025	1.052	1.079	1.107	1.135	1.165	1.196	1.227	1.260	1.294
	0.55	1.000	1.026	1.053	1.082	1.110	1.140	1.171	1.203	1.235	1.269	1.305
	0.60	1.000	1.027	1.055	1.084	1.114	1.145	1.176	1.209	1.244	1.279	1.315
	0.65	1.000	1.028	1.057	1.087	1.118	1.149	1.182	1.216	1.252	1.288	1.326
	0.70	1.000	1.029	1.059	1.090	1.122	1.154	1.188	1.224	1.260	1.298	1.337
	0.75	1.000	1.030	1.060	1.092	1.125	1.159	1.194	1.231	1.268	1.308	1.348
	0.80	1.000	1.031	1.062	1.095	1.129	1.164	1.200	1.238	1.277	1.317	1.359
	0.85	1.000	1.031	1.064	1.098	1.133	1.169	1.206	1.245	1.285	1.327	1.370
	0.90	1.000	1.032	1.066	1.101	1.137	1.173	1.212	1.252	1.294	1.337	1.382
	0.95	1.000	1.033	1.067	1.103	1.140	1.178	1.218	1.259	1.302	1.346	1.393
	1.00	1.000	1.034	1.069	1.106	1.144	1.183	1.224	1.266	1.310	1.356	1.404
	1.05	1.000	1.035	1.071	1.109	1.148	1.188	1.230	1.273	1.319	1.366	1.415
	1.10	1.000	1.036	1.073	1.112	1.151	1.193	1.236	1.281	1.328	1.376	1.427
	1.15	1.000	1.037	1.075	1.114	1.155	1.198	1.242	1.288	1.336	1.386	1.438
	1.20	1.000	1.037	1.076	1.117	1.159	1.203	1.248	1.295	1.345	1.396	1.450
	1.25	1.000	1.038	1.078	1.120	1.163	1.207	1.254	1.303	1.358	1.406	1.461
For I	$\frac{x}{c}$	0.000	0.033	0.067	0.101	0.134	0.168	0.202	0.236	0.270	0.304	0.338
Correction		0.000	0.000	0.000	0.000	0.000	0.000	0.000	0.001	0.001	0.001	0.001

For B	$\frac{x}{c}$	0.50	0.55	0.60	0.65	0.70	0.75	0.80	0.85	0.90	0.95	1.00
or $\frac{aV_1}{r}$	0.00	1.190	1.212	1.234	1.257	1.281	1.305	1.330	1.355	1.382	1.409	1.437
	0.05	1.200	1.223	1.247	1.271	1.296	1.322	1.348	1.375	1.403	1.43	1.461
	0.10	1.210	1.234	1.259	1.285	1.311	1.339	1.366	1.395	1.425	1.455	1.486
	0.15	1.220	1.246	1.272	1.299	1.327	1.356	1.385	1.415	1.447	1.479	1.512
	0.20	1.231	1.258	1.285	1.314	1.343	1.373	1.404	1.436	1.469	1.503	1.538
	0.25	1.241	1.269	1.298	1.328	1.359	1.390	1.423	1.457	1.491	1.527	1.563
	0.30	1.252	1.281	1.311	1.343	1.375	1.408	1.442	1.477	1.514	1.551	1.590
	0.35	1.262	1.293	1.325	1.357	1.391	1.425	1.461	1.499	1.536	1.576	1.616
	0.40	1.273	1.305	1.338	1.372	1.407	1.443	1.48	1.520	1.559	1.601	1.643
	0.45	1.283	1.317	1.351	1.387	1.423	1.461	1.500	1.541	1.583	1.626	1.670
	0.50	1.294	1.329	1.365	1.402	1.440	1.479	1.520	1.563	1.606	1.651	1.697
	0.55	1.305	1.341	1.378	1.417	1.457	1.498	1.540	1.584	1.630	1.677	1.725
	0.60	1.315	1.353	1.392	1.432	1.473	1.516	1.560	1.606	1.654	1.703	1.753
	0.65	1.326	1.365	1.406	1.447	1.490	1.535	1.581	1.629	1.678	1.729	1.781
	0.70	1.337	1.378	1.420	1.463	1.507	1.553	1.601	1.651	1.704	1.755	1.810
	0.75	1.348	1.390	1.433	1.478	1.524	1.572	1.622	1.674	1.727	1.782	1.839
	0.80	1.359	1.403	1.447	1.494	1.542	1.591	1.643	1.696	1.751	1.809	1.868
	0.85	1.370	1.415	1.462	1.509	1.559	1.610	1.664	1.719	1.776	1.836	1.897
	0.90	1.382	1.428	1.476	1.525	1.577	1.630	1.685	1.743	1.802	1.863	1.927
	0.95	1.393	1.440	1.490	1.541	1.594	1.649	1.706	1.766	1.827	1.891	1.957
	1.00	1.404	1.453	1.504	1.557	1.612	1.669	1.728	1.789	1.853	1.919	1.987
	1.05	1.415	1.466	1.519	1.573	1.630	1.688	1.749	1.813	1.879	1.947	2.017
	1.10	1.427	1.479	1.533	1.590	1.648	1.708	1.771	1.837	1.905	1.975	2.048
	1.15	1.438	1.492	1.548	1.606	1.666	1.728	1.793	1.861	1.931	2.004	2.079
	1.20	1.450	1.505	1.563	1.623	1.684	1.749	1.816	1.886	1.958	2.033	2.111
	1.25	1.461	1.518	1.578	1.639	1.703	1.769	1.838	1.910	1.985	2.062	2.142
For I	$\frac{x}{c}$	0.338	0.372	0.407	0.441	0.476	0.511	0.545	0.580	0.615	0.650	0.685
Correction		0.001	0.002	0.002	0.002	0.003	0.003	0.004	0.004	0.005	0.005	0.006

Values of *B* and *I*.—(*Continued.*)

For *B*	$\frac{x}{c}$	1.00	1.05	1.10	1.15	1.20	1.25	1.30	1.35	1.40	1.45	1.50
For $\frac{aV_1}{r}$	0.00	1.437	1.465	1.494	1.525	1.556	1.588	1.621	1.654	1.689	1.725	1.762
	0.05	1.461	1.492	1.523	1.555	1.588	1.622	1.657	1.693	1.730	1.768	1.808
	0.10	1.486	1.519	1.552	1.586	1.621	1.657	1.694	1.732	1.772	1.812	1.854
	0.15	1.512	1.546	1.581	1.620	1.654	1.692	1.732	1.772	1.814	1.857	1.902
	0.20	1.538	1.573	1.610	1.649	1.688	1.728	1.770	1.813	1.857	1.903	1.950
	0.25	1.563	1.601	1.640	1.681	1.722	1.765	1.809	1.854	1.901	1.949	1.999
	0.30	1.590	1.629	1.670	1.713	1.757	1.802	1.848	1.896	1.945	1.996	2.049
	0.35	1.616	1.658	1.701	1.746	1.792	1.839	1.888	1.938	1.990	2.044	2.100
	0.40	1.643	1.687	1.732	1.779	1.827	1.877	1.928	1.981	2.036	2.093	2.151
	0.45	1.670	1.716	1.763	1.812	1.863	1.915	1.969	2.025	2.083	2.142	2.203
	0.50	1.697	1.745	1.795	1.846	1.899	1.954	2.011	2.069	2.129	2.192	2.256
	0.55	1.725	1.775	1.827	1.881	1.936	1.993	2.053	2.114	2.177	2.242	2.310
	0.60	1.753	1.805	1.859	1.915	1.973	2.033	2.095	2.159	2.225	2.293	2.364
	0.65	1.781	1.836	1.892	1.950	2.011	2.073	2.138	2.205	2.274	2.345	2.419
	0.70	1.810	1.866	1.925	1.986	2.049	2.114	2.182	2.251	2.323	2.398	2.475
	0.75	1.839	1.897	1.958	2.022	2.088	2.155	2.226	2.298	2.373	2.451	2.532
	0.80	1.868	1.929	1.992	2.058	2.127	2.197	2.270	2.346	2.424	2.505	2.589
	0.85	1.897	1.960	2.026	2.095	2.166	2.239	2.315	2.394	2.475	2.560	2.648
	0.90	1.927	1.992	2.061	2.132	2.206	2.282	2.361	2.443	2.527	2.616	2.707
	0.95	1.957	2.025	2.096	2.169	2.246	2.325	2.407	2.492	2.580	2.672	2.766
	1.00	1.987	2.057	2.131	2.207	2.287	2.369	2.454	2.542	2.633	2.728	2.827
	1.05	2.017	2.090	2.167	2.246	2.328	2.413	2.501	2.593	2.687	2.786	2.888
	1.10	2.048	2.127	2.203	2.284	2.370	2.458	2.549	2.644	2.742	2.844	2.950
	1.15	2.079	2.157	2.240	2.323	2.412	2.503	2.597	2.695	2.797	2.903	3.013
	1.20	2.111	2.191	2.276	2.363	2.454	2.548	2.646	2.748	2.853	2.963	3.076
	1.25	2.142	2.225	2.313	2.403	2.497	2.594	2.696	2.801	2.909	3.023	3.141
For *I*	$\frac{x}{c}$	0.685	0.721	0.756	0.791	0.827	0.863	0.899	0.934	0.970	1.006	1.043
Correction ...		0.006	0.007	0.007	0.008	0.009	0.010	0.011	0.012	0.013	0.014	0.015

For *B*	$\frac{x}{c}$	1.50	1.55	1.60	1.65	1.70	1.75	1.80	1.85	1.90	1.95	2.00
For $\frac{aV_1}{r}$	0.00	1.762	1.799	1.838	1.878	1.920	1.962	2.006	2.051	2.097	2.145	2.194
	0.05	1.808	1.848	1.890	1.933	1.977	2.022	2.069	2.117	2.167	2.218	2.271
	0.10	1.854	1.897	1.942	1.988	2.035	2.083	2.133	2.185	2.238	2.293	2.349
	0.15	1.902	1.949	1.995	2.044	2.094	2.145	2.199	2.254	2.310	2.369	2.429
	0.20	1.950	1.999	2.049	2.101	2.154	2.209	2.265	2.324	2.384	2.446	2.511
	0.25	1.999	2.051	2.104	2.158	2.215	2.273	2.333	2.395	2.459	2.525	2.594
	0.30	2.049	2.103	2.159	2.217	2.277	2.339	2.402	2.468	2.536	2.606	2.678
	0.35	2.100	2.157	2.216	2.277	2.340	2.405	2.473	2.542	2.614	2.688	2.765
	0.40	2.151	2.211	2.274	2.339	2.405	2.473	2.544	2.617	2.693	2.771	2.852
	0.45	2.203	2.267	2.332	2.400	2.470	2.542	2.617	2.694	2.774	2.857	2.942
	0.50	2.256	2.323	2.391	2.463	2.536	2.61[illegible]	2.691	2.772	2.856	2.943	3.033
	0.55	2.310	2.380	2.452	2.526	2.604	2.683	2.766	2.851	2.940	3.031	3.126
	0.60	2.364	2.437	2.513	2.591	2.672	2.756	2.842	2.932	3.025	3.121	3.220
	0.65	2.419	2.496	2.575	2.657	2.742	2.829	2.920	3.014	3.111	3.212	3.316
	0.70	2.475	2.555	2.638	2.723	2.812	2.904	2.999	3.097	3.199	3.304	3.413
	0.75	2.532	2.615	2.702	2.791	2.884	2.979	3.079	3.181	3.288	3.398	3.512
	0.80	2.589	2.676	2.766	2.860	2.956	3.056	3.160	3.267	3.379	3.494	3.613
	0.85	2.648	2.738	2.832	2.929	3.030	3.134	3.242	3.354	3.471	3.591	3.715
	0.90	2.707	2.801	2.898	3.000	3.105	3.213	3.326	3.443	3.564	3.689	3.819
	0.95	2.766	2.864	2.966	3.071	3.180	3.293	3.411	3.532	3.659	3.790	3.925
	1.00	2.827	2.928	3.034	3.144	3.257	3.375	3.497	3.623	3.755	3.891	4.032
	1.05	2.888	2.993	3.103	3.217	3.335	3.457	3.584	3.716	3.852	3.994	4.141
	1.10	2.950	3.059	3.173	3.291	3.414	3.541	3.673	3.809	3.951	4.099	4.251
	1.15	3.013	3.126	3.244	3.367	3.494	3.625	3.762	3.904	4.052	4.205	4.363
	1.20	3.076	3.194	3.316	3.443	3.575	3.711	3.853	4.000	4.153	4.312	4.477
	1.25	3.141	3.262	3.389	3.520	3.657	3.798	3.945	4.098	4.257	4.421	4.592
For *I*	$\frac{x}{c}$	1.043	1.079	1.115	1.151	1.188	1.225	1.261	1.298	1.335	1.372	1.409
Correction		0.015	0.017	0.018	0.019	0.021	0.022	0.024	0.025	0.027	0.029	0.031

TABLE 3.—Values of U for velocities and D for times.

For U	$\frac{x}{c}$	0.00	0.10	0.20	0.30	0.40	0.50	0.60	0.70	0.80	0.90	1.00
For $\frac{aV_{\prime}}{r}$	0.00	1.000	1.051	1.105	1.162	1.221	1.284	1.350	1.419	1.492	1.568	1.649
	0.05	1.000	1.054	1.110	1.170	1.233	1.298	1.367	1.440	1.516	1.597	1.681
	0.10	1.000	1.056	1.116	1.178	1.244	1.312	1.385	1.461	1.541	1.625	1.714
	0.15	1.000	1.059	1.121	1.186	1.255	1.327	1.402	1.482	1.566	1.654	1.746
	0.20	1.000	1.062	1.126	1.194	1.266	1.341	1.420	1.503	1.590	1.682	1.779
	0.25	1.000	1.064	1.132	1.202	1.277	1.355	1.437	1.524	1.615	1.710	1.811
	0.30	1.000	1.067	1.137	1.210	1.288	1.369	1.455	1.545	1.639	1.739	1.843
	0.35	1.000	1.069	1.142	1.219	1.299	1.383	1.472	1.566	1.664	1.767	1.876
	0.40	1.000	1.072	1.147	1.227	1.310	1.398	1.490	1.587	1.689	1.796	1.908
	0.45	1.000	1.074	1.153	1.235	1.321	1.412	1.507	1.608	1.713	1.824	1.941
	0.50	1.000	1.077	1.158	1.243	1.332	1.426	1.525	1.629	1.738	1.853	1.973
	0.55	1.000	1.080	1.163	1.251	1.343	1.440	1.542	1.650	1.762	1.881	2.006
	0.60	1.000	1.082	1.168	1.259	1.354	1.454	1.560	1.671	1.787	1.909	2.038
	0.65	1.000	1.085	1.174	1.267	1.365	1.469	1.577	1.692	1.812	1.938	2.070
	0.70	1.000	1.087	1.179	1.275	1.376	1.483	1.595	1.712	1.836	1.966	2.103
	0.75	1.000	1.090	1.184	1.283	1.388	1.497	1.612	1.733	1.861	1.995	2.135
	0.80	1.000	1.092	1.189	1.291	1.399	1.511	1.630	1.754	1.885	2.023	2.168
	0.85	1.000	1.095	1.195	1.299	1.410	1.525	1.647	1.775	1.910	2.051	2.200
	0.90	1.000	1.097	1.200	1.308	1.421	1.540	1.665	1.796	1.935	2.080	2.233
	0.95	1.000	1.100	1.205	1.316	1.432	1.554	1.682	1.817	1.959	2.108	2.265
	1.00	1.000	1.103	1.210	1.324	1.443	1.568	1.700	1.838	1.984	2.137	2.297
	1.05	1.000	1.105	1.216	1.332	1.454	1.582	1.717	1.859	2.008	2.165	2.330
	1.10	1.000	1.108	1.221	1.340	1.465	1.597	1.735	1.880	2.033	2.194	2.362
	1.15	1.000	1.110	1.226	1.348	1.476	1.611	1.752	1.901	2.057	2.222	2.395
	1.20	1.000	1.113	1.231	1.356	1.487	1.625	1.770	1.922	2.082	2.250	2.427
	1.25	1.000	1.115	1.237	1.364	1.498	1.629	1.787	1.943	2.107	2.279	2.460
For D	$\frac{x}{c}$	0.000	0.198	0.393	0.585	0.775	0.962	1.146	1.327	1.506	1.683	1.858

For U	$\frac{x}{c}$	1.00	1.10	1.20	1.30	1.40	1.50	1.60	1.70	1.80	1.90	2.00
For $\frac{aV_{\prime}}{r}$	0.00	1.649	1.733	1.822	1.916	2.014	2.117	2.226	2.340	2.460	2.586	2.718
	0.05	1.681	1.770	1.863	1.961	2.064	2.173	2.287	2.407	2.533	2.665	2.804
	0.10	1.714	1.807	1.904	2.007	2.115	2.229	2.348	2.474	2.606	2.744	2.890
	0.15	1.746	1.843	1.945	2.053	2.166	2.285	2.409	2.541	2.679	2.824	2.976
	0.20	1.779	1.880	1.987	2.099	2.217	2.340	2.471	2.608	2.752	2.903	3.062
	0.25	1.811	1.917	2.028	2.144	2.267	2.396	2.532	2.675	2.825	2.982	3.148
	0.30	1.843	1.953	2.069	2.190	2.318	2.452	2.593	2.742	2.898	3.061	3.234
	0.35	1.876	1.990	2.110	2.236	2.369	2.508	2.655	2.809	2.971	3.141	3.320
	0.40	1.908	2.027	2.151	2.282	2.419	2.564	2.716	2.876	3.043	3.220	3.406
	0.45	1.941	2.063	2.192	2.328	2.470	2.620	2.777	2.943	3.116	3.299	3.492
	0.50	1.973	2.100	2.233	2.373	2.521	2.676	2.838	3.010	3.189	3.379	3.577
	0.55	2.006	2.137	2.274	2.419	2.571	2.731	2.900	3.077	3.262	3.458	3.663
	0.60	2.038	2.173	2.315	2.465	2.622	2.787	2.961	3.143	3.335	3.537	3.749
	0.65	2.070	2.210	2.357	2.511	2.673	2.843	3.022	3.210	3.408	3.616	3.835
	0.70	2.103	2.247	2.398	2.556	2.723	2.899	3.083	3.277	3.481	3.696	3.921
	0.75	2.135	2.283	2.439	2.602	2.774	2.955	3.145	3.344	3.554	3.775	4.007
	0.80	2.168	2.320	2.480	2.648	2.825	3.011	3.206	3.411	3.627	3.854	4.093
	0.85	2.200	2.357	2.521	2.694	2.875	3.066	3.267	3.478	3.700	3.933	4.179
	0.90	2.233	2.393	2.562	2.740	2.926	3.122	3.329	3.545	3.773	4.013	4.265
	0.95	2.265	2.430	2.603	2.785	2.977	3.178	3.390	3.612	3.846	4.092	4.351
	1.00	2.297	2.467	2.644	2.831	3.028	3.234	3.451	3.679	3.919	4.171	4.437
	1.05	2.330	2.503	2.685	2.877	3.078	3.290	3.512	3.746	3.992	4.251	4.523
	1.10	2.362	2.540	2.726	2.923	3.129	3.346	3.574	3.813	4.065	4.330	4.608
	1.15	2.395	2.577	2.768	2.968	3.180	3.402	3.635	3.880	4.138	4.409	4.694
	1.20	2.427	2.613	2.809	3.014	3.230	3.457	3.696	3.947	4.211	4.489	4.780
	1.25	2.460	2.650	2.850	3.060	3.281	3.513	3.758	4.014	4.284	4.568	4.866
For D	$\frac{x}{c}$	1.858	2.030	2.199	2.369	2.535	2.701	2.864	3.026	3.186	3.344	3.501

TABLE 4.—Values of $\frac{x}{c}B$ for ranges.

	$\frac{x}{c}$	0.00	0.05	0.10	0.15	0.20	0.25	0.30	0.35	0.40	0.45	0.50
For $\frac{aV_{\prime}}{r}$	0.00	0.000	0.0508	0.1034	0.1578	0.2140	0.2722	0.3324	0.3947	0.4591	0.5258	0.5949
	0.05	0.000	0.0509	0.1036	0.1582	0.2148	0.2734	0.3341	0.3970	0.4622	0.5298	0.6000
	0.10	0.000	0.0509	0.1038	0.1586	0.2155	0.2745	0.3357	0.3993	0.4654	0.5339	0.6051
	0.15	0.000	0.0510	0.1039	0.1590	0.2162	0.2757	0.3374	0.4017	0.4685	0.5379	0.6102
	0.20	0.000	0.0510	0.1041	0.1594	0.2169	0.2768	0.3391	0.4040	0.4716	0.5420	0.6154
	0.25	0.000	0.0511	0.1043	0.1598	0.2177	0.2780	0.3408	0.4064	0.4748	0.5461	0.6206
	0.30	0.000	0.0511	0.1045	0.1602	0.2184	0.2791	0.3425	0.4088	0.4780	0.5503	0.6258
	0.35	0.000	0.0511	0.1046	0.1606	0.2191	0.2803	0.3443	0.4112	0.4812	0.5544	0.6310
	0.40	0.000	0.0512	0.1048	0.1610	0.2199	0.2815	0.3460	0.4136	0.4844	0.5586	0.6363
	0.45	0.000	0.0512	0.1050	0.1614	0.2206	0.2826	0.3477	0.4160	0.4877	0.5628	0.6416
	0.50	0.000	0.0513	0.1052	0.1618	0.2213	0.2838	0.3494	0.4184	0.4909	0.5670	0.6470
	0.55	0.000	0.0513	0.1053	0.1622	0.2221	0.2850	0.3512	0.4209	0.4942	0.5712	0.6523
	0.60	0.000	0.0514	0.1055	0.1626	0.2228	0.2862	0.3529	0.4233	0.4974	0.5755	0.6577
	0.65	0.000	0.0514	0.1057	0.1630	0.2236	0.2874	0.3547	0.4257	0.5007	0.5797	0.6632
	0.70	0.000	0.0514	0.1059	0.1634	0.2243	0.2886	0.3564	0.4282	0.5040	0.5841	0.6686
	0.75	0.000	0.0515	0.1060	0.1638	0.2250	0.2897	0.3582	0.4307	0.5074	0.5884	0.6741
	0.80	0.000	0.0515	0.1062	0.1643	0.2258	0.2909	0.3600	0.4332	0.5107	0.5927	0.6796
	0.85	0.000	0.0516	0.1064	0.1647	0.2265	0.2921	0.3617	0.4356	0.5140	0.5971	0.6852
	0.90	0.000	0.0516	0.1066	0.1651	0.2273	0.2933	0.3635	0.4381	0.5174	0.6015	0.6908
	0.95	0.000	0.0517	0.1067	0.1655	0.2280	0.2946	0.3653	0.4407	0.5208	0.6059	0.6964
	1.00	0.000	0.0517	0.1069	0.1659	0.2288	0.2958	0.3671	0.4432	0.5242	0.6103	0.7020
	1.05	0.000	0.0517	0.1071	0.1663	0.2295	0.2970	0.3689	0.4457	0.5276	0.6147	0.7076
	1.10	0.000	0.0518	0.1073	0.1667	0.2303	0.2982	0.3707	0.4482	0.5310	0.6192	0.7133
	1.15	0.000	0.0518	0.1075	0.1671	0.2310	0.2994	0.3725	0.4508	0.5344	0.6237	0.7191
	1.20	0.000	0.0519	0.1076	0.1676	0.2318	0.3006	0.3743	0.4533	0.5379	0.6282	0.7248
	1.25	0.000	0.0519	0.1078	0.1680	0.2326	0.3019	0.3761	0.4559	0.5414	0.6327	0.7306

	$\frac{x}{c}$	0.50	0.55	0.60	0.65	0.70	0.75	0.80	0.85	0.90	0.95	1.00
For $\frac{aV_{\prime}}{r}$	0.00	0.5949	0.6664	0.7404	0.8171	0.8964	0.9786	1.0638	1.1521	1.2435	1.3384	1.4365
	0.05	0.6000	0.6727	0.7480	0.8262	0.9072	0.9912	1.0784	1.1689	1.2628	1.3603	1.4613
	0.10	0.6051	0.6789	0.7556	0.8353	0.9180	1.0039	1.0931	1.1860	1.2823	1.3825	1.4864
	0.15	0.6102	0.6852	0.7633	0.8445	0.9290	1.0107	1.1080	1.2031	1.3020	1.4050	1.5118
	0.20	0.6154	0.6916	0.7711	0.8538	0.9400	1.0296	1.1230	1.2205	1.3219	1.4276	1.5375
	0.25	0.6206	0.6981	0.7789	0.8632	0.9511	1.0427	1.1382	1.2380	1.3420	1.4505	1.5634
	0.30	0.6258	0.7045	0.7868	0.8726	0.9623	1.0558	1.1535	1.2558	1.3622	1.4736	1.5896
	0.35	0.6310	0.7110	0.7947	0.8822	0.9736	1.0691	1.1690	1.2737	1.3828	1.4970	1.6162
	0.40	0.6363	0.7175	0.8027	0.8917	0.9850	1.0824	1.1846	1.2917	1.4035	1.5207	1.6430
	0.45	0.6416	0.7241	0.8107	0.9014	0.9964	1.0959	1.2002	1.3099	1.4243	1.5445	1.6701
	0.50	0.6470	0.7307	0.8188	0.9111	1.0080	1.1095	1.2161	1.3282	1.4455	1.5686	1.6974
	0.55	0.6523	0.7374	0.8269	0.9209	1.0196	1.1232	1.2321	1.3467	1.4667	1.5930	1.7251
	0.60	0.6577	0.7441	0.8351	0.9307	1.0318	1.1370	1.2482	1.3654	1.4882	1.6176	1.7530
	0.65	0.6632	0.7509	0.8434	0.9407	1.0431	1.1510	1.2645	1.3853	1.5100	1.6424	1.7813
	0.70	0.6686	0.7577	0.8517	0.9507	1.0551	1.1650	1.2809	1.4034	1.5319	1.6674	1.8098
	0.75	0.6741	0.7645	0.8600	0.9608	1.0671	1.1791	1.2974	1.4225	1.5539	1.6927	1.8386
	0.80	0.6796	0.7714	0.8684	0.9709	1.0792	1.1933	1.8141	1.4419	1.5763	1.7183	1.8676
	0.85	0.6852	0.7783	0.8769	0.9811	1.0914	1.2077	1.3309	1.4614	1.5988	1.7440	1.8970
	0.90	0.6908	0.7852	0.8854	0.9914	1.1036	1.2222	1.3478	1.4811	1.6215	1.7700	1.9267
	0.95	0.6964	0.7922	0.8940	1.0018	1.1159	1.2368	1.3649	1.5009	1.6444	1.7964	1.9566
	1.00	0.7020	0.7993	0.9026	1.0122	1.1284	1.2515	1.3821	1.5210	1.6675	1.8229	1.9868
	1.05	0.7076	0.8064	0.9113	1.0226	1.1409	1.2663	1.3995	1.5411	1.6909	1.8495	2.0173
	1.10	0.7133	0.8135	0.9200	1.0331	1.1535	1.2812	1.4170	1.5615	1.7144	1.8764	2.0481
	1.15	0.7191	0.8206	0.9289	1.0439	1.1663	1.2963	1.4346	1.5819	1.7382	1.9037	2.0792
	1.20	0.7248	0.8278	0.9377	1.0546	1.1791	1.3115	1.4524	1.6027	1.7621	1.9312	2.1105
	1.25	0.7306	0.8350	0.9466	1.0654	1.1920	1.3268	1.4702	1.6236	1.7861	1.9588	2.1422

Table 5.—Values of $\dfrac{\dfrac{V_{\prime}}{r}}{\sqrt{B}}$ for initial velocities.

For $\dfrac{aV_{\prime}}{r}$

$\frac{x}{c}$	0.00	0.05	0.10	0.15	0.20	0.25	0.30	0.35	0.40	0.45	0.50
		0.	0.	0.	0.	0.	0.	0.	0.	0.	0.
0.05	0.0000	.04956	.04913	.04869	.04825	.04782	.04738	.04695	.04651	.04608	.04564
0.10	0.0000	.09908	.09816	.09725	.09634	.09543	.09452	.09362	.09271	.09181	.09091
0.15	0.0000	.14857	.14713	.14569	.14426	.14284	.14143	.14001	.13860	.13719	.13778
0.20	0.0000	.19800	.19601	.19402	.19204	.19007	.18811	.18614	.18418	.18223	.18028
0.25	0.0000	.24739	.24480	.24221	.23963	.23708	.23454	.23200	.22945	.22692	.22440
0.30	0.0000	.29675	.29351	.29029	.28710	.28392	.28075	.27759	.27444	.27130	.26818
0.35	0.0000	0.3461	0.3422	0.3383	0.3344	0.3306	0.3267	0.3229	0.3191	0.3153	0.3116
0.40	0.0000	0.3953	0.3907	0.3861	0.3815	0.3770	0.3725	0.3680	0.3635	0.3590	0.3546
0.45	0.0000	0.4446	0.4392	0.4338	0.4285	0.4232	0.4180	0.4128	0.4075	0.4024	0.3973
0.50	0.0000	0.4938	0.4876	0.4815	0.4754	0.4693	0.4633	0.4573	0.4513	0.4454	0.4396
0.55	0.0000	0.5429	0.5359	0.5289	0.5220	0.5152	0.5084	0.5016	0.4948	0.4881	0.4815
0.60	0.0000	0.5920	0.5841	0.5762	0.5685	0.5608	0.5532	0.5456	0.5380	0.5305	0.5231
0.65	0.0000	0.6411	0.6323	0.6235	0.6148	0.6063	0.5978	0.5894	0.5810	0.5727	0.5645
0.70	0.0000	0.6901	0.6803	0.6706	0.6610	0.6516	0.6422	0.6329	0.6236	0.6144	0.6053
0.75	0.0000	0.7391	0.7283	0.7176	0.7071	0.6967	0.6864	0.6761	0.6659	0.6558	0.6459
0.80	0.0000	0.7880	0.7762	0.7645	0.7530	0.7416	0.7303	0.7191	0.7080	0.6971	0.6862
0.85	0.0000	0.8370	0.8241	0.8113	0.7987	0.7863	0.7741	0.7619	0.7498	0.7379	0.7261
0.90	0.0000	0.8857	0.8717	0.8579	0.8443	0.8309	0.8176	0.8044	0.7913	0.7784	0.7657
0.95	0.0000	0.9347	0.9195	0.9045	0.8897	0.8752	0.8609	0.8467	0.8327	0.8188	0.8051
1.00	0.0000	0.9834	0.9671	0.9509	0.9349	0.9194	0.9040	0.8887	0.8736	0.8587	0.8439
1.05	0.0000	1.0322	1.0145	0.9971	0.9801	0.9634	0.9469	0.9305	0.9143	0.8984	0.8826
1.10	0.0000	1.0809	1.0620	1.0434	1.0251	1.0072	0.9895	0.9720	0.9547	0.9377	0.9210
1.15	0.0000	1.1295	1.1094	1.0895	1.0700	1.0508	1.0320	1.0134	0.9949	0.9768	0.9590
1.20	0.0000	1.1781	1.1566	1.1354	1.1146	1.0942	1.0741	1.0543	1.0348	1.0156	0.9967
1.25	0.0000	1.2267	1.2088	1.1813	1.1592	1.1375	1.1162	1.0952	1.0745	1.0541	1.0841

For $\dfrac{aV_{\prime}}{r}$

$\frac{x}{c}$	0.50	0.55	0.60	0.65	0.70	0.75	0.80	0.85	0.90	0.95	1.00
	0.	0.	0.	0.	0.	0.	0.	0.	0.	0.	0.
0.05	.04564	.04521	.04478	.04435	.04392	.04349	.04306	.04263	.04221	.04178	.04136
0.10	.09091	.09001	.08911	.08821	.08732	.08644	.08555	.08466	.08378	.08289	.08202
0.15	.13778	.13438	.13299	.13160	.13021	.12883	.12746	.12608	.12471	.12334	.12198
0.20	.18028	.17835	.17643	.17451	.17259	.17069	.16880	.16691	.16503	.16315	.16128
0.25	.22440	.22190	.21942	.21694	.21448	.21203	.20958	.20715	.20473	.20233	.19994
0.30	.26818	.26507	.26198	.25890	.25588	.25285	.24983	.24682	.24383	.24087	.23793
0.35	0.3116	0.3078	0.3041	0.3004	0.2968	0.2932	0.2895	0.2859	0.2824	0.2788	0.2753
0.40	0.3546	0.3502	0.3458	0.3415	0.3372	0.3330	0.3287	0.3245	0.3203	0.3162	0.3121
0.45	0.3973	0.3922	0.3871	0.3821	0.3772	0.3723	0.3674	0.3625	0.3577	0.3529	0.3482
0.50	0.4396	0.4338	0.4280	0.4223	0.4167	0.4111	0.4055	0.4000	0.3945	0.3891	0.3838
0.55	0.4815	0.4750	0.4685	0.4621	0.4557	0.4494	0.4432	0.4370	0.4308	0.4247	0.4187
0.60	0.5231	0.5158	0.5086	0.5014	0.4943	0.4873	0.4804	0.4735	0.4666	0.4598	0.4531
0.65	0.5645	0.5563	0.5483	0.5403	0.5325	0.5247	0.5170	0.5093	0.5018	0.4944	0.4870
0.70	0.6053	0.5964	0.5875	0.5788	0.5702	0.5616	0.5532	0.5448	0.5365	0.5284	0.5203
0.75	0.6459	0.6361	0.6264	0.6169	0.6075	0.5981	0.5889	0.5798	0.5708	0.5619	0.5531
0.80	0.6862	0.6755	0.6650	0.6546	0.6443	0.6342	0.6242	0.6143	0.6045	0.5949	0.5854
0.85	0.7261	0.7145	0.7031	0.6919	0.6808	0.6698	0.6589	0.6482	0.6377	0.6273	0.6171
0.90	0.7657	0.7532	0.7409	0.7288	0.7168	0.7050	0.6934	0.6819	0.6706	0.6594	0.6484
0.95	0.8051	0.7916	0.7783	0.7652	0.7524	0.7398	0.7273	0.7149	0.7028	0.6909	0.6792
1.00	0.8439	0.8295	0.8153	0.8014	0.7876	0.7741	0.7608	0.7476	0.7341	0.7219	0.7094
1.05	0.8826	0.8671	0.8520	0.8371	0.8225	0.8081	0.7939	0.7798	0.7660	0.7525	0.7393
1.10	0.9210	0.9045	0.8883	0.8725	0.8570	0.8417	0.8265	0.8116	0.7970	0.7827	0.7686
1.15	0.9590	0.9415	0.9243	0.9075	0.8909	0.8747	0.8588	0.8430	0.8275	0.8124	0.7975
1.20	0.9967	0.9781	0.9599	0.9421	0.9246	0.9075	0.8906	0.8739	0.8576	0.8416	0.8260
1.25	1.0841	1.0145	0.9958	0.9764	0.9579	0.9398	0.9220	0.9045	0.8873	0.8705	0.8540

TABLE *of Times, calculated for the* WEST POINT BALLISTIC MACHINE.

Length of simple pendulum, 5.769 in.; and $\frac{2\pi l}{360\sqrt{2gl}}=0.001509''$.

Degrees.	Time of passage for each degree.	Sum of Times.	Degrees.	Time of passage for each degree.	Sum of Times.
1	.00151		26	.00159	.03987
2	.00151	.00302	27	.00159	.04146
3	.00151	.00453	28	.00160	.04306
4	.00151	.00604	29	.00161	.04467
5	.00151	.00755	30	.00162	.04629
6	.00151	.00906	31	.00163	.04792
7	.00151	.01057	32	.00163	.04955
8	.00151	.01208	33	.00164	.05119
9	.00151	.01359	34	.00165	.05284
10	.00152	.01511	35	.00166	.05450
11	.00152	.01663	36	.00167	.05617
12	.00152	.01815	37	.00168	.05785
13	.00152	.01967	38	.00170	.05955
14	.00153	.02120	39	.00171	.06126
15	.00153	.02273	40	.00172	.06298
16	.00153	.02426	41	.00173	.06471
17	.00154	.02580	42	.00175	.06646
18	.00154	.02734	43	.00176	.06822
19	.00155	.02889	44	.00178	.07000
20	.00155	.03044	45	.00179	.07179
21	.00156	.03200	46	.00181	.07360
22	.00156	.03356	47	.00182	.07542
23	.00157	.03513	48	.00184	.07726
24	.00157	.03670	49	.00186	.07912
25	.00158	.03828	50	.00188	.08100

EXAMPLE.—What is the velocity of a projectile when the time of its passage between two targets, 100 feet apart, corresponds to 20.5 degrees of the graduated arc?

Time of 20°	= 0.03044	Log. of 100	= 2.000000
Add for 0.5°	0.00077	Log. 0.06242	= $\bar{2}$.795324
Time of 20°.5	= 0.03121	Log. 1602.	= 3.204676
	2		
Double arc	= 0.06242	Velocity	= 1602. feet.

TABLES OF FIRE.

WITH the exception of those for mortars, the following tables of fire were calculated by Bvt. Brig.-Gen'l J. A. HASKIN, U. S. Artillery, by aid of the equations of the movement of projectiles in air given in Chapter VIII., and so far as they have been verified they agree remarkably well with those obtained in practice.

The "range" is understood to be the distance in yards at which the projectile first strikes the horizontal plane drawn through the centre of the muzzle of the piece. When the object is situated below this plane the angle of elevation of the *piece*, if measured by the quadrant, should be corrected by the table on page 532. In aiming at the object direct, with the breech-sight, no correction for the elevation of the piece is considered necessary, if the angle of elevation or depression of the *object* does not exceed 15°.

The weight of each projectile, the value (c), and the initial velocity (I. V.), are given in the tables in each case.

Kind of Ordnance.	Charge.	Projectile.	Elevation.	Range.	Time.	Remarks.
			° ′	Yards.	Seconds	
6 Pdr. Field	Lbs.	SHOT.	.30	298		
Gun.	1.25		1.	507		
		Wt.=6.15 lbs.	1. 30	670		
		c.=2660	2.	804		
		I.V.=1440	2. 30	918		
			3.	1018		
			3. 30	1106		
			4.	1184		
			5.	1323		
			6.	1442		
			7.	1546		
			8.	1638		
			9.	1721		
			10.	1796		
			.30	328	.77	
	1.25	SPHERICAL.	1.	537	1.44	
		CASE.	1. 30	697	2.07	
			2.	827	2.65	
		Wt.=5.5lbs.	2. 30	936	3.18	
			3.	1030	3.68	
		c.=2379	3. 30	1112	4.17	
		I. V.=1550	4.	1186	4.62	
			5.	1315	5.52	
			6.	1425	6.36	
			7.	1521	7.14	
			.15	181		Ranges obtained
12 Pdr. Field	2.5	SHOT.	.30	320		at Washington
Gun.			.45	452		Arsenal,
		Wt.=12.3 lbs.	1.	562		March, 1865.
		c.=3338	1. 15	662		° ′ Yards
		I.V.=1486	1. 30	750		1. 599
			1. 45	832		1. 30 750
			2.	906		2. 875
			2. 30	1040		2. 30 1050
			3.	1158		3. 1185
			3. 30	1261		3.30 1280
			4.	1355		4. 1360
			5.	1521		5. 1520
			6.	1663		
			7.	1788		
			8.	1899		
			9.	1999		
			10.	2090		
			.30	324	.77	
	2.5	SPHERICAL	1.	554	1.45	
		CASE.	1. 30	733	2.09	
			2.	882	2.65	
		Wt.=11.25	2. 30	1010	3.25	
			3.	1120	3.8	
		c.=3053	3. 30	1218	4.3	
		I. V.=14.95	4.	1307	4.81	
			5.	1462	5.73	
			6.	1595	6.62	
			7.	1711	7.46	
			8.	1815	8.27	
			9.	1907	9.05	
			10.	1992	9.82	

Kind of Ordnance.	Charge.	Projectile.	Elevation.	Range.	Time.	Remarks.
			° ′	Yards.	Seconds	
12 Pdr. Field Gun.	Lbs. 2.5	SHELL.	.30	367	.82	
			1.	590	1.52	
		Wt.=9 lbs.	1.30	765	2.18	
		c.=2442	2.	900	2.78	
		I.V.=1680	2.30	1014	3.36	
			3.	1112	3.88	
			3.30	1197	4.36	
			4.	1274	4.84	
			5.	1408	5.75	
			6.	1521	6.57	
			7.	1619	7.42	
			8.	1706	8.19	
			9.	1784	8.92	
			10.	1855	9.64	
			11.	1919	10.34	
			12.	1979	11.98	
12 Pdr. Field Howitzer.	1.	SHELL.	.30	232	.64	
			1.	405	1.23	
			1.30	545	1.78	
		Wt.=9 lbs.	2.	662	2.30	
		c.=2442	2.30	763	2.77	
		I. V.=1239	3.	850	3.26	
			3.30	930	3.72	
			4.	1002	4.15	
			5.	1127	5.	
			6.	1235	5.79	
			7.	1329	6.54	
			8.	1413	7.26	
			9.	1488	7.95	
			10.	1556	8.61	
	1.	SPHERICAL CASE.	.30	214	.61	
			1.	385	1.18	
			1.30	529	1.72	
			2.	653	2.24	
		Wt.=11.25	2.30	763	2.73	
		c.=3053	3.	859	3.21	
		I. V.=1160	3.30	947	3.67	
			4.	1028	4.12	
			5.	1171	5.	
			6.	1295	5.79	
24 Pdr. Field Howitzer.	2.5	SHELL.	.30	233	.64	
			1.	415	1.23	
			1. 30	566	1.79	
		Wt.=17.5lbs.	2.	695	2.32	
		c.=3007	2. 30	807	2.83	
		I.V.=1220	3.	906	3.32	
			3. 30	996	3.79	
			4.	1078	4.24	
			5.	1222	5.11	
			6.	1348	5.95	
			7.	1457	6.75	
			8.	1556	7.5	
			9.	1646	8.24	
			10.	1726	8.95	

Kind of Ordnance.	Charge.	Projectile.	Elevation.	Range.	Time.	Remarks.
			° ′	Yards.	Seconds	
24 Pdr. Field Howitzer.	Lbs. 2.5	SPHERICAL CASE.	.30	215	.61	
			1.	396	1.19	
			1.30	551	1.74	
		Wt.=22.46	2.	688	2.27	
		c.=3860	2.30	810	2.78	
			3.	920	3.28	
		I.V.=1150	3.30	1020	3.76	
			4.	1112	4.23	
			5.	1278	5.13	
			6.	1423	5.99	
			7.	1552	6.81	
			8.	1669	7.61	
			9.	1775	8.39	
			10.	1871	9.13	
32 Pdr. Field Howitzer.	3.25	SHELL.	.30	223	.63	
			1.	403	1.21	
			1.30	554	1.76	
		Wt.=23.03	2.	685	2.29	
		c.=3307	2.30	801	2.8	
			3.	903	3.29	
		I.V.=1182	3.30	997	3.75	
			4.	1083	4.21	
			5.	1235	5.08	
			6.	1367	5.94	
			7.	1484	6.75	
			8.	1589	7.52	
			9.	1684	8.26	
			10.	1770	9.	
	3.25	SPHERICAL CASE.	.30	205	.59	
			1.	381	1.15	
			1.30	536	1.7	
			2.	674	2.23	
		Wt.=30.75	2.30	799	2.74	
		c.=4365	3.	913	3.24	
			3.30	1016	3.71	
		I.V.=1110	4.	1113	4.17	
			5.	1288	5.07	
			6.	1442	5.93	
			7.	1580	6.75	
			8.	1704	7.53	
			9.	1818	8.29	
			10.	1922	9.02	
			15.	2344	12.03	
24 Pdr. Siege Gun.	6.	SHOT.	.30	411		
			1.	704		
			1.30	934		
		Wt.=24.3 lbs.	2.	1124		
		c.=4167	2.30	1283		
			3.	1426		
		I.V.=1680.	3.30	1555		
			4.	1672		
			5.	1874		
			6.	2046		
			7.	2195		
			8.	2329		
			9.	2450		
			10.	2562		

Kind of Ordnance.	Charge.	Projectile.	Elevation.	Range.	Time.	Remarks.
			° ′	Yards.	Seconds	
24 Pdr. Siege	Lbs.	SPHERICAL.	.30	418	.87	
Gun.	6	CASE.	1.	707	1.65	
			1. 30	930	2.37	
		Wt.=22.46	2.	1114	3.03	
		c.=3860	2. 30	1271	3.66	
			3.	1408	4.26	
		I.V.=1715	3. 30	1529	4.33	
			4.	1637	5.38	
			5.	1828	6.43	
			6.	1993	7.43	
			7.	2134	8.36	
			8.	2261	9.26	
			9.	2377	10.13	
			10.	2480	10.97	
			.30	419	.88	
	6	SHELL.	1.	684	1.64	
			1. 30	863	2.34	
		Wt.=17.5 lbs.	2.	1043	2.98	
		c.=3007	2. 30	1176	3.58	
			3.	1292	4.15	
		I.V.=1782	3. 30	1393	4.68	
			4.	1484	5.20	
			5.	1654	6.18	
			6.	1777	7.09	
			7.	1895	7.94	
			8.	1997	8.73	
			9.			
			10.			
			.30	402		
32 Pdr. Sea Coast Gun.	8	SHOT.	1.	699		
			1. 30	935		
		Wt.=32.3 lbs.	2.	1135		
		c.=4584	2. 30	1306		
			3.	1457		
		I.V.=1640	3. 30	1591		
			4.	1712		
			5.	1927		
			6.	2112		
			7.	2274		
			8.	2419		
			9.	2549		
			10.	2664		
			.30	413	.86	
	8	SPHERICAL	1.	709	1.64	
		CASE.	1. 30	943	2.36	
			2.	1140	3.04	
		Wt.=130.75	2. 30	1308	3.68	
			3.	1454	4.29	
		c.=4365	3. 30	1584	4.87	
		I. V.=16.75	4.	1702	5.44	
			5.	1910	6.52	
			6.	2089	7.54	
			7.	2246	8.51	
			8.	2387	9.43	
			9.	2511	10.33	
			10.	2626	11.20	

Kind of Ordnance.	Charge.	Projectile.	Elevation.	Range.	Time.	Remarks.
			° ′	Yards.	Seconds	
32 Pdr. Sea Coast Gun.	Lbs. 8.	SHELL.	30	410	.86	
			1.	693	1.65	
		Wt.=23.3 lbs.	1. 30	901	2.35	
		c.=3307	2.	1069	3.	
		I.V.=1754.	2. 30	1213	3.6	
			3.	1337	4.19	
			3. 30	1443	4.74	
			4.	1542	5.28	
			5.	1712	6.28	
			6.	1857	7.24	
			7.	1984	8.14	
			8.	2096	9.	
			9.	2197	9.82	
			10.			
42 Pdr. Sea Coast Gun.	10.5	SHOT.	.30	407		
			1.	712		
			1. 30	960		
		Wt.=42.5 lbs.	2.	1170		
		c.=5036	2. 30	1350		
		I.V.=1638	3.	1510		
			3. 30	1653		
			4.	1781		
			5.	2012		
			6.	2208		
			7.	2382		
			8.	2537		
			9.	2675		
			10.	2805		
8 Inch Siege Howitzer.	4.	SHELL.	.30	187	.56	
			1.	350	1.1	
			1. 30	498	1.63	
		Wt.=46.6	2.	623	2.14	
		c.=4161	2. 30	740	2.62	
		I. V.= 1060	3.	847	3.1	
			3. 30	944	3.56	
			4.	1036	4.02	
			5.	1200	4.86	
			6.	1346	5.7	
			7.	1477	6.51	
			8.	1597	7.3	
			9.	1704	8.05	
			10.	1805	8.8	
			12.	1986	10.3	
8 In. Sea Coast Howitzer.	6.	SHELL.	.30	214	.6	
			1.	397	1.17	
			1. 30	558	1.73	
		Wt.=49.75	2.	700	2.27	
		c.=4442	2. 30	829	2.79	
		I.V.=1137	3.	946	3.3	
			3. 30	1053	3.79	
			4.	1153	4.27	
			5.	1333	5.19	
			6.	1489	6.07	
			7.	1630	6.91	
			8.	1758	7.74	
			9.	1874	8.55	
			10.	1984	9.33	

Kind of Ordnance.	Charge.	Projectile.	Elevation.	Range.	Time.	Remarks.	
			° ′	Yards.	Seconds		
10 Inch Sea Coast Howitzer.	Lbs. 10	SHELL.	.30	313	.73		
			1.	593	1.45		
		Wt.=101.75	1 30	821	2.13		
		c.=5786	2.	1020	2.77		
		I.V.=1420.	2. 30	1197	3.4		
			3.	1355	3.99		
			3. 30	1500	4.57		
			4.	1634	5.13		
			5.	1870	6.21		
			6.	2079	7.23		
			7.	2263	8.2		
			8.	2428	9.13	Final Velo-city.	Angle of Fall.
			9.	2579	10.03		
			10.	2718	10.9		
			.30	298		Feet.	° ′
8 Inch Rodman.	10.	SHOT	1.	549		1065.	1. 12
			1. 30	765			
		Wt.=65 lbs.	2.	966		885.	2. 42
		c.=5893	2. 30	1127			
		I.V =1350	3.	1283		775.	4. 23
			3. 30	1428			
			4.	1558		695.	6. 19
			5.	1794		635.	8. 24
			6.	2004		589.	10. 31
			7.	2191		545.	12. 48
			8.	2363		517.	15. 10
			9.	2599		490.	17. 34
			10.	2665		466.	20. 8
			15.	3269		381.	
			20.	3746		327.	
			.30	318		If the gun is above the object, the angle of fall must be increased by the angle of depression corresponding to the range and height of the gun.	
	10.	SHOT.	1.	582			
			1. 30	807			
		Wt.=65 lbs	2.	1005			
		c.=5893	2. 30	1182			
		I. V.= 1400	3.	1342			
			3. 30	1488			
			4.	1622			
			5.	1865			
			6.	2077			
			7.	2267			
			8.	2441			
			9.	2599			
			10.	2747			
			15.	3354			
			20.	3833			
			.30	355	.79		
	10.	SHELL.	1.	628	1.52		
			1. 30	852	2.21		
		Wt.=51.5	2.	1042	2.85		
		c.=4669	2. 30	1207	3.47		
		I.V.=1517	3.	1354	4.07		
			3. 30	1487	4.65		
			4.	1608	5.21		
			5.	1822	6.28		
			6.	2008	7.29		
			7.	2174	8.25		
			8.	2322	9.19		
			9.	2457	10.1		
			10.	2581	10.97		
			15.	3093	15.1		

Kind of Ordnance.	Charge.	Projectile.	Elevation.	Range.	Time.	Remarks. Final velocity.	Angle of Fall.
			° ′	Yards.	Seconds		
8 inch Rodman.	Lbs. 10	SHELL.	1.	669	1.58		
			2.	1088	2.95		
		Wt.=49.75	3.	1399	4.17		
		c.=4442	4.	1647	5.3		
			5.	1855	6.36		
		I.V.=1598	6.	2034	7.35		
			7.	2191	8.29		
			8.	2332	9.19		
			9.	2458	10.05		
			10.	2574	10.87		
			11.	2680	11.65		
			12.	2779	12.4		
10 Inch Rodman.	15	SHOT.	.30	272		Feet.	° ′
			1.	511		1146.	
			1. 30	724		1047.	1. 8
		Wt.=123.5	2.	916		968.	
		c.=7028	2. 30	1090		905.	2. 30
			3.	1251		853.	
		I.V.=1275	3. 30	1401		808.	4. 3
			4.	1539		769.	
			5.	1793		735.	5. 44
			6.	2019		677.	7. 32
			7.	2225		631.	9. 23
			8.	2414		593.	11. 22
			9.	2587		560.	13. 24
			10.	2749		532.	15. 27
			15.	3429		508.	17. 34
			20.	3976		419.	
	10	SHELL.	.30	265	.68	372.	
			1.	504	1.33		
			1. 30	708	1.95		
		Wt.=101.75	2.	886	2.56		
		c.=5786	2. 30	1048	3.15		
		I.V.=1284	3.	1195	3.71		
			3. 30	1330	4.25		
			4.	1455	4.79		
			5.	1680	5.83		
			6.	1879	6.82		
			7.	2057	7.78		
			8.	2217	8.71		
			9.	2363	9.6		
			10.	2498	10.46		
			11.	2621	11.31		
			12.	2737	12.14		

Kind of Ordnance.	Charge.	Projectile.	Elevation.	Range.	Time.	Final Velocity.	Angle of Fall.
			° ′	Yards.	Seconds	Feet.	° ′
15 Inch Rodman Gun.	Lbs.	SHOT.	1.	360		944.	1. 7
	40.		2.	684		877.	2. 14
		Wt.=428 lbs.	3.	976		822.	3. 30
		c.=10760.	4.	1243		776.	4. 49
		I.V.=1028.	5.	1490		736.	6. 14
			6.	1718		702.	7. 41
			7.	1931		672.	9. 12
			8.	2129		646.	10. 49
			9.	2314		622.	12. 25
			10.	2489		603.	14. 4
			15.	3225		525.	22. 34
			20.	3787		475.	30. 51
			30	253			
	40.	SHELL.	1.	482		1048.	
			1. 30	692			
		Wt.=344 lbs.	2.	885		932.	2. 23
		c.=8648.3	3.	1230		845.	3. 49
			4.	1534		778.	5. 22
		I.V.=1218.	5.	1806		724.	7.
			6.	2053		679.	8. 44
			7.	2280		641.	10. 34
			8.	2488		609.	12. 24
			9.	2682		580.	14. 20
			10.	2865		555.	16. 18
			15.	3641		463.	26. 26
			20.	4271		401.	
			20. 26.	4322			
			1.	556	1.4		
	40.	SHELL.	2.	1090	2.7		
			3.	1369	3.92		
		Wt.=328 lbs	4.	1686	5.1		
		c.=8187.	5.	1965	6.22		
			6.	2214	7.29		
		I. V.=1325.	7.	2439	8.35		
			8.	2644	9.37		
			9.	2831	10.36		
			10.	3003	11.31		
			12.	3312	13.13		
			15.	3707	16.		
			20. 26	4165	19.25		
			1.	406	1,12		
	40.	SHELL.	2.	752	2.3		
			3.	1053	3.4		
		Wt.=314 lbs	4.	1314	4.43		
		c.=7918.	5.	1551	5.4		
			6.	1765	6.35		
		I.V.=1110.	7.	1963	7.29		
			8.	2141	8.18		
			9.	2309	9.05		
			10.	2464	9.88		
			12.	2745	11.49		
			15.	3105	13.69		

Kind of Ordnance.	Charge.	Projectile.	Elevation.	Range.	Time.	Remarks.
			° ′	Yards.	Seconds	
3 Inch Rifle Gun.	Lbs. 1.	DYER SHELL.	.30	258	.66	Ranges at Washington Arsenal, March 8th, 1865.
			1.	489	1.29	
		Wt.=9 lbs.	1.30	700	1.91	° Yards.
		c.=8167	2.	892	2.53	3 1236.
		I.V.=1232	2.30	1070	3.12	4 1510.
			3.	1234	3.7	5 1785.
			3.30	1386	4.25	10 2790.
			4.	1530	4.8	
			5.	1794	5.87	
			6.	2031	6.9	
			7.	2244	7.87	
			8.	2441	8.83	
			9.	2621	9.75	
			10.	2788	10.67	
			12.	3114	12.45	
			20.	3972	17.88	
3 In. Parrott Gun.	1.	SHELL.	.30	259	.66	
			1.	494	1.3	
		Wt.=10 lbs.	1.30	710	1.93	
		c.=9075	2.	909	2.54	
		I.V.=1232.	2.30	1092	3.14	
			3.	1262	3.72	
			3.30	1422	4.29	
			4.	1573	4.85	
			5.	1851	5.96	
			6.	2101	7.02	
			7.	2328	8.04	
			8.	2536	9.04	
			9.	2729	10.02	
			10.	2907	10.96	
			12.	3212	12.7	
			15.	3558	15.09	
4½ In. Rifle Gun.	3.25	DYER SHELL.	.30	290	.69	
			1.	553	1.37	
		Wt.=25.5	1.30	799	2.05	
		c.=10491.4	2.	1017	2.69	
		I.V.=1303	2.30	1224	3.32	
			3.	1414	3.94	
			3.30	1593	4.54	
			4.	1762	5.14	
			5.	2071	6.3	
			6.	2354	7.42	
			7.	2610	8.51	
			8.	2844	9.57	
			9.	3061	10.6	
			10.	3265	11.59	

Kind of Ordnance.	Charge.	Projectile.	Elevation.	Range.	Time.	Remarks.
			° ′	Yards.	Seconds	
24 Pdr. Gun Rifled.	Lbs. 4.5	Shot.	1.	495		
		Wt.=52.2 lbs.	2.	929		
		c.=12587	3.	1317		
		I.V.=1215	4.	1663		
			5.	1976		
			6.	2266		
			7.	2530		
			8.	2777		
			9.	3007		
			10.	3221		
			12.	3611		
			15.	4118		
	4.5	Shell.	1.	497	1.3	
		Wt.=45.2 lbs.	2.	925	2.5	
		c.=10899	3.	1300	3.7	
		I.V.=1224	4.	1629	4.9	
			5.	1927	6.	
			6.	2200	7.	
			7.	2449	8.1	
			8.	2682	9.1	
			9.	2897	10.1	
			10.	3092	11.	
			12.	3452	12.8	
			15.	3918	15.3	
32 Pdr. Gun Rifled.	6.	Shot.	1.	513		
		Wt.=60.4 lbs.	2.	961		
		c.=12286.5	3.	1353		
		I.V.=1241	4.	1705		
			5.	2024		
			6.	2314		
			7.	2582		
			8.	2830		
			9.	3061		
			10.	3276		
			12.	3665		
			15.	4170		
	6.	Shell.	1.	482	1.28	
		Wt.=55.5 lbs.	2.	905	2.52	
		c.=11289.7	3.	1274	3.7	
		I.V.=1205.	4.	1603	4.84	
			5.	1902	5.94	
			6.	2175	7.	
			7.	2424	8.03	
			8.	2657	9.03	
			9.	2872	10.	
			10.	3067	10.95	
			12.	3437	12.75	

Kind of Ordnance.	Charge.	Projectile.	Elevation.	Range.	Time.	Remarks.
			′ °	Yards.	Seconds	
30 Pdr. Parrot Gun.	Lbs.	SHOT.	.30	289	.09	
	3.25		1.	557	1.37	
		Wt.=29.2 lbs.	1. 30	808	2.03	
		c.=13698.	2.	1043	2.69	
		I.V.=1293.	2. 30	1264	3.34	
			3.	1471	3.97	
			3. 30	1667	4.59	
			4.	1854	5.21	
			5.	2201	6.41	
			6.	2519	7.58	
			7.	2811	8.71	
			8.	3082	9.82	
			9.	3334	10.9	
			10.	3568	11.96	
			.30	287	.69	
	3.25	CASE.	1.	554	1.37	
			1. 30	804	2.03	
		Wt.=29.2 lbs.	2.	1039	2.69	
		c.=13792	2. 30	1259	3.33	
			3.	1465	3.96	
		I.V.=1289	3. 30	1662	4.58	
			4.	1849	5.2	
			5.	2196	6.4	
			6.	2513	7.56	
			7.	2805	8.69	
			8.	3076	9.8	
			9.	3328	10.87	
			10.	3562	11.93	
			.30	303	.7	
	3.25	SHELL.	1.	584	1.4	
			1. 30	843	2.1	
		Wt.=27.5	2.	1085	2.8	
		c.=12989	2. 30	1310	3.5	
		I. V.=1328	3.	1520	4.1	
			3. 30	1719	4.7	
			4.	1907	5.3	
			5.	2254	6.5	
			6.	2573	7.7	
			7.	2862	8.8	
			8.	3130	9.9	
			9.	3379	11.	
			10.	3610	12.1	
			.30	313		
	3.25	SHELL.	1.	603		
			1. 30	867		
		Wt.=26.5	2.	1111		
		c.=12517	2. 30	1338		
			3.	1551		
		I. V.=1353	3. 30	1752		
			4.	1941		
			5.	2293		
			6.	2606		
			7.	2898		
			8.	3162		
			9.	3408		
			10.	3636		

Kind of Ordnance.	Charge.	Projectile.	Elevation.	Range.	Time.	Remarks.
			° ′	Yards.	Seconds	
42 Pdr. Gun Rifled.	Lbs. 8.	JAMES SHOT.	.30	305		
			1.	587		
		Wt.=84 lbs.	1.30	851		
			2.	1098		
		c.=14283.5	2.30	1330		
		I.V.=1328	3.	1549		
			3.30	1756		
			4.	1953		
			5.	2322		
			6.	2660		
			7.	2973		
			8.	3266		
			9.	3543		
			10.	3803		
			.30	343	.76	
	8.	JAMES SHELL.	1.	650	1.49	
			1.30	930	2.21	
		Wt.=68 lbs.	2.	1187	2.91	
		c.=11563	2.30	1422	3.59	
			3.	1642	4.26	
		I.V.=1420	3.30	1848	4.92	
			4.	2041	5.56	
			5.	2396	6.81	
			6.	2718	8.02	
			7.	3013	9.21	
			8	3283	10.35	
			9	3535	11.46	
			10.	3771	12.57	

Kind of Ordnance.	Charge.	Projectile.	Elevation.	Range.	Time.	Remarks.
			° ′	Yards.	Seconds	
100 Pdr. Parrot Gun.	Lbs. 10.	SHOT.	1.	515		
			2.	986		
		Wt.=99.5 lbs.	3.	1420		
		c.=20240	4.	1824		
		I.V.=1222	5.	2200		
			6.	2554		
			7.	2888		
			8.	3205		
			9.	3506		
			10.	3792		
			15.	5062		
			20.	6127		
			25.	7068		
			30.	7906		
			1.	486	1.27	
	10.	SHELL.	2.	937	2.52	
			3.	1356	3.74	
		Wt.=101 lbs.	4.	1745	4.94	
		c.=20545	5.	2110	6.11	
		I V.=1188	6.	2454	7.26	
			7.	2780	8.4	
			8.	3089	9.52	
			9.	3384	10.63	
			10.	3664	11.72	
			15.	4932	17.28	
			20.	5918	22.1	
			25.	6899	27.55	
			30.	7740	33.03	
			1.	599		
	10.	SHOT.	2.	1126		
			3.	1594		
		Wt.=80 lbs.	4.	2015		
		c.=16273	5.	2401		
		I. V.=1335	6.	2752		
			7.	3077		
			8.	3378		
			9.	3660		
			10.	3924		
			15.	5022		
			20.	5830		
			25.	6416		
			30.	6831		
			.30	327	.74	
	8.	DYER SHELL.	1.	627	1.46	
			1. 30	900	2.17	
		Wt.=61.5	2.	1153	2.86	
		c.=12510.2	2. 30	1388	3.53	
		I.V.=1384	3.	1606	4.19	
			3. 30	1810	4.83	
			4.	2003	5.47	
			5.	2359	6.71	
			6.	2682	7.91	
			7.	2976	9.07	
			8.	3245	10.20	
			9.	3494	11.33	
			10.	3726	12.42	

The following ranges are determined by practice, and lie in the plane of the platform on which the bed stands.

Kind of Ordnance.	Charge.	Projectile.	Elevation.	Range.	Time.			Remarks.
			° '	Yards.	Seconds			
8 Inch Siege	lbs. oz.	SHELL.	45.	360	8.0			Ranges obtain-
Mortar.	0. 8	Wt.=46 lbs.	"	703	12.5			ed from experi-
(Model 1861.)	0. 12		"	1082	15.0			ments made near
	1. 0		"	1412	17.0			Petersburgh, Va.,
	1. 4		"	1741	18.5			by the 1st Conn.
	1. 8		"	1985	20.0			Artillery, in Sept.
	1. 12		"	2225	21.0			'64.
	2. 0							
				Model 1861.		Old Model.		
						Range.	Time.	
10 Inch Siege	0. 8	SHELL.	"	189	6.4	123		
Mortar.	1. 0	Wt.=90 lbs.	"	245	10.4	276	6.9	
	1. 8		"	854	14.2	522		
	2. 0		"	1122	17.2	774	11.5	
	2. 8		"	1410	18.4	1144	14.6	
	3. 0		"	1676	19.8	1466	17.5	
	3. 5		"	1848	20.9	1811		
	4. 0		"	2064	21.9	2028		
8 Inch Siege	0. 4	SHELL.	"	314		300		
Howitzer.	0. 8	Wt.=46 lbs.	"	620		553		
As a Mortar.	0. 12		"	1082				
	1. 0		"	1440		1332		
	1. 4		"	1925		1695		

RANGES WITH SEA-COAST 13 INCH MORTARS, 20° ELEVATION.

CHARGE.	Mean time of Flight.	Least Range.	Greatest Range.	Mean Range.
Lbs.	Seconds.	Yards.	Yards.	Yards.
4	8	840	877	869
6	9.5	1209	1317	1263
8	11.66	1653	1840	1744
10	12.50	2010	2128	2066
12	14.25	2369	2688	2528
14	15.25	2664	2780	2722

RANGES WITH 13 INCH MORTARS, AT 45° ELEVATION.

13 INCH MORTAR.	Powder.	Shell.	Elevation.	Range.
	Lbs.	Lbs.		Yards.
	20	200	45°	4325

RANGES WITH 13 INCH MORTARS AT 45° ELEVATION.

Charge.		Flight.	Fuze.		Range.
Lbs.	oz.	Seconds.	Inches.	10ths.	Yards.
7		21.4	4	2⅔	2190
7	8	22.4	4	4	2346
8		23.2	4	6	2480
8	8	23.8	4	7½	2600
9		24.4	4	8¾	2734
9	8	24.9	4	9¾	2853
10		25.4	5	1	2958
10	8	25.9	5	1¾	3026
11		26.3	5	2½	3150
11	8	26.7	5	3½	3246
12		27.0	5	4	3327
12	8	27.4	5	4¾	3404
13		27.7	5	5½	3470
13	8	28.0	5	6	3552
14		28.3	5	6½	3617
14	8	28.5	5	7	3681
15		29.0	5	8	3739
15	8	29.1	5	8¼	3797
16		29.2	5	8½	3849
16	8	29.4	5	8¾	3901
17		29.6	5	9	3949
17	8	29.8	5	9½	3997
18		29.8	5	9¾	4040
18	8	30.0	6	0	4085
19		30.2	6	0¼	4123
19	8	30.3	6	0½	4160
20		30.5	6	1	4200

Table showing the angles of depression of an object for different distances and heights of a gun above the water.

DISTANCE.	HEIGHT.							
	1 Foot.	2 Feet.	4 Feet.	8 Feet.	16 Feet.	32 Feet.	64 Feet.	96 Feet.
Yards.	° ′	° ′	° ′	° ′	° ′	° ′	° ′	° ′
200	5.7	11.5	23.	45.8	1 31.6	3 3.2	6 5.5	9 5.5
250	4.6	9.2	18.3	36.7	1 13.3	2 26.6	4 53.	7 17.6
300	3.8	7.6	15.3	30.6	1 1.1	2 2.2	4 4.	6 5.4
350	3.3	6.5	13.1	26.2	52.4	1 44.7	3 29.4	5 13.5
400	2.9	5.7	11.4	22.9	45.8	1 31.6	3 3.2	4 34.5
450	2.5	5.1	10.2	20.4	40.7	1 21.5	2 42.8	4 4.
500	2.3	4.6	9.1	18.3	36.7	1 13.3	2 26.6	3 39.7
550	2.1	4.2	8.3	16.7	33.3	1 6.6	2 13.3	3 19.8
600	1.9	3.8	7.6	15.3	30.6	1 1.1	2 2.2	3 3.3
650	1.7	3.5	7.	14.1	28.2	56.4	1 52.8	2 49.1
700	1.6	3.3	6.5	13.1	26.2	52.4	1 44.7	2 37.
750	1.5	3.	6.1	12.2	24.4	48.9	1 37.7	2 26.6
800	1.4	2.8	5.7	11.4	22.9	45.8	1 31.7	2 17.4
850	1.3	2.7	5.4	10.8	21.6	43.1	1 26.3	2 9.4
900	1.3	2.5	5.1	10.2	20.4	40.7	1 21.5	2 2.2
950	1.2	2.4	4.8	9.6	19.3	38.6	1 17.2	1 55.8
1000	1.1	2.3	4.6	9.2	18.3	36.7	1 13.3	1 50.
1100	1.	2.1	4.2	8.3	16.7	33.3	1 6.7	1 40.
1200	.9	1.9	3.8	7.6	15.3	30.6	1 1.1	1 31.7
1300	.9	1.8	3.5	7.	14.1	28.2	56.4	1 24.6
1400	.8	1.6	3.3	6.5	13.1	26.2	52.4	1 18.6
1500	.8	1.5	3.	6.1	12.2	24.4	48.9	1 13.3
1600	.7	1.4	2.9	5.7	11.4	22.9	45.8	1 8.7
1700	.7	1.3	2.7	5.4	10.8	21.6	43.1	1 4.7
1800	.6	1.3	2.5	5.1	10.2	20.4	40.7	1 1.1
1900	.6	1.2	2.4	4.8	9.6	19.3	38.6	57.9
2000	.6	1.2	2.3	4.6	9.2	18.3	36.7	55.
2100	.5	1.1	2.2	4.3	8.7	17.5	34.9	52.4
2200	.5	1.	2.1	4.2	8.4	16.7	33.3	50.
2300	.5	1.	2.	4.	7.9	15.9	31.9	47.8
2400	.5	1.	1.9	3.8	7.6	15.3	30.6	45.8
2500	.4	.9	1.8	3.6	7.3	14.7	29.3	44.
3000	.4	.8	1.5	3.	6.1	12.2	24.4	36.7
3500	.3	.7	1.3	2.6	5.2	10.4	21.	31.4
4000	.3	.6	1.1	2.3	4.6	9.2	18.3	27.5
4500	.3	.5	1.	2.	4.1	8.1	16.3	24.4
5000	.2	.5	.9	1.8	3.7	7.3	14.7	22.

When the height of the piece above the water or horizontal plane is known, the angle of depression for different distances can be found by using the foregoing table, Find the angle for any height not given in the table, as follows: divide the given height into parts, which are found in the table, using the largest number possible; and add the angles corresponding to those parts, for the required distance. Example: required the angle for distance 1000 yards and height 23 feet. 23 feet gives the parts 16, 4, 2, & 1.; the sum of the angles for these heights is 18.3′+4.6′+2.3′+1.1′=26.3′.

In using the quadrant, or giving the elevation from the horizontal plane of the piece, if the picce is higher than the object fired at, the angle of depression should be deducted from the "elevation" in the table of ranges. Example: the 10 inch Columbiad at 6° gives about 2000 yards range; if the piece is 32 ft. above the object the angle of depression is 18.3′ and the quadrant should be set at 51° 41.7′.

Remaining velocities for Sea Coast Projectiles, calculated by Capt. Stockton, Ord. Dept. by formula on page 417.

Calibre.	Initial Velocity.	Distance for which remaining velocity is calculated.	Remaining Velocity.	Loss of Velocity.
	Feet.	Feet.	Feet.	Feet.
15 inch Shot.	1200	1500	1066	134
	"	3000	961	239
	"	6000	773	427
	1500	1500	1315	185
	"	3000	1165	335
	"	6000	929	571
13 inch Shot.	1200	1500	1045	155
	"	3000	924	276
	"	6000	695	505
	1500	1500	1293	207
	"	3000	1118	382
	"	6000	865	635
10 inch Shot.	1200	1500	1007	193
	"	3000	854	346
	"	6000	632	568
	1500	1500	1251	249
	"	3000	1032	468
	"	6000	741	753
8 inch Shot.	1200	1500	964	236
	"	3000	792	408
	"	6000	550	650
	1500	1500	1177	323
	"	3000	948	552
	"	6000	643	857
12 inch Rifle Shot. (*c.*=34920.)	1200	1500	1154	46.
	"	3000	1110	90.
	"	6000	1028	172.
10 inch Rifle Shot. (*c.*=21056.)	"	1500	1125	75
	"	3000	1056	144
	"	6000	939	261
8 inch Rifle Shot. (*c.*=19636.)	"	1500	1115	85
	"	3000	1048	152
	"	6000	927	273

RESULTS OF EXPERIMENTAL FIRING with 15-inch Columbiad, from barbette platform No. 120 of Fort Monroe, Va., made between Oct. 24th, 1865, and Nov. 10th, 1865, to determine the penetration of a Solid Shot and Shell into a Sand target (well rammed) 30 feet thick, 15 feet high, and fifteen feet front, located 400 yards from the Gun.

No. of Discharge.	Charge.	Depresssion.	Recoil.	Reference of Lodgment.	Penetration.	Remarks.
	Lbs.	° ′	Ft. In.	Ft. In.	Ft. In.	
1	50	0 30	6 5	2 0	15 8	Solid Shot.
2	"	0 15	5 0	2 6	10 6	" " Struck ground before entering target.
3	"	0 10	5 4	8 0	20 0	" "
4	"	0 15	7 4	8 0	16 0	" "
5	"	0 15		6 0	17 0	" " Carriage struck c-heurter of Chassis and rebounded 3 inches.
1	"	0 15	5 10	10 0	14 3	Shell.
2	"	0 30	5 8½	1 6	11 0	"
3	"	0 30	6 2	7 0	10 0	"
4	"	0 30	7 8	7 6	9 8	"
5	"	0 30	3 10	10 0	15 0	" Chassis rails sanded.
1	45	0 15	7 4	8 0	18 0	Solid Shot.
2	"	0 15		10 0	19 0	" " Carriage struck c-heurter of Chassis and rebounded 1½ inch. Carriage struck c-heurter of Chassis and rebounded 1½ inches.
3	"	0 8	6 7	10 0	18 0	" "
4	"	0 15		7 0	15 0	" "
5	"	0 15	6 10	7 0	16 0	" "
6	"	0 15	6 6	6 0	12 0	" "
1	"	0 30	5 6	6 0	9 0	Shell.
2	"	0 30	4 4	3 0	8 0	"
3	"	0 30	5 1	7 0	9 0	"
4	"	0 30	4 6	6 0	7 0	"
5	"	0 30	4 8	7 0	12 0	"
6	"	0 30	4 6	3 0	7 0	"
1	40	0 23	5 0	6 0	11 0	Solid Shot. Struck ground before entering target.
2	"	0 8	6 3	4 0	12 0	" "
3	"	0 8	6 2½	4 0	12 6	" "
4	"	0 8	6 11	6 0	17 0	" "
5	"	0 8	5 11	5 0	7 0	" " Same as above.
6	"	0 8	5 10	5 0	12 0	" "
1	"	0 30	2 9	3 0	6 0	Shell. Chassis rails sanded. Struck ground before entering target.
2	"	0 23	3 4	7 0	13 0	"
3	"	0 30	3 0	5 0	9 6	"
4	"	0 23	3 6	4 0	8 0	"
5	"	0 23	3 6	4 0	6 6	"
6	"	0 23	3 7	7 0	7 0	"

Reference of base of Target=0 ft. 0 inch. Reference of axis of trunnions of Gun =30 ft. 9 inches. Average weight of Solid Shot=434 lbs., ditto of Shell=320 lbs.

Results of Experiments made at West Point, by Captain Benet, Ordnance Corps, upon the Penetration and Explosive Effects of the Projectiles of the Parrot Gun, into a mixture of sand and clay, free from gravel, and compactly rammed. The sides were confined by a strong vertical boarding. The front and rear had the natural slope. Distance, 24 yards.

Gun.	Projectile, lbs.	Powder, lbs.	Penetrat'n Feet.	Crater. Diam. Ft.	Crater. Depth. Ft.	
20 Pdr.	——	3	9½	——	——	Greatest Penetrat'n
20 Pdr.	——	2	8½	——	——	do.
20 Pdr.	——	2	6	4 & 8	1 & 2	Per'n Shell, Av. Pen
30 Pdr.	25¾	3¼ to 4	4½ to 13½	10 to 12	2½ to 3	Percussion Shells.
30 Pdr.	29½	3¼ to 4	10	——	——	Average Penetration
100 Pdr.	80 to 96½	10	14	——	——	Hazwell's No. 5 & 7 p
100 Pdr.	81	12	15½	——	——	do. No. 7.
100 Pdr.	81	12	18½	——	——	Mammouth powder.
100 Pdr.	32	12	10½	——	——	Hazwell's No. 7 do.
100 Pdr.	100	10	12½ to 14½	12	2	Hazwell's No. 5 do.
200 Pdr.	150	18	12½ to 18½	——	——	
200 Pdr.	146	16 to 18	8½ to 10½	12 to 17	1½ to 5	

Experiments were also been made for penetration with the same guns into a but of well-rammed gravel, at a distance of 1,000 yards, with the following results, viz.

For the solid projectiles of the 20-pounder and 30-pounder guns, the average penetration was three feet.

For the solid projectiles of the 100-pounder gun, weighing eighty pounds, the average penetration was seven feet.

For the solid projectiles of the 200-pounder gun, the average penetration was eight feet.

RESULTS OF EXPERIMENTS ON THE PENETRATION OF RIFLE BULLETS INTO VARIOUS SUBSTANCES.

See Professional Papers of the Corps of Royal Engineers, Vol. VII., New Series.

The arms used were the Enfield Rifle and the Lancaster Rifled Carbine.

RANGE		INCHES.
20 Yards.	With Lancaster Rifle, the mean penetration into a parapet of light, sandy soil, rammed lightly, was..	15.
	With Enfield Rifle, circumstances the same.	15.
	Lancaster Rifle, mean penetration into solid oak	2.8
	Enfield " " " "	2.75
	Lancaster Rifle, mean penetration into two thicknesses of three inch oak	3.
	Enfield Rifle, mean penetration into two thicknesses of three inch oak........................	3.58
	Enfield Rifle, penetration into solid pine, with the grain	9.83
	Lancaster Rifle, penetration into sand bags, filled with light sandy earth, and formed as a parapet	All stopped
	Enfield Rifle, one bullet passed through a stretcher.	
	Lancaster Rifle, penetration into a sap roller four feet exterior diameter, 2.5 feet interior diameter, nine inches thickness, fascine stuffing	INCHES. 9.
	Enfield Rifle, same penetration as Lancaster Rifle.	
	Lancaster Rifle, penetration into fascines, nine inches diameter, of green wood, backed by earth. Through the fascines, and into the earth...............	7.
	Enfield Rifle, same circumstances	8.
	With gabions of both brushwood and sheet iron, filled fresh with light sandy earth, the bullets of neither arm passed through.	
50 Yards.	It required a rope mantelet, like those used at the siege of Sebastapol, consisting of three vertical and two horizontal layers of four and a half inch rope, weighing twenty-seven and a half pounds to the square foot, to give full protection against the bullets of each arm.	

COST OF GUNS.—(HOLLEY'S ORD. AND ARMOR.)

Name of Gun.	Material.	Bore.	Weight	Cost per pound.	Total Cost.
		Inches.	Lbs.	Cts.	$ Cts.
13.3 inch Armstrong.	Wrought iron coils in hoops...............	13.3	51351	37.	19000.00
10½ inch do.	Do.	10.5	26880	33.6	9000.00
110 pdr. do.	Do.	7.	9184	23.9	2195.75
Horsfall gun........	Wrought iron forged solid	13.	53846	23.2	12000.00
Alfred gun..........	Wrought iron forged hollow	10.	24094	20.7	5000.00
Krupp's 15 inch gun.	Cast Steel forged solid...	15.	33600	87.5	29400.00
Krupp's 9 inch gun..	Do.	9.	18000	56.5	10125.00
Bessemer forging.....	Do.	7 to 8	11200	13.0	1466.00
Blakely 12 inch gun..	Cast Steel hooped with steel................	12.	40000	87.5	35000.00
Blakely 11 inch gun..	Do.	11.	35000	78.5	27500.00
Blakely 10 inch gun..	Do.	10.	30000	58.3	17500.00
Blakely 120 pdr gun.	Do.	7.	9600	62.5	6000.00
Whitworth 120 pdr. gun	Do.	7.	13440	37.2	5000.00
Parrott 100 pdr. gun.	Cast Iron hooped with wrought iron........	6.4	9700	12.4	1300.00
Parrott 8-inch gun...	Do.	8.	16300	14.1	2200.00
Parrott 10 inch gun...	Do.	10.	26500	17.0	4700.00
Rodman 15 inch....	Cast Iron—Cast hollow.	15.	49100	13.2	6500.00
Rodman 10 inch....	Do.	10.	15059	11.0	1665.00
Rodman 8 inch.....	Do.	8.	8465	11.8	1000.00
Ames gun..........	Wrought Iron, built up.	7.	19400	100.0	19400.00
3 inch field gun.....	Wrought Iron, solid....	3.	820	55.0	450.00

NOTE.—The cost of bronze field guns is about fifteen cents per pound more than the market price of copper.

APPENDIX.

APPENDIX.

It is proposed in this Appendix to give a short description of some of the most noted modern cannon and projectiles, and in doing so the author has freely consulted the valuable work of Mr. Holley, on Ordnance and Armor.

1. **Armstrong gun.** Armstrong guns are of two kinds,—breech-loaders and muzzle-loaders. In consequence of not possessing the requisite strength and endurance, the breech-loading apparatus of the Armstrong guns is now only applied to the smaller calibres.

Construction of breech-loaders. The body of the piece is made by welding together several wrought-iron tubes at their ends; each tube is from two to three feet long, and is formed by twisting a square bar of iron around a mandrel and welding the edges together. Thus far the piece resembles the barrel of a fowling-piece. To strengthen the part in rear of the trunnions, it is enveloped with two additional thicknesses, or tubes. The outer tube, like the inner one, consists of spiral coils; but the intermediate tube is formed of an iron slab, bent into a circular shape and welded at the edges. The reason for this distinction is, that the intermediate layer has chiefly to sustain the pressure on the bottom of the bore, while the other layers are formed to resist the tangential strain.

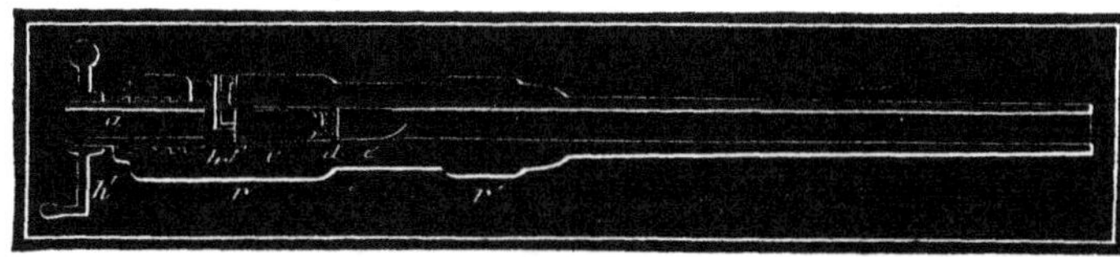

Fig. 155.

Breech. The breech is closed with a vent-piece (*b*), fig. 155, which is slipped with the hand into a slot cut in the breech of the piece, and held in its place by a breech-screw (*a*), which supports it from behind. This *screw* is made in the form of a tube, so that its hollow forms a part of the bore prolonged, when the vent-piece is withdrawn. The object of this hollow is to allow the charge to be passed into the chamber.

Bore. The bore of the field-gun, which is represented in the drawing, is three inches in diameter, and is rifled with thirty-four narrow grooves. *Twist,* one turn in 9 feet. The diameter of the chamber is one-eighth of an inch larger than that of the bore.

Vent. The vent is formed in the breech-piece in order that when it becomes enlarged it may be easily replaced. For this purpose, spare vent-pieces are carried with each battery. The diameter of the vent is one-half an inch, and is primed by filling it with a small paper cartridge of fine powder, which is ignited by an ordinary friction-tube.

Muzzle-loading. The most approved method of making the large muzzle-loading Armstrong guns, as practised at the Royal Gun Factory at Woolwich Arsenal in 1864, is shown in fig. 156.*

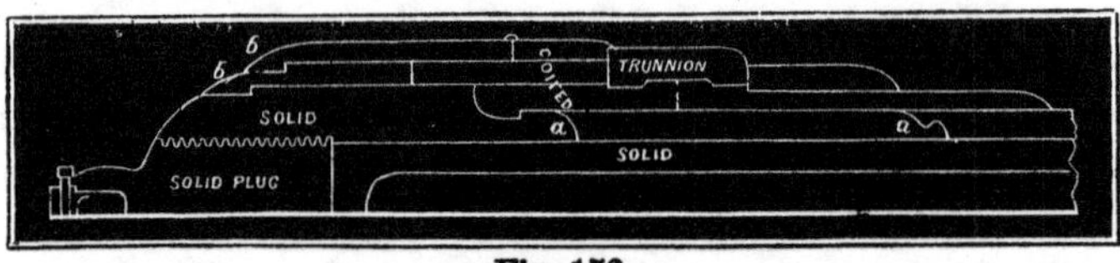

Fig. 156.

The *barrel,* or part surrounding the bore, is made solid of steel tempered in oil, by which its brittleness is not only diminished, but its tenacity is increased. That part of the barrel at and in rear of the trunnions is enveloped by three layers of wrought-iron tubes. These tubes are not joined at their ends by welding, as in the breech-loading guns of this system, but are hooked on to each other by a system of shoulders and recesses, as shown at *a a* in the accom-

* **Fraser gun.** This mode of manufacture has been somewhat changed at the suggestion of Mr. Fraser, Manager and Assistant Inspector of Machinery at the Woolwich Arsenal, and the guns now made at that establishment are known as Fraser guns, but in fact are nothing more than modified Armstrong guns. The modifications consist in reducing the number of coils; using a cheaper iron for the outer coils; shrinking on of the outer coils and trunnion piece together; and an improved arrangement of shoulders and recesses to prevent separation of the parts. It is understood that while these guns have nearly, if not quite, the strength of the Armstrong guns proper, their cost has been reduced from £100 per ton to £40 per ton for the coiled inner tube, and to £55 per ton for the solid steel inner tube,—a cost per pound not much exceeding that of cast-iron guns in this country. Three sizes of these guns are now being made, viz.: 7-inch, or 115-pdr.; 8-inch, or 180-pdr., and 9-inch, or 250-pdr.; and to these will shortly be added a 10-inch, or 350-pdr., and a 12-inch, or 500-pdr. It is stated that guns of this system have undergone very severe tests, and that they are now being made in large numbers.

panying figure. When one end of a tube is expanded by heat, the shoulders are slipped over each other, and the contraction which afterwards takes place in cooling, causes them to fit into the corresponding recesses. There are also projections fitting into corresponding recesses, as at *b b*, which serve to prevent the tubes from slipping within each other. The tube which immediately surrounds the barrel opposite to the seat of the charge is called the *breech-piece.* It is not made of spiral bars as the other tubes are, but it is formed with its fibres and welds running longitudinally so as to resist the recoil of the barrel against the head of the *breech-plug* which is screwed into the breech-piece. The object of making the breech-plug separately is to insure a better quality of iron than would usually be at the bottom of the breech-piece if made solid.

Mr. Anderson recommends that the tubes next to the barrel have a tension of about 6 tons to the inch, in which state they are in a condition to bear an enlargement of .004 of their diameter without going beyond their limit of elasticity; a little more tension than this should be given to the next layer of tubes, while the outer layer should have a tension of 8 tons per inch.

The Armstrong gun is formed very strong to resist tangential strains; and in consequence of the great ductility of the wrought-iron used for the tubes, this gun is not liable to burst without previously showing considerable enlargement. It is stated that of 3,000 guns built on this principle, not one has burst explosively.

The number of grooves of the muzzle-loaders varies from 3 to 10, and the twist varies from one turn in 30 to one turn in 38 calibres.

Dimensions and Weight of the Armstrong Guns of the latest Elswich Patterns.

Name of Gun.	Length of Gun.	Length of Bore.	Diam. of Bore.	Diam. of Powder Chamb.	Diam. at Muzzle.	Weight.	Prepon-derance.
	Inch.	Inch.	Inch.	Inch.	Inch.	Lbs.	Lbs.
12-pdr. Breech-Loader.......	83.	73.5	3.	9.75	5.75	918	
12-pdr. Muzzle-Loader.......	76.	67.75	3.	10.9	5.6	996	60
25-pdr. Breech-Loader.......	96.	93.	3.75	12.75	6.	1,882	123
40-pdr. Breech-Loader.......	121.	106.5	4.75	16.4	7.75	3,696	392
70-pdr. Muzzle-Loader.......	126.5	103.	6.4	25.3	12.4	9,016	548
150-pdr. Muzzle-Loader.......	129.75	102.25	8.5	31.	15.4	14,896	504
300-pdr. Muzzle-Loader.......	156.	124.	10.	38.3	19.	26,880	
600-pdr. Muzzle-Loader.......	183.	145.25	13.3	51.5	21.5	51,296	952

Armstrong guns are proved by firing them twice with service charge of powder and shot, and three times with service shot and a charge of powder equal to one-sixth of the weight of the service shot. The service charges do not vary much from one-seventh to one-eighth of the weight of the projectile.

Rifling. The muzzle-loading Armstrong guns proper are rifled to "center" the projectile by what is known as the *shunt* system.* In this system the grooves are considerably wider than the buttons on the projectile, except near the bottom of the bore, where they taper down to the ordinary width. That part, or side, of the groove passed over by the projectile in going down the bore, is made deep enough to admit the buttons freely. The other side is so shallow as to press against the ends of the buttons as the projectile comes out, and thereby force its center towards the center of the bore. The changing, or "shunting" from one side of the grooves to the other, takes place at the seat of the charge, and follows from the narrowing of the grooves, and from an inclined plane which leads from the deep to the shallow side of the groove.

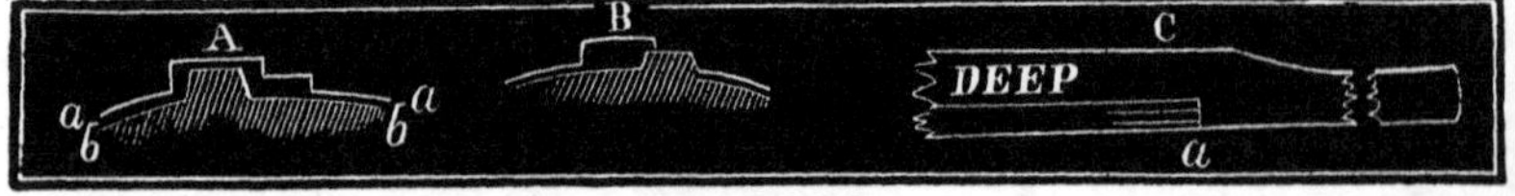

Fig. 156 (*a*).

Diagram *A* shows the position of the button in the deep part of the groove when entering the bore. *C* shows the manner of shunting it to the shallow side ; *a* is the point at which the inclined plane commences. *B* shows the button on the driving or shallow side.

2. **Whitworth gun.** The Whitworth guns are made of a substance called "homogeneous iron," a species of low steel, which is said to be made by melting short bars of Swedish iron and adding a small quantity of carbonaceous matter, after which it is cast into round ingots.

The smaller Whitworth guns are forged solid; the larger are built up with coils or hoops, after the manner shown in fig. 157.

In the Armstrong guns the coils are shrunk on by the aid of heat; in the Whitworth guns the hoops are forced on by hydraulic

* The Fraser, or "Woolwich guns," it is said, are rifled on the French button system as modified by Captain Palliser. The twist of the grooves is increasing with a view to diminish the strain on the breech of the gun, and to increase the accuracy of fire.

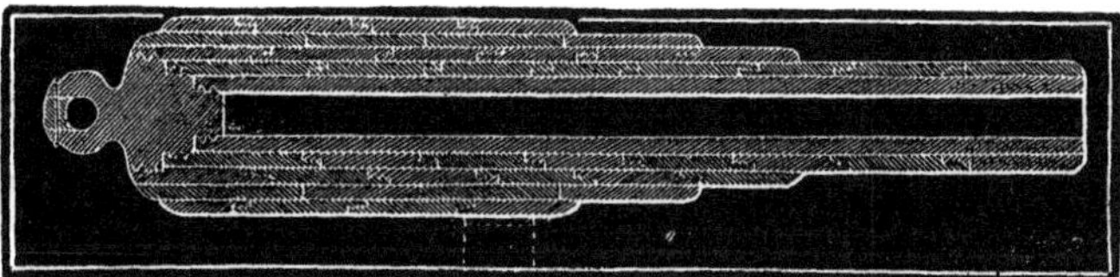

Fig. 157.

pressure, and for this purpose they are made with a slight taper and with the design to secure initial tension. The ends of the hoops are joined by screw-threads. The hoops are first cast hollow and then hammered out over a steel mandrel, or rolled out in a machine like that used for forming wheel-tire. Before receiving their final finish, they are subject to an annealing process for some three or four weeks, which makes the metal very ductile, but, at the same time, slightly impairs its tenacity. The breech-pin is made with offsets in such a way as to screw into the end of the barrel and the next two surrounding hoops.

The breech in the case of the large guns is hooped with a harder and higher steel than that used for the barrel. The 70-pdr. gun has one hoop; the 120-pdr. proposed by Mr. Whitworth was to have four tiers of hoops.

The cross-section of the bore of the Whitworth guns is a hexagon with rounded corners. The twist is very rapid, and the projectiles are made very long.

Dimensions, &c., of Whitworth Guns.

	Diam. of Bore across flats.	Length.	Weight.	Twist. Inches for one turn.	Weight of Powder.	Weight of Projectile.
	Inches.	Inches.	Lbs.		Lbs.	Lbs.
120-pounder........	6.4	144	16,000	130	27.	151
70-pounder........	5.	118	8,582	100	13.	81
12-pounder........	2.75	104	1,092	55	1.75	12

The proof consists in firing once with a service charge, with the powder increased one-fourth—and once with a service charge of powder and a six-calibre projectile.

3. **Blakely gun.** The most approved pattern of the Blakely gun combines in its construction the principles of "initial tension" and "varying elasticity," the object of which is to bring the strength of all the metal of the piece into simultaneous play, to resist explosion. (See page 150.)

35

Fig. 158 represents a 9-inch Blakely gun of this kind.

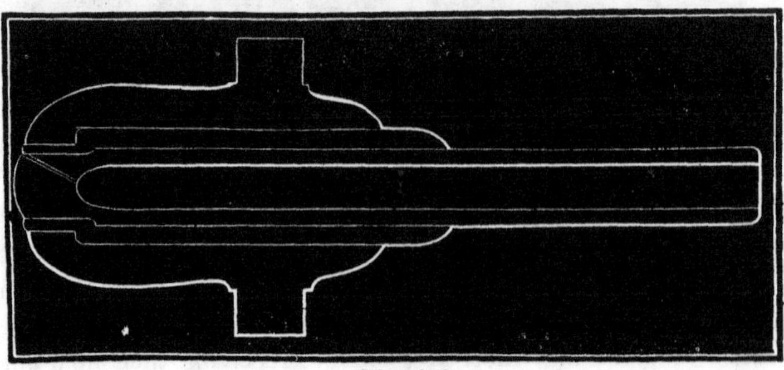

Fig. 158.

The inner tube, or barrel, is made of low steel, having considerable but not quite enough elasticity. The next tube is made of high steel with less elasticity, and is shrunk on to the barrel with just sufficient tension to compensate for the insufficient difference of elasticity between the two tubes. The outer cast-iron jacket, to which the trunnions are attached, is the least elastic of all, and is put on with only the shrinkage attained by warming it over a fire. The steel tubes are cast hollow and hammered over steel mandrels, under steam hammers; by this process they are elongated about 130 per cent., at the same time the tenacity of the metal is increased. All the steel parts are annealed.

Captain Blakely uses other combinations of these metals, the simplest of which is a cast-iron gun with hoops of steel surrounding the reinforce. He objects to the use of wrought-iron on account of its tenacity to stretch permanently. Blakely guns are rifled with one-sided grooves, and are fired with expanding projectiles.

Dimensions, &c., of Blakely All Steel Guns.

Gun.	Weight.	Diam. of Bore.	Length of Bore.	Number of Grooves.	Twist.	Weight of Projectile.	Weight of Powder.
	Lbs.	Inches.	Inches.		1 turn in Calibres.	Lbs.	Lbs.
100-pdr...	8,000	6.4	96	8	48	100	10
120-pdr..........	9,000	7.	100	8	48	120	12
200-pdr..........	17,000	8.	144 to 156	12	48	200	20
250-pdr..........	24,000	9.	do.	12	48	250	25
350-pdr..........	30,000	10.	do.	15	48	350	35
350-pdr..........	35,000	11.	do.	12	36	550	55
700-pdr..........	40,000	12.	do.	12	36	700	70

4. **Palliser gun.** Captain Palliser, of the British service, describes his manner of making a gun to consist in introducing into a cast-iron gun a barrel or hollow cylinder of coiled wrought-iron, of such thickness in proportion to its calibre, that the residual strain borne by the tube shall have a relation to the strain it transmits to the surrounding cast-iron, which shall be most suitably proportioned to their respective elasticities.

The precise proportions will depend on various circumstances, viz.: the excessive expansion of wrought-iron due to heat, also the greater range between the limits of elasticity and rupture of this metal, and that the cast-iron will have to do nearly all the longitudinal work. By varying the thickness of the tube, the transmitted strains can be regulated with the greatest nicety.

The mechanical method, says Captain Palliser, by which I propose to insert the tube, is by making it very slightly taper, and placing it in the gun, the bore of which is tapered correspondingly; as soon as the tube comes in contact with the gun throughout its length, a screw washer around the muzzle will screw it home into its place. See fig. 159.

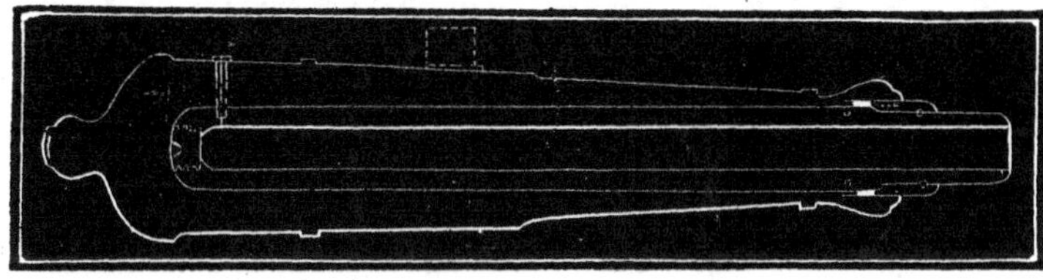

Fig. 159.

Since the amount of taper, as well as the distance the tube is driven by the washer, is known, and the increment or decrement in cast or wrought-iron due to any pressure is also known, we shall in this manner be able to measure most accurately the strain placed on the cast-iron outer gun.

In the larger guns Captain Palliser proposes to use two or more concentric tubes. In the very largest guns he proposes three tubes, the inner one to be of the softest and most ductile wrought-iron; the next may be of a stronger and harsher nature; the third of steel for some distance in front of the chamber.

Old smooth-bored guns have been reamed out and strengthened on Captain Palliser's plan, and have shown remarkable strength. The guns tested were chiefly 68-pounders. A 9-inch gun has lately been tested with severe charges, some as high as 45 pounds powder and 250 pounds projectile.

5. **Parsons gun.** The principle upon which Mr. Parsons makes his gun would seem to be similar to that of Captain Palliser's, *i. e.*, by varying elasticity. As applied to strengthening a 68-pounder cast-iron gun, his method consists of boring into the breech of the gun, coincident with its axis, reaming out the bore into a slightly conical shape as far as the front of the trunnions, and then inserting into this space a reinforced wrought-iron tube, which is secured in its place by a breech-plug. The exterior of this compound tube is turned to fit the conical space easily, its length being cut so that it will be compressed longitudinally by screwing up the breech-plug, thus communicating to the outer cast-iron portion the entire longitudinal strain of the powder.

Mr. Parsons bases his method on the fact, as stated by him, that wrought-iron may be stretched three times as much as cast-iron, and will offer from three and a half to six times the resistance within the limit of its elasticity.

6. **Krupp gun.** Mr. Krupp, of Prussia, makes his guns out of solid cast-steel of low quality. The steel is formed in crucibles in the usual way, and is then run into a large ingot, which constitutes the mass of the gun. This ingot is worked under powerful steam hammers, to give the requisite texture to the metal. In this way 9-inch muzzle-loading rifle cannon, weighing 16,800 pounds, have been made. The success of this manufacture is said to be owing to the very heavy machinery employed, the skillful heating of the large masses to the centre without burning the outside, and the presence of manganese in certain proportions in the iron from which the steel is made.

7. **Parrott gun.** The Parrott rifle-gun is a cast-iron piece of about the usual dimensions, strengthened by shrinking a coiled band or barrel of wrought-iron over that portion of the reinforce which surrounds the charge. See fig. 160.

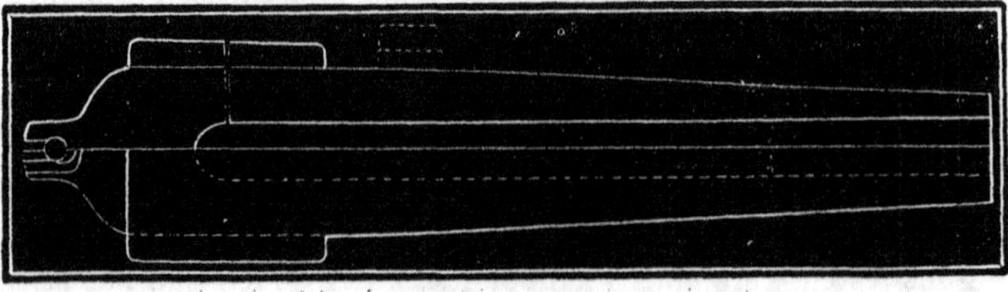

Fig. 160.

The body of the larger Parrott guns are cast hollow, and cooled from the interior on the Rodman plan. The barrel is formed by bending a rectangular bar of wrought-iron spirally around a mandrel and then welding the mass together by hammering it in a strong cast-iron cylinder, or tube. In bending the bar, the outer side being more elongated than the inner one, is diminished in thickness, giving the cross section of the bar a wedge shape, which possesses the advantage of allowing the cinder to escape through the opening, thereby securing a more perfect weld.

The barrel is shrunk on by the aid of heat, and for this purpose the reinforce of the gun is carefully turned to a cylindrical shape, and about one-sixteenth of an inch to the foot larger than the interior diameter of the barrel in a cold state. To prevent the cast-iron from expanding when the barrel is slipped on to its place, a stream of cold water is allowed to run through the bore. At the same time, and while the band hangs loosely upon it, the body of the gun is rotated around its axis to render the cooling uniform over the whole surface of the barrel.

Dimensions, &c., of Parrott Guns.

Gun.	Length of Bore.	Diam. of Bore.	Diam. of Barrel	Weight.	Number of Grooves.	Depth of Grooves.	Twist Increasing.	Charge.	Weight of Projectile.
	Inches.	Inches.	Inches.	Lbs.		Inches.	1 turn in ft. at Muzzle.	Lbs.	Lbs.
10-pdr....	70	3.	11.3	890	3	0.1	10	1	10
20-pdr....	79	3.67	14.5	1,750	5	0.1	10	2	19
30-pdr....	120	4.20	18.3	4,200	7	0.1	12	3	28
100-pdr....	130	6.4	25.9	9,700	9	0.1	18	10	86
8-inch...	136	8.	32.	16,300	11	0.1	23	16	150
10-inch...	144	10.	40.	26,500	15	0.1	30	25	250

The proof of the Parrott guns consists in firing each piece *ten* rounds with service charges.

*A large number of these guns were used in the late war, both on sea and land; and the amount of work done by them, especially in breaching masonry, is probably not exceeded by the rifle-guns of any other system. While a few of them have failed in the service, others have shown very great endurance. The cause of this

* The number of Parrott guns procured by the War Department alone was upwards of 1,700, besides 3,000,000 of Parrott projectiles.

failure has been attributed to the bursting of shells in the bore, the presence of sand in the bore, etc., but late investigations show that the Parrott projectiles were frequently broken at the bottom by the force of the powder in such a manner as to wedge the body against the bore. It is quite probable that this cause had much to do with the bursting of the guns. The inventor thinks he has corrected this evil.

8. **Brooke gun.** The gun made after the plan of Captain Brooke, for the Confederate service, resembles Parrott's in shape and construction, except that the reinforcing band is made up of wrought-iron rings not welded together. In some cases two layers of rings are used. The rifling appears to be similar to that used in the Blakely guns.

9. **Ames gun.** The rifle-guns of Mr. Horatio Ames, of Falls Village, Conn., are made of wrought-iron on the built-up principle. Rings of a certain size are made by bending a bar of wrought-iron around a mandrel and welding it together at the ends. Two or more of these rings are carefully turned in a lathe, and fitted one within the other to form a disk. A number of these disks are welded together, two by two, commencing at the end of a bar of iron, which forms the breech of the gun, and is used for handling it in manufacture. The mass is then reamed out and turned to the proper shape. The trunnions are attached by being screwed into the body of the gun. It is understood that Mr. Ames now proposes to try the experiment of lining his guns with a steel tube.

10. **Dahlgren gun.** The Dahlgren guns of large calibre are made of cast-iron, solid, and cooled from the exterior. To produce uniformity in the cooling, the piece is cast nearly cylindrical, and then turned down to the required shape, which is shown in the annexed figure.

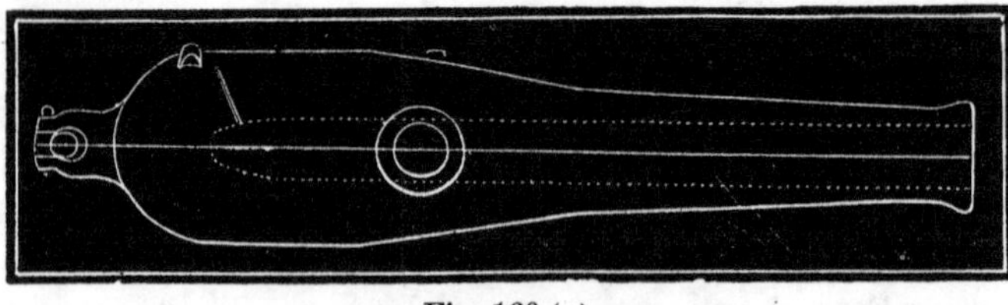

Fig. 160 (*a*).

The thickness of metal around the seat of the charge is a little more than the diameter of the bore, which rule holds good for nearly all cast-iron guns. The chase, however, tapers more rapidly than in other cast-iron guns, which gives the appearance of greater thickness of metal at the reinforce. The chamber is of the Gomer form.

The principal guns of this system are of 9 and 11-inch calibre. A piece of 10-inch calibre has, however, been introduced into the Navy, on Admiral Dahlgren's plan, for firing solid shot with 40 lbs. of powder. The 15-inch and 20-inch Naval guns are shaped exteriorly after the Dahlgren pattern, but are cast hollow and have the elliptical chamber of the Rodman system.

Dimensions, &c., of Dahlgren guns.

Gun.	Length of Bore.	Maximum Diam.	Weight.	Service Charge.	Maximum Charge.	Weight of Shot.	Weight of Shell.
	Inch.	Inch.	Lbs.	Lbs.	Lbs.	Lbs.	Lbs.
20-inch	163	64.	100,000	100		1080	
15-inch	130	48.	42,000	35	60	400	330
13-inch	130	44.7	36,000	40		280	224
11-inch	132	32.	16,000	15	20	170	130
10-inch	119⅓	29.1	12,000	12½	16	125	100
9-inch	107	72.2	9,200	10	13	93	70
125-pdr	117¾	33.25	16,500	40		125	100

11. **Rodman gun.** The principal difficulty formerly experienced in manufacturing very large cast-iron cannon was the injurious strains produced by cooling the casting from the exterior. As far back as the year 1844, General Rodman, of the Ordnance Department, sought to discover the means to overcome this difficulty. After much observation and study, he developed his theory of the strains produced by cooling a casting like that of a cannon, and as a remedy for them he proposed that cannon should be cast on a hollow core, and cooled by a stream of water, or air, passing through it; see page 136. After an elaborate series of experiments the truth of his theory was established, and his new mode of casting was adopted by the War Department. As a result of General Rodman's theory, he claimed that he could cast cannon of any practicable size, and asked that a 15-inch cast-iron gun might be made. This was done in 1860, and the gun was successfully tested shortly afterwards. General Rodman then projected a 20-inch gun, which was made at the Fort Pitt foundry in 1863, under his directions.

Formerly it was customary to use but one kind or size of grain of powder for all cannon, whatever their size. General Rodman proposed for his large cannon that there should be a proportional increase in the size of the grain, expecting thereby to get as high a velocity for the projectile without a corresponding increase in the strain on the breech or weak part of the piece; this led to the introduction of our present Mammoth powder. He also thought that the powder which would produce the least strain on the gun, giving certain initial velocity to the projectile, would be that which should develop its gas as the space behind the projectile increased; or in other words, that the powder should burn on an increasing instead of a decreasing surface. With this object in view he proposed to compress the substance of the powder into short hexagonal prisms, which could be easily fitted together without loss of space. These prisms were perforated with longitudinal holes, from which the combustion of the powder spread. While this idea has to a certain extent been confirmed by experiment, this powder has not been officially adopted in this country; it is understood that it has been to a certain extent in Russia for service in heavy rifle-guns.

The trial, or No. 1 15-inch Army gun, has been fired 509 times with charges varying from 35 to 50 lbs. of powder. The effect on the bore is hardly perceptible. The Navy 15-inch trial gun was fired 900 times with charges varying from 35 to 70 lbs., mostly Mortar or Navy cannon powder, when it burst.

Within a short time another Army 15-inch gun has been fired without injury 250 times, with charges varying from 40 to 100 lbs. of Mammoth powder; *one hundred* of these were with 100 lbs. of powder and projectiles of 450 lbs. each. 15-inch gun No. 105 has likewise been fired as follows, viz.:

No. times fired.	Charge.	Weight of projectile.	Initial velocity.
4	60 lbs.	430 lbs.	1191 ft.
3	70 lbs.	431 lbs.	1278 ft.
3	80 lbs.	433 lbs.	1355 ft.
3	90 lbs.	453 lbs.	1433 ft.
2	100 lbs.	453 lbs.	1509 ft.

The Chief of Ordnance reports that 286 15-inch Rodman guns have been purchased for the use of the Army; besides a large number have been procured for the Navy, for use on the Monitors, where they rendered good service against the enemy's iron-clads and fortifications during the late war. Only, some two or three of the Navy 15-inch guns have split at the muzzle where they had been turned down very thin to fit the port-holes originally intended for the 13-inch Dahlgren gun.

A 12-inch rifle-gun, having the exterior form and dimensions of the 15-inch gun, has been made and tested by firing it 420 times without injury, with charges of powder varying from 40 to 85 lbs., and projectiles varying from 475 to 620 lbs.; an initial velocity of 1121 feet was obtained. Further experiments in large rifle-cannon are to be made at Fort Monroe, under the direction of the Chief of Ordnance.

20-inch gun. The 20-inch gun made in 1863 has been thus far fired only eight times, a delay having been occasioned by a failure to get a suitable iron target against which to test its destructive powers. Soon after it was mounted at Fort Hamilton, New York Harbor, it was fired four times with 50, 75, 100, and 125 lbs. of mammoth powder, and solid shot weighing 1080 lbs. In March of this year (1867), it was again fired as follows:

1st fire,	125 lbs. powder,	25° elevation,	6144 yds. range.
2d "	150 " "	25° "	6860 " "
3d "	175 " "	25° "	6828 " "
4th "	200 " "	25° "	8001 " "

The maximum pressure on the bore was 25,000 lbs. The form of the 15-inch and 20-inch Rodman gun is shown in fig. 46.

Particulars and Charges of Rodman guns.

Name of Gun.	Length.	Length of Bore.	Maximum Diameter.	Weight.	Service Charge.	Bursting Charge. Shell.	Weight of Shot.	Weight of Shell.
Smooth-Bores.	In.	In.	In.	Lbs.	Lbs.	Lbs.	Lbs.	Lbs.
20-in. gun.	243.5	210.	64.	115200	100		1080	
15-in. do.	190.	165.	48.	49100	50 mammoth.	17	440 / 425	330
13-in. do.	177.6	155.94	41.6	32731	30 cannon.	7	300 / 280	224
10-in. do. of 1860	136.66	105.5	32.	15059	15 for shell. / 18 for shot.	3	127	100
8-in. do.	123.5	110.	25.6	8465	10	1	68	48

12. **Gatling gun.** The Gatling gun is a machine gun composed of six barrels made to revolve around a central axis parallel to their bores, by means of a hand-crank. As each barrel comes opposite to a certain point a self-primed metal case cartridge, falling from a hopper, is pushed into the breech by a plunger, and held there until it is exploded by the firing-pin.

This gun is capable of firing as many as 200 shots a minute, with great range and precision. The machinery is simple, and little liable to get out of order. A number of these guns of 1-inch and ½-inch calibre have lately been procured by the Government for use on the Plains, in flanking ditches, defending block-houses, etc. In addition to the solid ½ lb. bullet projected by the gun of 1-inch calibre, this piece fires a cartridge containing 16 smaller projectiles, giving a very destructive fire for short distances. As the weight of this gun—about 1000 lbs.—is very great compared to that of the charge, the aim is not disturbed by recoil.

13. **Various guns.** The continental nations of Europe have done but little in the manufacture of rifle-cannon throwing projectiles weighing over 100 lbs. In France and Spain the cast-iron sea-coast cannon, corresponding to our 32-pounders, have been banded with steel hoops and rifled. In Russia 9-inch rifle-guns, made of Krupp steel, have been used to a certain extent, but it is understood that one or more of them have failed in service.

In the matter of field artillery, most of these nations have rifled their bronze guns, thus making use of the material on hand. The Prussian field guns, however, are made of Krupp steel, and are breech-loaders. In Russia the Broadwell system of breech-loading has been introduced into the field service.

A Table of Foreign Cannon, as given by Major Owen, R. A.

Nature of Gun.	Weight.	Charge of Powder in terms of Projectile.	Projectile.
	Cwt.		Lbs.
French 4-pounder (bronze)................	6½	1-10	8¾
French 12-pounder (bronze)...............	12	1-10	25¼
French 30-pounder (cast-iron).............	56	1-9 to 1-13	60 to 100
Prussian 12-pounder (Krupp steel).........		1-10	13
Prussian —-pounder (cast-iron)............	53½	1-10	57
Spanish 32-pounder (cast-iron)............	62	1-9	61

PROJECTILES.

14. **Armstrong projectile.** But one kind of projectile is used in the Armstrong breech-loading guns for the field service, and this is so constructed as to act as a shot, shell or case-shot, at pleasure.

It consists—fig. 162—of a very thin cast-iron shell (*A A*), enclosing forty-two segment-shaped pieces of cast-iron (*B B*), built up so as to form a cylindrical cavity in the centre (*D*), which contains the bursting charge and the concussion fuze. The exterior of the shell is thinly coated with lead (*C C*), which is applied by placing the shell in a mould and pouring it in a melted state. The lead is also allowed to percolate among the segments, so as to fill up the interstices, the central cavity being kept open by the insertion of a steel core.

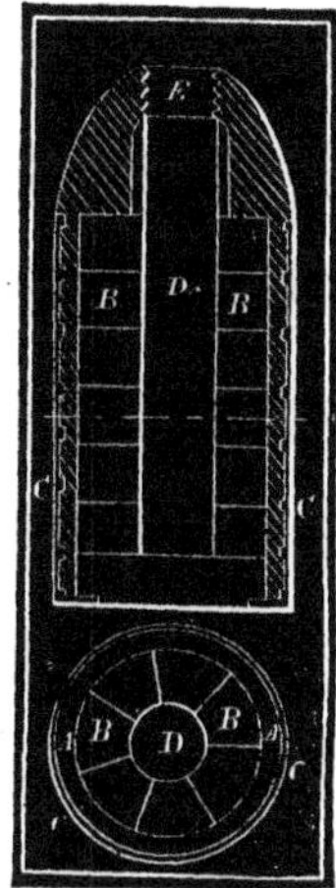

Fig. 162.

In this state the projectile is so compact that it may be fired without injury; while its resistance to a bursting charge is so small that less than one ounce of powder is required to burst it. When the projectile is to be fired as a shot, it requires no preparation; but the expediency of using it otherwise than as a shell is doubted.

To make it available as a shell, the bursting tube, the concussion and time fuzes, are all to be inserted; the bursting tube entering first and the time-fuze being screwed in at the apex. If the time-fuze be correctly adjusted, the shell will burst when it reaches within a few yards of the object; or, failing in this, it will burst by the concussion-fuze when it strikes the object, or grazes the ground near it. If it be required to act as a canister-shot upon an enemy close to the gun, the regulation of the time-fuze must be turned to the zero of the scale, and then the shell will burst on leaving the gun.

The explosion of one of these shells in a closed chamber, where the pieces could be collected, resulted in the following number of fragments:—106 pieces of cast-iron, 90 pieces of lead, and 12 pieces of fuze, etc.—making in all 217 pieces.

The Armstrong projectiles for the muzzle-loading guns have

rows of brass or copper studs projecting from their sides to fit into the grooves of the gun, which are constructed on the *shunt* principle. Fig. 163 represents a 10-inch Armstrong shell for penetrating armor plates. It is made of wrought-iron, or low steel, with very thick sides. There is no fuze, the explosion resulting from the heat generated by the impact, and the crushing in of the thin cap which closes the mouth of the powder chamber. The sides and bottom of the shell being thick enough to resist crushing by the impact and also to resist the explosive force of the bursting charge, its effect will, after penetration, be expended on the backing of the armor, or the decks which the armor is intended to screen.

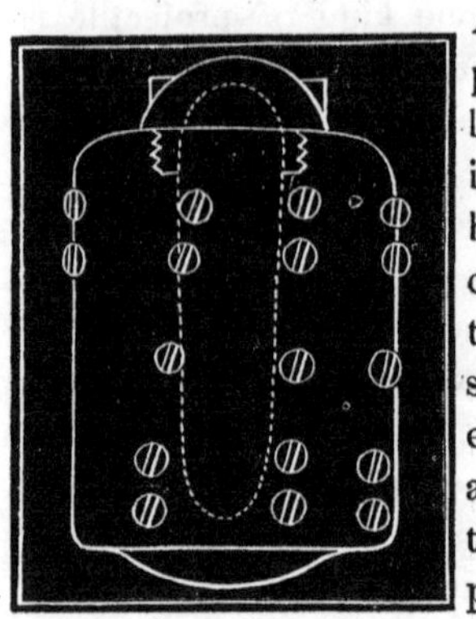

Fig. 163.

Such projectiles are called "blind shells."

15. **Whitworth projectile.** The cross-section of the bore of the Whitworth gun is a hexagon with the corners slightly rounded. The projectile is first formed so that its cross-section is a circle, and its sides taper towards both ends. The middle portion is then carefully planed off to fit the bore of the gun. Fig. 164 represents a Whitworth blind shell for firing against armor plates. It is made of tempered steel, and each end is closed with a screw. To prevent the heat of impact from acting too soon on the bursting charge, it is surrounded by one or more thicknesses of flannel.

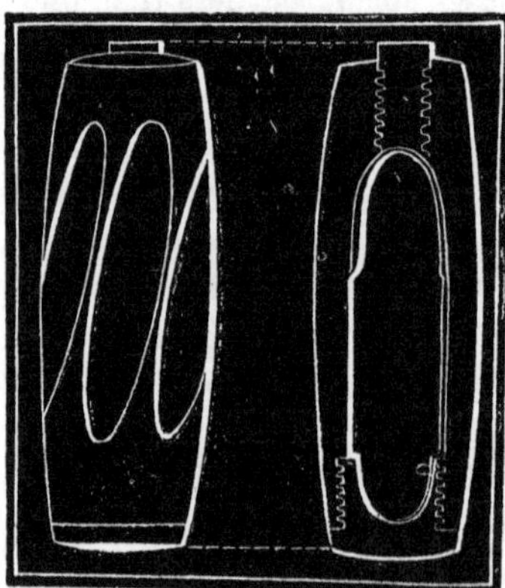

Fig. 164.

A 7-inch shell of this kind has been found to have sufficient strength and stiffness to penetrate five inches of wrought-iron before bursting.

16. **French projectile.** The projectile used in the French field service is made of cast iron, and has twelve zinc studs on its sides, arranged in pairs, so as to fit the six grooves of the gun. See fig. 165.

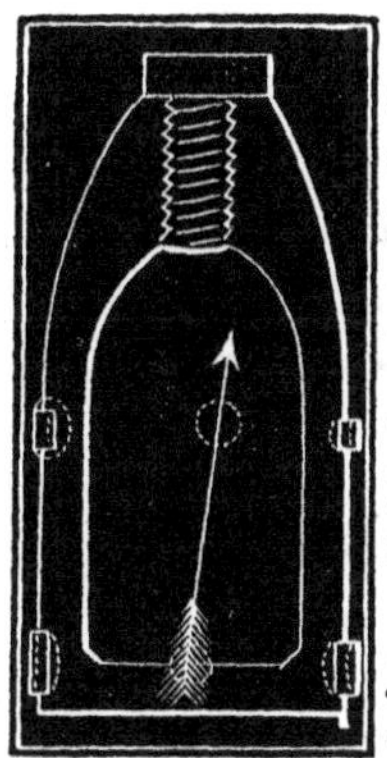
Fig. 165.

For the larger cannon projectiles but three studs are used, and these are cast on the projectile, nearly opposite to its centre of gravity; the bearing sides of the studs are faced with white metal to diminish friction against the grooves of the bore. The shape of the grooves is such as to centre the projectile. The latter projectile is used with increasing, the former with grooves of uniform twist.

Russian, Austrian and Spanish artillery projectiles belong to the studded, or button class, but differ from each other in the details of their construction.

17. **Blakely projectile.** Captain Blakely's projectile has an expanding copper cup attached to its base by means of a single tap-bolt in the centre (see fig. 166). It is prevented from turning by radial grooves cast on the surface of the bottom of the projectile, into which the cup is pressed by the charge. The angle between the curved sides of the cup and the bottom of the projectile is filled with a lubricating material. On the forward part of the body are soft metal studs, more numerous than the grooves of the bore of the piece, that some of them may always form a bearing surface for the projectile against the lands. The driving sides of the grooves are deeper than the other.

Fig. 166.

18. **Scott projectile.** The shell devised by Commander Scott, of the British Navy, for firing molten iron, is shown in fig. 167.

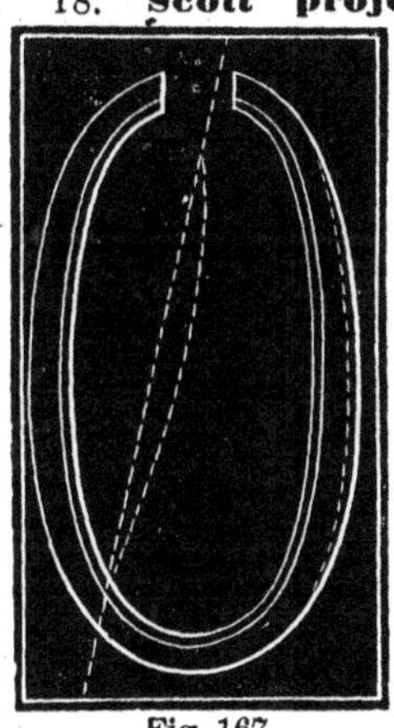
Fig. 167.

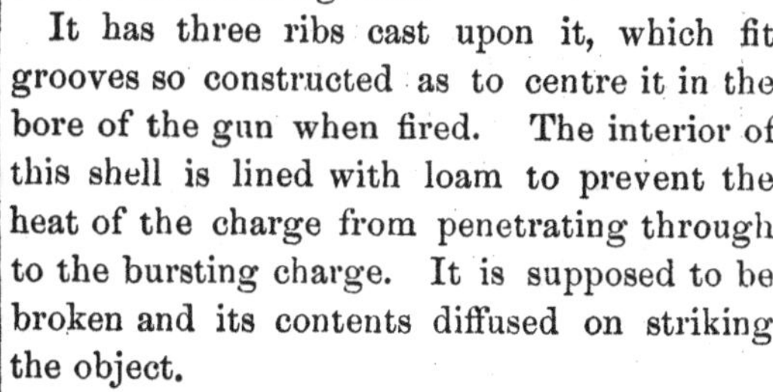

It has three ribs cast upon it, which fit grooves so constructed as to centre it in the bore of the gun when fired. The interior of this shell is lined with loam to prevent the heat of the charge from penetrating through to the bursting charge. It is supposed to be broken and its contents diffused on striking the object.

19. **Parrott projectile.** Capt. Parrott's projectile is composed of a cast-iron body and a brass ring cast into a rabbet formed around

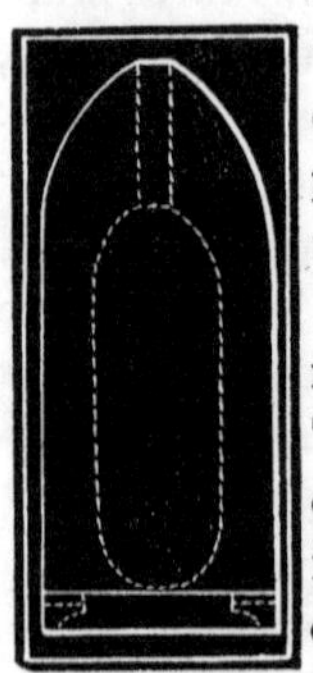

Fig. 168.

its base (see fig. 168). The flame presses against the bottom of the ring and underneath it so as to expand it into the grooves of the gun. To prevent the ring from turning in the rabbet, the latter is recessed at several points of its circumference.

Parrott's incendiary shell has two compartments formed by a partition at right angles to its length. The lower and larger space is filled with a burning composition; the upper one is filled with a bursting charge of powder, which is fired by a time or concussion-fuse. The burning composition is introduced through a hole in the bottom of the shell, which is stopped up with a screw plug.

20. **Schenkle projectile.** Schenkle's projectile is shown in fig. 169. It is composed of a cast-iron body (*a*), the posterior portion of which is a cone. The expanding portion is a *papier mache* sabot or ring (*b*), which is expanded into the rifling of the bore by being forced on to the cone by the action of the charge. On issuing from the bore the wad is blown to pieces, leaving the projectile unencumbered in its flight. A great difficulty has been found in practice in always getting a proper quality of material for the sabot, and in consequence, these projectiles have not been found to be reliable.

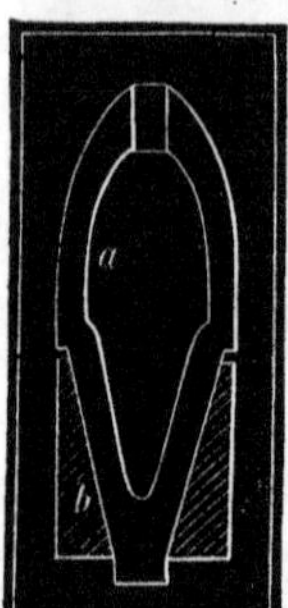

Fig. 169.

21. **Hotchkiss projectile.** The Hotchkiss projectile is composed of three parts: the body (*a*),—fig. 170—the expanding ring of lead (*b*), and the cast-iron cup (*c*). The action of the charge is to crowd the cup against the soft metal ring, thereby expanding it into the rifling of the gun. The time-fuse projectile has deep longitudinal grooves cut on its sides to allow the flame to pass over and ignite the fuse.

The last rifle-projectile submitted by Mr. Hotchkiss has an expanding cup of brass attached to its base in a peculiar manner. The cup is divided into four parts by thin projections on the base of the projectile. This arrangement is intended to facilitate the expansion of the cup and to allow the flame to pass over to ignite the fuse.

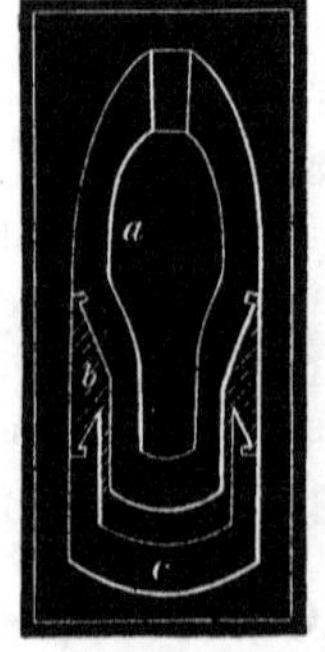

Fig. 170.

22. **Sawyer projectile.** The Sawyer projectile has upon its sides six rectangular flanges or ribs to fit into corresponding grooves of the bore. To soften the contact with the surface of the bore, the entire surface of the projectile is covered with a coating of lead and brass foil. The soft metal at the corner of the base is made thicker than at the sides to admit of being expanded into the grooves, and thereby closing the windage. In the latest pattern of Sawyer projectiles the flanges are omitted, and the projectiles are made to take the grooves by the expansion of the soft metal at the base, which is peculiarly shaped for this purpose.

23. **James projectile.** The expanding part of James's projectile consists of a hollow (*c*),—fig. 171—formed in the base of the projectile, and eight radial openings (*b*), which extend from this hollow to the surface for the passage of the flame of the charge, which presses against and expands into the grooves of the bore, an envelope or patch (*e*), composed of paper, canvas and lead. (*a*) represents the body of the projectile, which in this case is a solid shot. (*d*) is a partition between two of the openings.

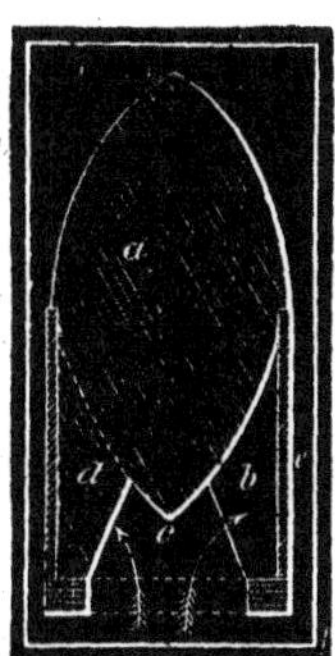

Fig. 171.

In a later pattern of this projectile, the internal cavity and radial openings are omitted, and the outside is furrowed with longitudinal grooves which increase in depth towards the base of the projectile, forming inclined planes, up which the outer covering of lead and canvas is moved by the force of the charge and expanded into the rifling of the piece.

24. **Dyer projectile.** The Dyer projectile is composed of a cast-iron body (*a*),—see fig. 172—and a soft metal expanding cup (*b*), attached to its base. The adhesion of the cup is effected by tinning the bottom of the projectile, and then casting the cup on to it. The cup is composed of an alloy of lead, tin and copper in certain proportions. This projectile, as improved by Mr. Taylor at the Washington Arsenal, gives good results for even as large a calibre as 12 inches.

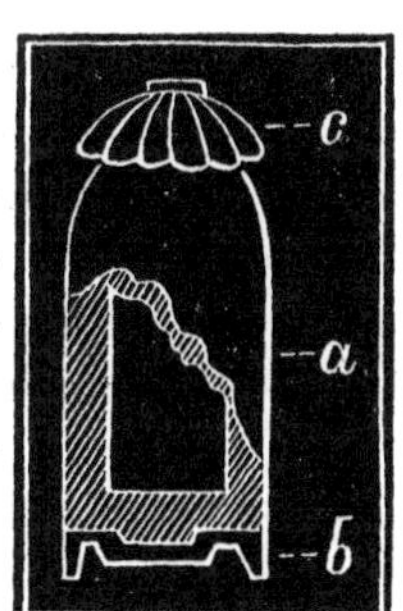

Fig. 172.

(*c*) represents a corrugated cap of tinned sheet-iron, used with 3-in. projectiles to catch and direct that portion of the flame of the charge which escapes over the projectile on to the fuze to ignite it.

Fig. 173.

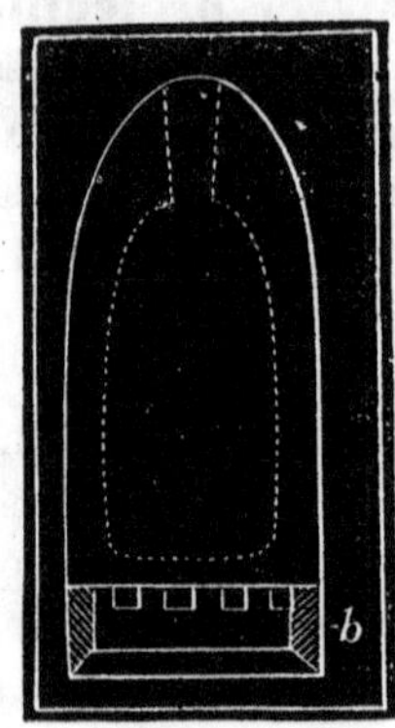

Fig. 174.

25. **Confederate projectiles.** The rifle-projectiles used by the Confederates in the late war, belonged, with a few exceptions, to the expanding class. Fig. 173 represents a Confederate wrought-iron solid shot, for use against iron-clads. For the larger sizes this shot is formed by welding rings around a bar of iron and then turning the mass to the proper size. It has an annular-shaped cup at the base for the purpose of expansion.

Fig. 174 represents a shell with a copper ring (*b*) fitting into a rabbet formed around its base in casting. This projectile would seem to resemble the Parrott projectile in its construction. The lower edge of the band, however, projects below the bottom of the base, which in Parrott's it does not. Recesses are formed in the sides of the rabbet to prevent the ring from turning.

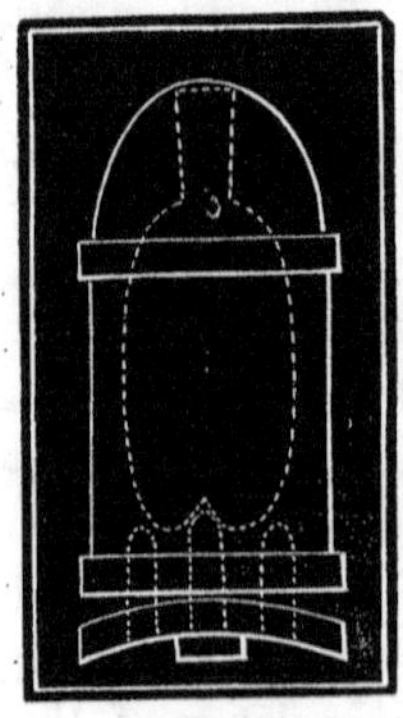
Fig. 175.

The projectile represented in fig. 175 has a thick circular plate of copper attached to its base by means of a screw-bolt at its centre. To prevent it from turning around this bolt there are three pins, or dowels, fastened into the base of the projectile, and projecting into corresponding holes in the circular plate. This plate is slightly cupped, and the angle between it and the bottom of the projectile is filled with a greased cord for lubricating the bore of the gun.

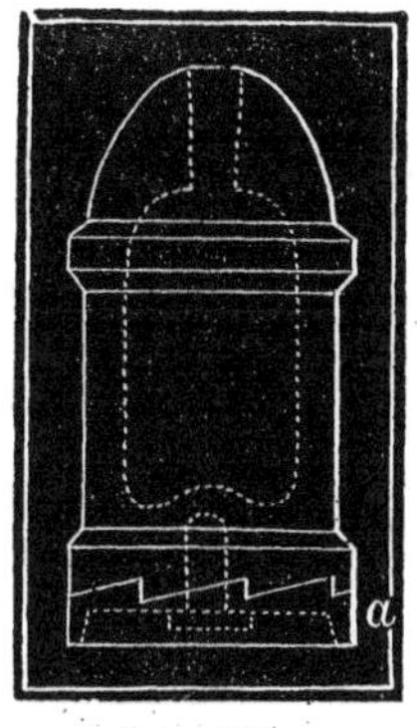

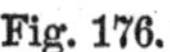

Fig. 176.

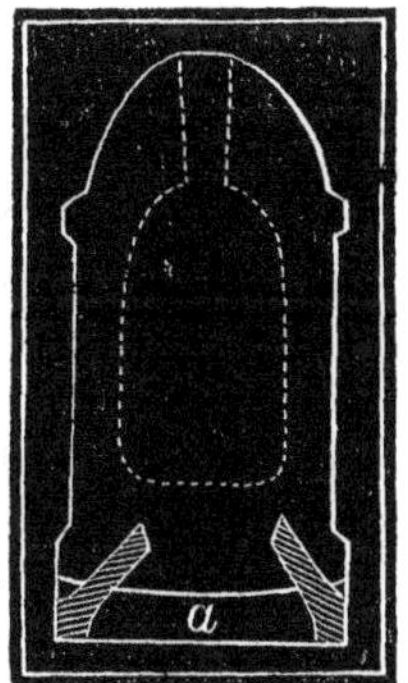

Fig. 177.

Fig. 176 represents a projectile of the Blakely class, with its expanding cup of copper (a). Instead of the soft metal studs which are placed on the forward part of the Blakely projectile, this projectile has a raised band carefully turned to fit the bore.

Fig. 177 represents a Reed projectile, in which the expanding cup is made of copper, as shown at (a). This cup is placed in mould, and the body of the projectile is cast upon it.

EXPERIMENTS ON ARMOR PLATES.

26. **Captain Noble's conclusions.** The following conclusions are drawn by Captain Noble, R. A., from the results obtained in England from various experiments carried out under the direction of the Ordnance Select Committee :

1. Where it is required to perforate the plate, the projectile should be of a hard material, such as steel or chilled iron.

2. The form of head best suited for the perforation of iron plates, whether direct or oblique, is the pointed ogeeval.

3. The best form of steel shell at present known is that in which the powder can act in a forward direction, and which is furnished with a solid steel head in the form of a pointed ogeeval.

4. When chilled iron can be made of the best quality, it is almost, if not quite, as good as ordinary steel for solid shot ; and when the projectile can perforate with ease, the chilled shot is more formidable than steel, as it enters the ship broken up, and would act as grape.*

* The introduction of chilled iron is due to Major Palliser, who has devoted much time and attention to the subject.

We have every reason to hope that chilled shells can be constructed, which will prove equal, if not superior, to steel.*

5. To attack well-built iron-clads effectively, the guns should be, if possible, not under twelve tons weight and nine inches calibre, firing an elongated projectile of two hundred and fifty pounds, with forty pounds of powder.

6. When the projectiles are of a hard material, such as steel, the perforation † is directly proportional to the "work" in the shot, and inversely proportional to the diameter of the projectile; and it is immaterial whether this "work" is made up of velocity or weight, within the usual limits which occur in practice.

7. The resistance of wrought-iron plates to perforation by steel projectiles varies as the squares of their thickness.

8. Hitting a plate at an angle diminishes the effect as regards the power of perforation in the proportion of the sine of the angle of incidence to unity.

9. The resistance of wrought-iron plates to perforation by steel shot is practically not much, if at all, increased, by backing simply of wood, within the usual limits of thickness; it is, however, much increased by a rigid backing either of iron combined with wood, or of granite, iron, brick, etc.‡

10. Iron-built ships in which the backing is composed of compact oak, or teak, offer much more resistance than similarly clad wooden ships.

11. The best form of backing seems to be that in which wood is combined with horizontal plates of iron, as in the "Chalmer," "Bellerophon," and "Hercules" targets.

12. An inner iron skin is of the greatest possible advantage; it not only has the effect of rendering the backing more compact, but it prevents the passage of many splinters, which would otherwise find their way into the ship.

No iron-clad, whether iron-built or wooden converted, should be without an inner iron skin.

* It is not meant by this that chilled iron would prove superior to the very best tool steel; but that it will be as effective as the ordinary steel hitherto used in the manufacture of projectiles.

† Or power of complete penetration.

‡ That is to say, as a shot which is capable of breaking a hole through a 4½-inch plate unbacked, will be also capable of doing so if the plate be only backed by wood, to the extent that, were the plate taken off the backing, the piece of iron where the shot had struck would fall out.

13. The bolts known as "Palliser's bolts" are, so far as known at present, the best for securing armor plates.

In these bolts the diameter of the shank is reduced to that which it is at the screwed end.

14. Laminated armor is much inferior to solid armor.

Captain Noble then proceeds to make the following remarks on the relative merits of rifled and smooth-bored guns against armored vessels, viz.:

There are two methods by which an iron-clad vessel can be destroyed by the fire of artillery:

1. *Racking* (American system), or the impact of heavy shot of large size moving at low velocities, and intended to shatter the vessel's armor, and by repeated shakes ultimately to knock the whole structure to pieces.

2. *Punching* (English system), or the penetration of the vessel's side, either by elongated shot or shell, intended to kill the crew, blow up the magazine, damage the machinery, and sink the vessel by holes made through her at or near the water-line.

Both these systems have their advocates, and there is undoubtedly a great deal to say on both sides.

All warlike operations tend to the crippling of your enemy; and that system is evidently the best which will cripple him in the shortest time, in the easiest manner, and at the least possible expense.

Now, time is an element which will largely enter into consideration in future actions with iron-clad vessels.

Suppose two opposing iron-clads to meet—one armed with guns on the "racking" system, the other with guns on the "punching" system, it is probable that the vessel which could send her shot clean through the side of her adversary would have the greatest *chance* of reaching a vital part *in a given time.* Besides which, a "punching" shot is usually an elongated rifle projectile animated by a moderately high velocity, and has consequently a *flatter trajectory* than the "racking" shot, which travels at a low velocity; and as accuracy and a flat trajectory are closely allied, the "punching" system would gain another chance, viz.: that of making the greatest number of hits for a given number of shots.

Suppose an iron-clad is desirous of running past a fort which defends an important harbor or roadstead. She would, if possible, probably pass at a rate of ten miles an hour. The fort in this case would only have time to fire a few rounds at her, and if the effect of those rounds was merely an external "racking," the vessel might receive no real injury at all—nothing, at least, which would in all likelihood stop her.

On the contrary, a happily directed "punching" shot would have the chance of destroying the machinery, blowing up the magazine, or establishing a leak at the water-line.

In attacking an iron-clad by the "racking" system, the whole effect is directed against the casing or armor-plating of the vessel, which, for all offensive purposes, is harmless; the enemy which we want to cripple are the men and guns behind the armor.

It appears from these considerations that an attack on the "punching" system will probably be attended by gain in time, as the vital parts of the vessel cannot be reached so quickly by an attack on the "racking" system. Even were an enemy's ship ultimately shattered and her offensive power destroyed by the effect of heavy blows, *this result might not be effected before she had accomplished her object, if not altogether.*

The attack on the "punching" system is carried on in an easier manner than that on the "racking" system. The former employs light rifle guns, from six to twelve tons; the latter, unwieldy heavy ordnance of from twelve to fifty tons. The "racking" projectiles are heavy cast-iron shot, fired with relatively small charges, and the loading and working of such projectiles and guns cannot be carried out as easily or expeditiously as in the case of a system which uses a lighter shot and relatively larger charge.*

The question of expense is one which, although it should come last in an inquiry of this nature, is too often made the most important consideration.

If, however, we compare the cost of the 9-inch rifled twelve-ton gun as fairly representing the *punching* system, and the American 15-inch twenty-ton smooth-bore gun as representing the *racking* system, we shall find that the total cost of gun, carriage and one hundred rounds of ammunition is very much the same for each.

On the one hand, the money will have procured a gun which can

* The present English system comprises a rifled gun throwing an elongated shot with a moderately high charge, with a view to *penetration.* If, however, it be desirable to adopt the opposite system, either altogether or in part, there will be no difficulty in employing heavy shot and low charges with our rifle-guns. The Americans, however, cannot apply our system to their smooth-bore guns, which is a point in our favor.

send a shot, and possibly a shell, through the strongest iron-clad yet afloat at one thousand yards' range.

On the other hand, a gun will be obtained which cannot pierce the above ship at any distance whatever; whose shot, at one thousand yards, would, if cast-iron, merely indent the armor and fall back broken into the water, and if steel, would merely lodge in the ship's side, and whose shell would be absolutely worthless against an iron-clad, and even against wooden ships or earth-works, inferior to the 9-inch rifle shell, both in accuracy and bursting power.

Remarks by the Author.—It is probable that Captain Noble bases his estimate of the power of the American 15-inch smooth-bored gun on the impression that it can only be fired with low charges of powder. It has been shown, however, that an initial velocity of 1200 feet can be easily obtained with this gun. A projectile, therefore, which weighs 450 lbs., fired from it, will have a striking velocity at 1000 yds. of 961 ft. equal to a total force of 3230 foot-tons, or 69.3 foot-tons per inch of the shot's circumference—a force, as shown by Captain Noble, more than sufficient to penetrate the Warrior target. It was with a force probably less than this, that the side of the iron-clad Atlanta was smashed in, and the ram Tennessee was disabled, each by a cored 15-inch shot made of ordinary cast-iron.

The power of the 15-inch gun, however, is not limited to 1200 feet initial velocity. Several of these guns of the army pattern have lately been fired with charges as high as 100 lbs. of mammoth powder, giving an initial velocity of 1500 feet. In one case, as many as 100 rounds have been fired with this charge with no material enlargement of the bore.

Within the limit of effective iron-clad warfare, the 15-inch gun has great accuracy of fire. In the trials lately made at Fort Monroe, it was shown that the accuracy of this gun at 1500 yds., with a charge of 100 lbs. of powder, was fully equal to that of the best rifle practice, and that while the least elevation of the rifle-guns was, for this distance, 3° 10′, that of the 15-inch gun was only 2° 25′. The latter, therefore, has a flatter trajectory within this distance, and the projectile being round, has greater accuracy in ricochet fire.

The advantage, however, which rifle-guns have of projecting loaded shells of peculiar construction (see Armstrong and Whitworth projectiles), so as to penetrate and explode in the sides of iron-clad vessels, is certainly an important one, and may, perhaps, outweigh, in actual service, the *smashing* power claimed by the ad-

TABLE of Results of Experiments with Armor Plates. Taken from Captain Noble's Report.

Target.	Gun.	Powder.		Projectile.			Striking Velocity.	$\frac{Wv^2}{2g}$ in Foot Tons on Impact.	Foot Tons per inch of Shot's Circumference.	Observed Effects.
		Weight.	Brand of Powder.	Nature and Length.	Weight.	Diameter.				
		lbs.		inches.	lbs.	inches.	feet.			
"Warrior," This target is composed of plates of 4½ inch wrought-iron, backed with 18 inches of teak, an inner skin 5/8 inch thick, and is stiffened with vertical iron ribs at the back.	13.3-inch M. L. rifle-gun of 23 tons.	70	Rifle L. G.	Steel shell 24.9	612.5	13.24	1046	4646	111.7	Target completely *penetrated*,* with much damage. Range 1000 yards.
	Horsfall's M. L. smooth-bore gun of 24 tons, calibre 13.014 inches.	74.4	L. G.	Spherical cast-iron shot.	285	12.88	1290	3288	81.3	Penetrated armor. Shot lodged in backing. Range 800 yards.
	10.5-in. wrought-iron M. L. gun of 12 tons.	50	"	"	150	10.37	1620	2730	83.8	Penetrated armor. Shot lodged in backing, bulging inner skin. Range 200 yards.
	9-inch M. L. rifle-gun of 12 tons.	23	"	Palliser's chilled shell, ogival, 1.5 diameters.	249	3.89	976	1645	58.9	Penetrated the target completely, bursting on passing through. Considerable damage in rear; the fragments of the shell passed many yards to the rear. Range 200 yards.
	7-inch M. L. gun, 130 cwt. (Woolwich.)	20	"	Steel shell, ogival, 1.5 diameters, screw base. 17.0	116.2	6.92	1397	1572	72.3	Struck one foot to the right of 1296; *penetrated* the target completely, bursting in the backing; base of shell blown back to the front. Head of shell perfect. Bursting charge 4.2 lbs. Range 70 yards. Palliser's chilled cast-iron shells fired under similar circumstances did not always explode, but were broken up impact. Range 200 yards.
	5-inch Whitworth gun.	13	"	Flat-headed steel shot. 19.0	81.0	{ 4.98 } { 5.42 }	1102	682	40.38	Penetrated armor and burst in backing. Range 600 yards.

* The word penetrated in this table means passed through.

TABLE—(continued).

Target.	Gun.	Powder.		Projectile.			Striking Velocity.	$\frac{Wv^2}{2g}$ in Foot Tons on Impact.	Foot Tons per inch of Shot's Circumference.	Observed Effects.
		Weight.	Brand of Powder.	Nature and Length.	Weight.	Diameter.				
		lbs.		inches.	lbs.	inches.	feet.			
8-in. plates on "Warrior" backing. Direct fire.	9-in. M. L. rifle-gun, 12 tons.	43	Rifle L. G.	Palliser's chilled shell, ogival, 20.	252.0	8.9	1323	3059	109.4	Penetrated target completely, bursting in backing; head uninjured. Struck fair on rib, which it broke. Bursting charge 2.37 lbs. Range 200 yards.
Same inclined at an angle of 30° with horizontal.	"	43	"	Steel shot. 17.2	251.0	8.92	1340	3125.1	111.5	Struck inclined target. Scooped out a piece of plate $14\frac{1}{2} \times 11$ inches; indent 3 inches. Slight buckle of plate; in rear nothing. Range 200 yards.
Same.	"	43	"	Palliser's chilled shell, ogival, 20.	250	8.9	1313	2989	106.9	Struck upper plate of inclined target 16 in. from 123 and 15 in. from 1241. Diameter of hole 11×11 in.; depth of indent 19.5 in. to head of shell. In rear a rib bent and one wood bolt broken. Bursting charge 2.12 lbs. Range 200 yards.
"Minotaur." 5.5 in. plate, 9 in. of teak, and an inner skin $\frac{5}{8}$ in. on ship's frame.	10.5 in. M. L. wrought-iron gun of 12 tons.	50	"	Spherical cast-iron shot.	150.0	10.37	1620	2730	83.8	Penetrated armor, and lodged in backing. Broke two ribs and seriously bulged the inner skin. Range 200 yards. The second shot passed clean through, carrying with it the piece of plate and many splinters.
"Bellerophon." 6-in. plate, 10-in. of teak and a $1\frac{1}{2}$ in. inner skin held by heavy ribs.	"	35	"	Spherical cast-iron shot.	150.0	10.36	1547	2489	76.5	Cracked and indented plate to a depth of 5 in. Skin slightly bulged. A steel shot (round) penetrated armor, and remained in backing. Inner skin bulged and cracked. Range 200 yards.

TABLE—(continued).

Target.	Gun.	Powder.		Projectile.			Striking Velocity.	$\frac{Wv^2}{2g}$ in Foot Tons on Impact.	Foot Tons per inch of Shot's Circumference.	Observed Effects.
		Weight.	Brand of Powder.	Nature and Length.	Weight.	Diameter.				
		lbs.		inches.	lbs.	inches.	feet.			
"Lord Warden." 1½-in. plate, 10 in. teak, 4½-in. plate, and 20 in. of timber	10.5 in. M. L. wrought-iron gun of 12 tons.	50	Rifle L. G.	Spherical cast-steel shot.	168.25	10.43	1576	2898	88.4	Penetrated armor. Hole 11 × 11.5 inch. Shot broke through inner skin and buried itself in ship's timbers. Back shaken. Range 200 yards. A shot from a 9.22 in. rifle-gun completely penetrated this target and went 500 yards out to sea.
"Hercules." 8-in. plate, 12 in. timber between ribs. 1½ in. plate, 18-in. of timber, and inner skin ¾ inch on iron ribs.	13-in. M. L. rifle-gun of 23 tons.	100.0	"	Cylindrical steel shot. 17.0	573.0	12.94	1268	6388	157.0	Struck 8-in. plate on rib; shot stuck in the plate, forcing the pieces into the backing; indent about 13 in. Inner skin slightly bulged; two ribs cracked. Range 700 yds.
Same.	"	100.0	"	Chilled shot, ogival head, 1.25 diam's. 21.2	577.5	12.95	1310	6872	169.0	Struck 8-in. plate just above a former shot, and passed through the target. This hit was hardly a fair one, as the rib was weakened. Shot broke up. Range 700 yards.
Same.	10.5-inch. M. L. rifle-gun of 12 tons.	45	"	Cylindrical steel shot. 14.1	299.8	10.42	1277	3390	103.6	Penetrated the plate, driving it into the backing, and slightly bulging the ¾ in. inner iron plate. Shot rebounded. Indent 13 in. Range 200 yards.
Same with a 9 instead of an 8-inch plate.	"	60	"	"	299.8	10.42	1425	4221	128.9	Struck fair on 9-in. plate, partly on rib. Indent 6.1 inches. Shot set up 2 inches and cracked. Diameter of indent 12.75 inches. Range 200 yards.

vocates of large smooth-bored guns. It shows the importance of not depending *entirely* on any one system, unless it combines the advantages of both.

It is understood to be the present intention of the Government to arm the sea-coast defences with equal proportions of smooth-bored and rifle-guns of the heaviest calibres.

The only extended experiments with armor plates in this country have been made by the Naval Ordnance Bureau, the most important details of which have not been published. A general summary of these results, as given in a report of the Chief of this Bureau, will be found on page 473.

27. **Improvements in Gun Carriages.**—The Ordnance Department has lately made and tested certain wrought-iron field and siege carriages. The principal improvements aimed at in one of the patterns of field-carriages on the wooden carriages now in use, are: 1. Lightness and cheapness. 2. Placing the pintle about two feet in rear of the limber axletree, so that the trail of the gun-carriage shall counterpoise the weight of the pole, and thereby relieve the shoulders of the wheel-horses which have now to support it. 3. Bringing the trunnion beds nearer to the axletree, thereby diminishing the liability of overturning the carriage in travelling. 4. Allowing no part of the carriage to project below the plane of the axletrees. This is found necessary to prevent the breaking of the implement fastenings in passing over stumps, stones, &c. 5. More convenient modes of carrying the rammers and trail handspike. 6. As the 12-pdr. carriage can be made of wrought-iron, so as to be little if any heavier than the wooden carriage used for mounting the 3-inch rifle-gun, but one carriage will be required for the field-service. In this case thimbles or washers will be required to fit the trunnions of the 3-inch gun into the trunnion beds of the 12-pdr. carriage.

The very large stock of wooden carriages now on hand may delay, on the score of economy, the introduction of wrought-iron carriages for some time to come; but that such carriages can be made superior to wooden ones scarcely admits of doubt.

INDEX.

Absolute force of gunpowder, 55.

Accidents, from the spontaneous combustion of charcoal, 17; with percussion locks, 297.

Aiming a fire-arm, 299.

Air, effect of the resistance of, on a rifle projectile, 177; resistance of the, 402; fall of a projectile in the, 404; trajectory in the, 412–424.

Alcohol in pyrotechny, 347.

Ames gun, 550.

Ammunition for small-arms, 348; for field and mountain cannon, 351; for siege and sea-coast cannon, 353; preparations for the service of, 358.

Analysis of gunpowder, 30.

Ancient arms, 270.

Ancient guns, 106.

Ancient howitzers, 107.

Ancient mortars, 106.

Ancient theory respecting length of cannon, 127.

Angle of arrival, in ricochet fire, 453.

Angle of fire, definition of, 437; in ricochet firing, 455.

Angle of sight, definition of, 437.

Antimony in pyrotechny, 345.

Appendix, 541.

Armament of sea-coast batteries, 503.

Armor, ancient, 272.

Armor, defensive, 286.

Armor plates, 471, 561.

Armorer, dismounting of fire-arms by an, 335.

Armories of the United States, where situated, 320.

Arms, package and storage of, 330; preservation and care of, when in service, 332.

Arms in service, inspection of, 336.

Armstrong's rifle-gun, 541; projectiles, 555.

Arquebuse, history and description of the, 273.

Arsenals, how classified, 268.

Artillery, material of, 108; system of, 108; first system of, in the 16th century, 109; second system of, in the reign of Louis XIV., 110; Valière's system of, 110; Gribeauval's system of, 111; Louis Napoleon's system of, 112; stock trail system of, 112; important recent improvements in, 113; General Paixhan's system of, 166; implements and machines for, 248.

Artillery carriages, classification of, 212; preservation and repairs of, 259; how to destroy, 259; materials for, 261–266; construction of, 267; painting of, 268.

Artillery harness, 215, 218.

Artillery wheel, 220.

Astragal of field-guns, 115.

Attachment of artillery harness, 216, 217.

Austrian army bullet, 313.

Axle-tree of the artillery carriage, 226; of the field-limber, 231.

Back-action lock, 297.

Bags of powder, explosive force of, 376.

Ballistic machine at West Point, 390, 395; table of times calculated for, 525.

Ballistic machine of Captain Navez, 389.

Ballistic pendulum, 383, 388.

Ballistics, history of, 382.

Balloting in the bore a cause of rotation, 425.

Bands of portable fire-arms, 300.

Barbette sea-coast carriages, 247.

Barrel of portable fire-arms, 290; length of, 316–319; inspection of, 328.

Barrel of the rifle-musket, how cleaned, 333.
Bar-shot, description of, 83.
Base of the breech of cannon, 115.
Base-ring of cannon, 115.
Battery-wagon, how employed, 236; description of, 237.
Battle-axe, ancient, 270.
Bayonet, origin and history of the, 276; description of, 280; inspection of, 329; when unserviceable, 337.
Beeswax and tallow, a lubricant for fire-arms, 316; in pyrotechny, 347.
Bill-hook, 259.
Bill of timber, 266.
Blakeley gun, 545; projectile, 557.
Blistered steel, how made, 140.
Blue color, how produced on the surface of iron and steel, 326.
Blue-light, ingredients for, 369.
Bombard, early cannon, 104; how constructed, 105.
Bomford, Colonel, plan of, for determining pressure of the charge in cannon, 152; columbiad invented by, 190.
Borda, experiments of, in relation to the forms of projectiles, 410.
Bore, influence of length of, on velocity of projectile, 127; effect of length of on maximum charge, 131; length of, in field cannon, 180; use of projectiles not suited to, 435.
Bore of cannon, 116; inspection of, 201; enlargement of, 208, 210.
Bore of fire-arms, length of, 127.
Bore of rockets, 96.
Boring cannon, 199; vent of cannon, 200; musket-barrels, 323.
Bormann fuze, description, 362.
Bow, the ancient, 271.
Breach, firing in, 496.
Breaching walls of fortifications, 481; with rifle-cannon, 485.
Breech of cannon, 115; best form of, for strength, 120; thickness of, 157.
Breeching of artillery horses, 219.
Breech-loading arms, 301; advantages of, 307.
Breech-loading cannon, early, 105; projectiles for, 170.
Breech-sight, how used, 255.
Breech-screw of portable fire-arms, 291.
British service bullet, 313.
Bronze as a material for cannon, 138; density and tenacity of, 139.
Bronze cannon, copper vent-pieces made for, 117; injured by the melting of the tin, 139; inspection of, 204; defects in, 205; how proved, 206; liable to injury from lodgement, 210.
Brooke gun, 550.
Browning musket-barrels, 327.
Buckshot cartridge, 349.
Budge-barrel, for carrying cartridges, 252.
Buildings for pyrotechny, how arranged, 342.
Bullets for small-arms, 76; present form of expanding, 312; in use in the United States and Europe, 312, 314; elongated, proper charge of powder for, 315; how made, 348; effects of, 486.
Butt of a musket, length and shape of, 298.
Butt-plate of portable fire-arms, 300.

Cadet-musket, description of, 317.
Caisson, description of the, 235; how to destroy, 259.
Cake powder made by compression of grain powder, 52; compared with grain powder, 154.
Calibre of cannon, 107; of siege-guns, 184; of portable fire-arms, 293.
Calibres in the American service (*note*), 107.
Canister fire, 468; of field artillery, 492.
Canister-shot, description of, 81.
Cannon, shape of the first, 104; early, construction of, 105; calibre of, 107; devices on, 108; construction of, 114; interior form of, 116; influence of length of bore of, on velocity of projectile, 127; materials best adapted for the construction of, 131, 132; strength of, how affected by cooling, 134; course of the fracture of, in bursting, 136; materials principally used in the fabrication of, 138; thickness of the metal of, 151; exterior form of, 151; nature of force to be restrained by, 153; various kinds of strain upon, 154; nomenclature of the exterior of, 157; peculiar form of, made in Sweden, 159; position of the centre of gravity of, 163; weight of, how determined, 165; different kinds of, 166; rifled, 167; uses to which applied, 180; mountain and prairie, 183; for sea-coast batteries, 188; where made, 194; proof of, 205; bronze, how proved, 206; inspection marks on, 207; how marked when rejected, 207; injuries to, caused by service, 207; how disabled, 260; cost of, 547.

Cannon-balls, preservation and piling of, 93.
Carbine, description of, 317.
Carcass, description of, 84; composition and preparation of, 371.
Carriage of Armstrong gun, 540.
Carriages, artillery, 212; materials for, 261–266.
Cartridge-bags, how made, 352, 353.
Cartridge of the rifle-musket, 349.
Cartridge with buckshot, 359.
Cartridge, metallic, 302.
Cartridges, materials for preparing, 348; for small-arms, how packed, 349; number expended in European wars, 469.
Cascable of cannon, parts of the, 114; knob of the, 163.
Case-hardening, the process of, 325.
Casemate carriages, 247.
Casemate howitzer, 193.
Casemate truck, for moving cannon, 250.
Case of a signal rocket, 366.
Cases for fireworks, how made, 357.
Case-shot, description of, 80; fabrication of spherical, 88; when to be fired in defence or attack of a work, 496.
Cast-iron, projectiles of, 72; effect of cooling on the strength of cannon made of, 134; better adapted for large than small cannon, 136; advantages and disadvantages of, as a material for cannon, 145; causes which affect the quality of, 145; how tested for cannon metal, 145; general properties of, 147; tenacity of, injured by the presence of sulphur, 149; endurance of, 211; injuries to cannon made of, 211; effect of projectiles on, 471.
Cast-steel, how made, and characteristics of, 141.
Cavalry, effect of field artillery against, 490.
Cavalry sabre, description of, 284.
Cavities in cannon, how produced, 208.
Centre of gravity of projectiles, 74; effect of the position of, 178.
Centre of gravity of cannon, position of, 163.
Chain-ball, proposed to be attached to projectiles, 75.
Chain-shot, 83.
Chambers of fire-arms, 121–123.
Charcoal, 15; properties of, 16; quality of, affected by temperature in manufacture, 16; spontaneous combustion of, 17; combustibility of various kinds of, 18, 19; influence of trituration on the combustibility of, 19; quantity of, in gunpowder, 31; pulverized, used in the process of casting cannon, 196; in pyrotechny, 345.
Charge, force of, influenced by the form of its seat, 120; influence of windage on the force of, 125; maximum, 130; the most suitable for smooth-bored cannon, 131; proportion of, to weight of projectile, 131; effect of length of bore on maximum, 131; to determine the pressure of, by calculation, 151; to determine the pressure of, by experiment, 152; for field-guns and howitzers, 181; for siege-guns, 185; for proving cannon, 206; in ricochet fire, to find, 455.
Charge of rupture of shells, 84.
Chase of cannon, 115; thickness of, 158.
Chassis for sea-coast gun-carriages, 245.
Chassis-rails, props of, 247,
Cheeks of artillery carriages, 226.
Chlorate of potassa, used in the manufacture of gunpowder, 14; in pyrotechny, 344.
Classification of cannon, 114; of artillery carriages, 212; of arms in service, 339; of fires, 450.
Closing the breech of breech-loading arms, 302.
Coehorn mortar, description of, 188.
Coke-wash, use of, in the process of casting cannon, 198.
Coloring materials in pyrotechny, 346.
Colt's pistol, description of, 318.
Columbiads, history and description of, 190; improved model of, adopted in 1860, 191; dimensions of the improved, 192; large, and iron-plated vessels (*note*), 473; ranges of, 532; the 16-inch, description of, 551.
Columbiad-shell, bursting charge of, 355.
Combined metals, as materials for cannon, 149.
Combustibility of various kinds of charcoal, 18, 19.
Combustion of gunpowder, velocity of, 39; nature of the products of, 52; gaseous products resulting from, 53.
Compositions for military fireworks, 356; for signal rockets, 366.
Compound projectiles, 73.
Compressible fluid, resistance of, 403.
Compression, strain by, upon cannon, 154.
Concave cutting-edge, action of, 283.

Concentric projectiles, effect of rotation on, 427.
Concussion fuze, description of, 364.
Cone of portable fire-arms, 292; defects in, 337.
Confederate projectiles, 560, 561.
Congreve rocket, how guided, 99.
Congreve, Sir William, improvements made by, in the construction of rockets, 102.
Conical chambers of fire-arms, 122.
Construction of artillery carriages, 267.
Construction of early cannon, 105.
Construction of cannon, 114.
Construction of rockets, 94.
Continuous effort exerted by draught-horses, 214.
Convex cutting-edge, action of, 283.
Cooling, effect of, on the strength of cannon, 134; unequal, effects of in cannon modified by time, 137; influence of, on the color and texture of cast-iron, 149.
Cooling cannon, 136, 198.
Corrosion, cannon metal should be able to resist, 138.
Counter and enfilading fires with siege cannon, 494.
Cracks in cannon, how produced, 208; on the exterior, 211.
Crossbow, the ancient, 272.
Cuirass and helmet, description of, 286.
Curving plates for musket-barrels, 321.
Cuts in cannon, 210.
Cutting-arms, 282.
Cutting and filing in the manufacture of small-arms, 325.
Culverins, 106; very long one cast during the reign of Charles V., 127.
Cylinder and cap in ammunition, 352.
Cylinder-gauge, 201.
Cylinder-staff, 200.
Cylindrical chambers, for fire-arms, 122; injurious effect of, on heavy cannon, 123.

Dahlgren, Captain, two vents placed by in his naval guns, 209; experiments of, in relation to deviation, 430; guns, 550.
Damascus steel, how made, 287.
Damask steel, appearance of, how produced, 142.
Dardanelles, defence of, by means of stone projectiles, 72.
Decorations of signal rockets, 367.
Defects in parts of fire-arms, 337–339.
Defects of timber trees, 265.
Defensive armor, 286.
Defensive weapons, ancient, 272.
Definition of velocities of projectiles, 384.
Deflection of the Armstrong gun, 540.
Delvigne's improvements in loading rifles, 309.
Density of a charge of gunpowder of cylindrical form, 62.
Density of gases developed in the combustion of gunpowder, 58.
Density of gunpowder, influence of, on the velocity of combustion, 43; in relation to force, 56.
Dérivation, or drift, causes of, 432.
Destruction of artillery, 259.
Determination of equations of motion, 395–400.
Deviation, 294, 424; conclusion respecting the causes of, 429; of oblong projectiles, 431; effect of wind in producing, 433; summary of the causes of, 433; measure of, 460; vertical and horizontal, 461; with the 15-inch columbiad, 565.
Devices on old cannon, 108.
Didion, Captain, on trajectory in the air, 412.
Different kinds of cannon, 166; small-arms, 316; fires, 450.
Dimensions of siege mortars, 186; of the siege howitzer, 186.
Direct fire, when used, 450.
Direct fire of field artillery, 490.
Disabling cannon, 260.
Discharging fire-arms, 435.
Dish of the artillery-carriage wheel, 221.
Dismounting arms by a soldier, 332; by an armorer, 335.
Dispart of cannon, 116.
Distance of objects, how estimated, 446.
Distillation, charcoal made by, 16.
Drag-rope, 258.
Draught-harness of the artillery horse, 218.
Draught-horses, force exerted by, 213; table relating to the force of, 214; ordinary load per day for, 215.
Drift, or *dérivation*, causes of, 432.
Drifts, for use in making fireworks, 358.
Drilling and tapping of musket-barrels, 323.
Driving the composition of rockets, 367.
Dry compositions for fireworks, 356.
Drying gunpowder, 25.
Ductility, a desirable quality in cannon metal, 133.
Ductility of wrought-iron, 144.

Dupont's gunpowder compared with Hazard's, 544.
Durability of the musket-barrel, 339.
Dyer projectiles, 559.

Early cannon, shape of, 104; construction of, 105.
Earth, effect of projectiles on, 485; penetrations of projectiles into, 489.
Eccentric handspike for sea-coast carriages, 257.
Eccentric projectiles, effect of rotation on, 427.
Eccentric turning, 324.
Eccentricity in projectiles a cause of rotation, 425.
Eccentrometer, uses of the, 426.
Effective range of field artillery, 488.
Effect of fire in general, 459.
Effects of gunpowder, 35; projectiles, 471.
Elasticity of material for cannon, 132.
Elasticity, absence of, in cast-iron cannon, 145.
Electro-ballistic machines, 389.
Electro-ballistic machine at West Point, 390–395.
Elevating arc of the sea-coast carriage, 245.
Elevating screw of the sea-coast carriage, 244.
Elongated bullets, proper charge of powder for, 315.
Elongated projectile, advantages of pointed out by Robins, 309.
Employment of field artillery, 488; siege cannon, 493; sea-coast cannon, 512.
Endurance of cast-iron cannon, 211; of the 15-inch columbiad, 552.
Enfilading and counter fires with siege cannon, 494.
Enlargement of bore, how produced, 208.
Equations of motion, 406–409; determination of, 395–400.
Equipments and implements for artillery, 251.
Escape of gas from breech-loading arms, 302.
Etching on a sword-blade, 289.
Expanding bullets, present form of, 312.
Expanding projectiles, 169.
Explosion of gunpowder, phenomenon of, 38.
Explosive force of gun-cotton, 70.
Exterior of cannon, 151; nomenclature of, 157.

Fabrication of projectiles, 88; of sword-blades, 287.
Face of cannon, 115.
Fascines, pitched, for incendiary purposes, 374.
Fall of a projectile in the air, 404, 422.
Felling timber for artillery purposes, 264.
Field ammunition, 351.
Field artillery, range of, 462; employment of, 488.
Field cannon, how classified, 180; weight of, 180; length of bore of, 180; charges of, 181; material for, 181; the Napoleon, 182; rapidity of fire of, 449.
Field carriages, turning of, 232; characteristics of, 234; varieties of, 234.
Field-guns, ranges of, 527.
Field howitzers, range of shells fired from, 492; ranges of, 527, 528.
Field-limber, construction of the, 230.
Field rifle-gun, Rodman's, 182.
Field-shells, uses and range of, 463; under what circumstances to be used, 493.
Fillets of field-guns, 115.
Filling-shells, 355.
Fire-arms, important recent improvements in, 113; shape of the chambers of, 121; length of bore of, 127; maximum charge of, 130; recoil of, 130; strongest near the bottom of the bore, 157; various modes of rifling, 168; portable, 273, 290; loading, pointing, and discharging, 435.
Fire-ball, for lighting up enemy's works, 373; use of, in attack or defence, 498.
Fire, preparations for communicating, 380.
Fire, tables of, 447; rapidity of, 448.
Fire in general, effect of, 459.
Fire of case-shot, in attack or defence, 496; of the siege howitzer, 497; of mortars, 499.
Fire-stone, description, preparation, and use of, 370–371.
Fireworks, proportions of nitre, charcoal, and sulphur for, 42; military, 356; for signals, 366; incendiary, 370; for light, 372–375; ornamental, 376; offensive and defensive, 376.
Firing at night, 444.
Firing in breach, 496.
Firing to effect a breach, rules for, 484.
First reinforce, thickness of, 157.
Fixed ammunition, 353.
Flame in ornamental fireworks, 378.
Flats of portable fire-arms, 292.
Flight of projectiles, to determine time of, 422.

Flint-lock, when introduced and discarded, 276.
Foot artillery sword, 285.
Foot-boards of the field-limber, 232.
Force, loss of, from windage, 124; nature of, to be restrained by cannon, 153.
Force of draught-horses, table relating to, 214.
Force of gunpowder, 55; relation between, and density, 56; when inflammation is instantaneous, 58; when inflammation is not instantaneous, 60; when inflammation is instantaneous or progressive, 66.
Forces acting on a gun-carriage, 228.
"Forcing" projectiles into rifle-barrels, 307.
Forge, travelling, description of, 236.
Forging a sword-blade, 287.
Fork of the field-limber, 231.
Form of a table of fire, 448.
Form of cannon, 151.
Form of projectile, theory in relation to, 409; results of experiments in relation to, 411.
Form of siege-guns, 185.
Formula for initial velocity, 385.
Fraser gun, 542.
French army bullets, 313.
French musket-barrel, durability of, 339; guns, 554; projectiles, 556.
Friction, the object of a carriage-wheel to diminish, 221.
Friction-powder, composition of, 360.
Friction-tube for firing cannon, 359.
Front sight of portable fire-arms, 299.
Funnel, for loading shells, 254.
Furnaces in laboratories, 343.
Furrows in cannon, how produced, 208.
Fuze instruments, 253.
Fuzes, percussion, concussion, and time, 360–366.

Galileo on the path of projectiles, 382.
Garrison and siege cannon, 184.
Gas, law of formation of, in the combustion of gunpowder, 44–46; table of quantities of, developed from various grains of gunpowder, 48; loss of, by the fuze-hole of a shell, 87; escape of, from breech-loading arms, 302.
Gases developed in the combustion of gunpowder, density of, 58.
Gatling gun, 554.
Gauges for inspecting fire-arms, 336.
Gerbe, in ornamental fireworks, 377.
Gin, an instrument for raising cannon, 248.
Glazing gunpowder, 25.
Globe and telescopic sights of fire-arms, 299.
Gomer chamber of fire-arms, advantages of, 123.
Graduation of rear-sights, 445.
Grain of various kinds of gunpowder, 27; results of experiments on six sizes of, 67.
Grained powder and caked powder, 154.
Granulating gunpowder, 24.
Grape and canister, inspection of, 91.
Grape and canister firing, 468.
Grape-shot, how composed, 80.
Gray iron, characteristics of, 147.
Greener's rifle projectile, 311.
Grenades, 79.
Gribeauval's system of artillery, 111; his method of attaching artillery horses, 217.
Grinding a sword-blade, 288; small-arms, 325.
Grommets or ring-wads, 355.
Grooved balls, 75.
Grooves, form of, in rifled fire-arms, 171, comparative advantages of variable and uniform, in rifled fire-arms, 173; method of cutting, in cannon, 173; number, width, and shape of, in cannon, 174; inclination of, in rifled fire-arms, 175–179; objects to be attained by, in rifles, 294; kind adopted by the United States Government, 295; rotation produced by, in gun-barrels, 294.
Guard-bow of portable fire-arms, 300.
Guard-plate of portable fire-arms, 300.
Gum-arabic in pyrotechny, 347.
Gun-carriages, classification of, 225; important requisites of, 225; forces acting on, 228; construction of, 234; sea-coast, 243; improvements, 569.
Gun-cotton, how prepared, 68; projectile force of, 69; explosive force of, 70.
Gun-metal, character of, 146; density and tenacity of, 147.
Gun-pendulum, for determining velocities, 388.
Gun-platform, how constructed, 242.
Gunner, implements carried by, 254.
Gunnery, science of, 382.
Gunpowder, general theory of, 7; composition of, 8; manufacture of, 21–23; pressing and granulating, 24; simple method of manufacturing, 25; glazing, drying, and dusting, 25; general qualities of good, 27; inspection and proof of, 27; size of grain of various kinds of, 27; specific gravity of, 28; mercury densimeter for, 28; initial velocity of, 29; how packed, 30; inspection report as to the qualities of, 30; quantity of charcoal in, 31; quantity of saltpetre in, 31; hygro-

metric qualities of, 32; quantity of sulphur in, 32; quickness of burning of, 33; unserviceable, how restored, 33; storage and preservation of, 34; history of, 35; transportation of, 35; effects of, 35; phenomenon of explosion of, 38; ignition of, 38; combustion of, progressive, 40; influence of purity and proportions of ingredients on combustion, 41; influence of density of, on velocity of combustion, 43; effects of trituration of ingredients, 43; velocity of combustion of, increased by moistening and subsequent drying, 44; law of formation of gaseous products from, 44; combustion of a spherical grain of, 45; combustion of a polyhedral grain of, 46; combustion of a grain of ordinary, 47; inflammation of, 49; influence of the size of the grain of, on combustion, 49; velocity of inflammation of, diminished by compression (*note*), 52; products of the combustion of, 52; gaseous products resulting from the combustion of, 53; for war purposes, proportions of, 53; temperature of the gaseous products of, 54; determination of the force of, 55; relation between the density and force of, 56; density of the gases developed in the combustion of, 58; force of, when inflammation is instantaneous, 58; force of, when inflammation is not instantaneous, 60; density of a cylindrical charge of, 62; expansive force of instantaneous or progressive inflammation of, 66; results on velocity, of various densities of, 67; for proving cannon, 205; injuries to cannon from, 208; in pyrotechny, 345.

Guns, ancient, 106; distinguishing characteristics of, 166; how proved, 206; how pointed, 439; for sea-coast service, 190; trajectory of projectiles of, 414–418.

Halberd, 271.

Hale's rocket, 98.

Hand-arms, ancient, 270; classification of, 277; general principles of, 277.

Hand-grenade, 79; description of Ketchum's (*note*), 80.

Handles of bronze field-pieces, 163.

Handling the sabre, 283.

Hardening and tempering steel, 326.

Hardness a necessary quality in cannon metal, 137.

Hardness of wrought-iron, 144.

Harness for artillery horses, 215.

Haversack, gunner's, for carrying cartridges, 252.

Head-gear of artillery horses, 218.

Heavy charges for cannon, 120.

Helmet and cuirass, description of, 286.

History of gunpowder, 35; of rockets, 102; of cannon, 104; of small-arms, 270; of ballistics, 382.

Hollow projectiles, cavities of, how made, 89; inspection of, 91, 92.

Hollow-shot, various denominations, 77.

Hotchkiss' rifle-projectile, 558.

Hot-shot, hay wads used for, 355; how prepared and used, 372.

Hounds of the field-limber, 231.

Howitzers, ancient, 107; characteristics and uses of the, 166; for mountain service, 183; for siege purposes, 186; weight of, 180; for sea-coast service, 193; how proved, 206; how pointed, 439; trajectory of projectiles of, 414–418; mountain, range of, 463.

Howitzer carriages, construction of, 234.

Hutton on the paths of projectiles, 383; experiments of, in relation to the forms of projectiles, 410.

Hygrometric qualities of gunpowder, 32.

Ignition of gunpowder, 38.

Implements and equipments for artillery, 251.

Incendiary fireworks, 370.

Incendiary match, how made, 372.

Inclination of grooves in rifled fire-arms, 175; limit of, 179.

Inclination most suitable for grooves in rifled fire-arms, 179.

Incompressible fluid, resistance of, 402.

Incorporating ingredients of gunpowder, 23.

Infantry, effect of field artillery against, 489.

Inflammation of gunpowder, 49; circumstances influencing the velocity of, 50; force developed by instantaneous, 58; force developed by, when not instantaneous, 60.

Ingredients of gunpowder, 21.

Initial velocity, of gunpowder, 29; of projectile from rifled fire-arms, 174; with American small-arms, 316–319; of projectiles, 384; causes affecting, 387; formula for, 385; practical rule for, 387; determination of, by experiment, 388; of a mortar-shell,

to find, 421; causes which affect, 424.
Initial velocities, table of, with service charges, 387; tables of values for, 508, 514.
Injuries to cannon, caused by service, 207; from the projectile, 209.
Inspection of gunpowder, 27; of projectiles, 90; of cannon, 200, 201; of bronze cannon, 204; of sword-blades, 289; of small-arms, 327; of arms in service, 336.
Inspection marks on cannon, 207.
Inspection report as to the qualities of gunpowder, 30.
Instruments for the inspection of hollow projectiles, 91; for the inspection of cannon, 200.
Interchange of parts in similar arms, 320, 339.
Interior form of cannon, 116.
Iron cannon, how proved, 206.
Iron ores used by the United States Government for the manufacture of cannon, 146.
Iron parts of a gun-carriage, 227.
Iron-plated ships, the smaller sea-coast guns useless against, 189; effects of projectiles on, 472.
Iron sling-cart, for moving cannon, 249.
Irreparable arms, 339.

James' rifle, dimensions of, 319; projectile for, 559.
Javelin, Roman, 271.
Jets, in movable fireworks, 379.

Knob of the cascable, 163.
Krupp gun, 548.

Laboratory, military, how it should be situated, 342; precautions to be used in, 343.
Lackering projectiles, 93.
Ladle, for withdrawing projectiles, 252.
Lampblack in pyrotechny, 346.
Lance, description of the, 279; advantages of the use of the, 280; mode of carrying on horseback, 280.
Lance, Macedonian, 270.
Lance, in ornamental fireworks, 377.
Land-batteries, advantages of, 502.
Large cannon for sea-coast batteries, 189.
Law of resistance, 402.
Lead, projectiles of, 72.
Length of barrel, of portable fire-arms, 294; influence of, on velocity of projectile, 294.
Length of bore, of fire-arms, influence of, on velocity of projectile, 127; effect of, on maximum charge, 131; experiments to determine the influence of, 128; of field cannon, 180; of siege-guns, 185.
Length of projectiles, effect of, 175.
Length of stock of field carriages, 233.
Level, gunner's, description of, 254.
Lever-jack, 250.
Lifting-jack, 250.
Light-artillery sabre, description of, 284.
Light-balls, 374.
Light-barrel, how constructed, 376.
Light charges for cannon, 121.
Limber, object of the, 230; for siege carriages, 240.
Limbering, difficulty of the old system of, 232.
Line of fire, definition of, 437.
Line of sight, definition of, 436.
Liquid compositions for fireworks, 356.
Load of pack-horses, 213; of artillery horses, 215; of field carriages, 233.
Loading cannon, implements for, 251.
Loading fire-arms, 435; improvements in, 276.
Loading rifles, various modes of, 307.
Lock of a portable fire-arm, 295; conditions to be fulfilled in the construction of, 296; inspection of, 330; defects in, 338.
Lock of a rifle-musket, how cleaned, 334; how taken apart, 335.
Locking wheels of artillery carriages, 233.
Lodgement, injury to cannon caused by, 209; means used to prevent, 210.
Longitudinal strain upon cannon, 154.
Long ranges of siege cannon, 493.
Loss of velocity by resistance of the air, 406.
Louis Napoleon's system of artillery, 112.
Lubricant for fire-arms, 315.

Macedonian lance, 270.
Machines and implements for artillery, 248.
Magnus, Prof., apparatus devised by, 428.
Manœuvre of artillery carriages, implements for, 256.
Manœuvring handspike for sea-coast carriages, 257.
Manufacture of cannon, 194; of small-arms, 320; of sword-blades, 287.
Marks, inspection, on cannon, 207.
Marrons in signal rockets, 369.
Martello tower, rapid destruction of one with rifle projectiles, 485.

Masonry, effect of projectiles on, 480.
Matchlock, description of the, 274.
Materials for the fabrication of cannon, 138; for field cannon, 181; for sea-coast carriages, 243; for artillery carriages, 261–266; for pyrotechny, 344–348.
Matériel of artillery, 108.
Maximum charge of fire-arms, 130.
Maynard's primer, how made, 350.
Maynard's self-priming percussion lock, 297.
"Mealed powder," 37; in pyrotechny, 345.
Measure of deviation, 460.
Measures, for charges of powder, 254.
Mechanical principles determining initial velocities, 384.
Men's harness, 258.
Mercury densimeter for gunpowder, 28.
Metal for cannon, qualities desirable in, 132; the various kinds used, 138.
Military fireworks, 356.
Mill-cake, pressing, 24.
Milling, in the manufacture of small-arms, 324.
Minié, form of projectile proposed by, 309; his rifle projectile, 311.
Model, wooden, for moulding cannon, 195.
Models for ordnance *matériel*, 268.
Molten iron used for incendiary purposes (*note*), 372.
Momentary effort exerted by draught-horses, 214.
Mordecai, Major, results of experiments of, with gun-cotton, 69; experiments of, on the influence of length of bore, 129.
Mortar, nitre obtained from, 12.
Mortar, a term applied to early cannon, 104; ancient, 106; characteristics and uses of, 167; for siege purposes, 186; form of, adopted in 1861, 187; spherical case-shot, for, 188; how proved, 206; inspection of, 204; how pointed, 442; fire of, in attack or defence, 499; case-shot, fire of, in attack or defence, 501; uncertainty of the fire of, from shipboard, 504.
Mortar-fuze, description of the, 361.
Mortar-platform, how constructed, 242.
Mortar-scraper, 254.
Mortar-shells, 79; bursting charge of, 355; table of times of flight of, 401; table of ranges of, 401; to find the initial velocity of, 421; how fired, 436; fire of, 464; tables of fire, 522, 523.
Mortar-wagon for siege purposes, 240.
Motion, equations of, 406, 409.
Motion of rockets, 95, 96; of a projectile in vacuo, 395.
Mottled iron, characteristics of, 148.
Mould for casting cannon, how formed, 196.
Moulding cannon, process of, 195.
Mountain ammunition, 351.
Mountain cannon, 183.
Mountain howitzer, 183.
Mountain howitzer carriage, requisites of, 237; description of, 238.
Mountain shells, range of, 463.
Mountings of portable fire-arms, 300; inspection of, 330.
Mountings of rifle-musket, how cleaned, 334.
Movable pieces of fireworks, 378.
Mules as pack-animals, 213.
Multipliers, tables of, 505.
Musket, introduced by Charles V., 274; boxes for packing, 330.
Musket-barrels, how made, 321; how browned, 327; defects in, 337; durability of, 339; strength of, 340.
Mutzig, trials made at, as to the strength of musket-barrels, 340.
Muzzle of cannon, 116; enlargement of, 210.

Nail-ball, 75.
Napoleon field-gun, description of, 182, advantages and disadvantages of, 183.
Natural angle of sight, 116, 181; of siege-mortars, 187.
Natural line of sight on cannon, how marked, 207.
Natural point-blank, definition of, 438.
Natural steel, 140.
Nave-box of the artillery carriage, 227.
Navez, Captain, ballistic machine of, 389.
Neck of cannon, 115.
Newton on the path of projectiles, 382.
Night-firing, 444.
Nitrate of soda, 15.
Nitre, preservation of, 12; refining of, 12; for laboratory use, 344.
Nitre-beds, how made, 11.
Nomenclature of artillery carriages, 225; of cannon of old pattern, 114; of artillery carriage-wheels, 220; of portable fire-arms, 291; of a percussion lock, 296.

Oak, effect of projectiles on, 474, 481.
Objects, distance of, how estimated, 446.
Oblong bullet, description of, used in the United States service, 77.

Oblong projectiles, superiority of, 74; trajectory of, 423; deviation of, 431.
Offensive and defensive fireworks, 376.
Ordnance and ordnance stores, what constitute, 108.
Ores of iron, used for the manufacture of cannon, 146.
Ornamental fireworks, 376.

Pack-horse, work of, to be regulated, 212.
Packing gunpowder, 30; arms, 330; small-arm cartridges, 349.
"Packing the joint" of breech-loading arms, 302.
Painting artillery carriages, 268.
Palliser gun, 547; projectiles, 473–561.
Paper, laboratory, classes of, 347.
Paper shells, how made, 380.
Parrott rifle-gun, how constructed, 548; ranges of, 527, 529.
Parrott's rifle projectile, 557.
Parsons gun, 548.
Pass-box, for carrying cartridges, 252.
Patch, mode of using with rifles, 308.
Pattern for moulding cannon, 195.
Pendulum hausse, how used, 255.
Penetration of projectiles, 476, 544, 545; of the rifle-musket bullet, 486, 546.
Percussion bullets, how made, 83.
Percussion-caps, description of, 350.
Percussion-fuze, description of, 365.
Percussion-lock, nomenclature of, 296.
Perriere, an early form of cannon, 105.
Petard, explosive, thrown aside, 376.
Petard, in ornamental fireworks, 377.
Petronel, description of the, 274.
Pig-iron, character of, 146; density and tenacity of, 147.
Pike, the ancient, 270.
Piling of cannon-balls, 93.
Pintle, the centre of the chassis of sea coast gun-carriages, 246.
Pintle-hook of the field-limber, 232.
Pistol, when and where invented, 274.
Pistol-carbine, description of, 318.
Pistol cartridge, how made, 350.
Pitch in pyrotechny, 347.
Pitched fascines, for incendiary purposes, 374.
Plane of fire, definition of, 437.
Plane of rupture of shells, 84.
Plane of sight, definition of, 437.
Platforms of siege pieces, how constructed, 241.
Plunging-fire, 459.
Point-blank, definition of, 437.
Pointing fire-arms, 435.
Pointing guns, implements for, 254.
Pointing small-arms, 445.
Pole of the field-limber, 231.
Polishing a sword-blade, 289; small-arms, 325.
Poplar, charcoal for gunpowder made from, 15.
Portable fire-arms, 290; early history of, 273; calibre of, 293.
Port-fires, composition of, 359.
Position of the centre of gravity of projectiles, 178.
Pot of signal-rockets, 367.
Potassa, chlorate of, used in the manufacture of gunpowder, 14; in pyrotechny, 344.
Powder, proper charge of, for elongated bullets, 314; see *Gunpowder*.
Powder, round, 25.
Powder-proof of cannon, 205.
Practical rule for initial velocity, 387.
Prairie artillery carriage, description of, 238.
Prairie-cannon, 183.
Precaution in using fire-arms, 336; against accidents in pyrotechny, 343; in loading cannon, 435.
Preponderance of cannon, 161; mortars and columbiads now made without (*note*), 162; of the siege-howitzer, 186.
Preservation of gunpowder, 34; of cannon-balls, 93; of artillery harness, 219; of artillery carriages, 259; of timber, 265; of arms in service, 332.
Pressure in the bore of the 15-inch columbiad, 542.
Priming-wire, for pricking cartridges, 253.
Pritchett bullet, how made, 313.
Projectile arms, ancient, 271.
Projectile force of gun-cotton, 69.
Projectiles, materials of, 71; advantages of spherical, 73; centre of gravity of, 74; oblong, superiority of the form of, 74; solid, for what purposes adapted, 75; fabrication of, 88; hollow, cavities of, how made, 89; inspection of, 90; maximum velocity of, on what dependent, 129; proportion of, to weight of charge, 131; with flanges, for rifles, 168; constructed on an expanding principle, 169; effect of length of, 175; influence of the resistance of the air upon, 177; effect of the position of the centre of gravity of, 178; shells and hot-shot the best for sea-coast batteries, 189; injuries to cannon from, 209; for small-arms, 307; of the Whitworth rifle, 311; initial velocity of, 383; fall of, in the air, 404; theory in relation to the form of, 409; oblong, trajectory of, 423; de-

viation of, 424; concentric, effect of rotation on, 427; eccentric, effect of rotation on, 427; oblong, deviation of, 431; smaller than the bore, how used, 435; effects of, 471; of the Armstrong gun, 528.
Projections, omitted on the surface of cannon of late construction, 160.
Prolonge, a stout hempen rope, 258.
Proof of gunpowder, 27; of cannon, 205; of sword-blades, 289; of scabbard, 290.
Properties of bronze, 139; of steel, 142.
Props of chassis rails, 247.
Puddled steel, how obtained, 142.
Pulverizing charcoal, 23.
Pyrotechny, definition of, 342; buildings for, 342; precautions against accidents in, 343; materials for 344–348.
Pyroxile, or gun-cotton, 68.

Quadrant, gunner's, 256.
Queen Anne's pocket-piece, 106.
Quick-match, description of, 359.

Rail platform, how constructed, 242.
Rammer-head used in the inspection of cannon, 201.
Rammer-head for loading cannon, 251.
Rampart grenade, 79.
Ramrods of portable fire-arms, 301; inspection of, 329.
Range, to determine, 422; definition of, 439.
Range of field artillery, 462, 463, 488; of siege cannon, 493; of the Armstrong projectile, 530; of artillery, 516–523.
Ranges, table of values for calculation of, 507, 513.
Ranges of mortar-shells, table of, 401.
Ranges obtained with Captain Rodman's 15-inch columbiad, 524.
Rapidity of fire, 448.
Rear-sight of portable fire-arms, 299; aiming without, 300.
Rear-sights, graduation of, 445.
Recoil, increase of, greater than increase of charge, 130; diminished by application of weights, 137.
Reinforce of cannon, 115.
Rejected cannon, how marked, 207.
Repairs of artillery carriages, 259.
Remaining velocities, table of, 543.
Resistance of the air, 402; effect of, on a rifle-projectile, 177; loss of velocity by, 406.
Retempering a sword-blade, 288.
Ricochet fire, 450; when used, 452; practical rules for, 453; of field artillery, 492; of sea-coast batteries, 504.
Ricochet firing, cartridge-bag for, 354.
Rifle-cannon, hardness of metal necessary to, 137; definition of, 167; breaching with, 485.
Rifled fire-arms, form of groove of, 171; variable groove of, 172; limit of inclination of grooves in, 179.
Rifle-grooves, points to be observed in constructing, 294.
Rifle-guns, ranges of, 522.
Rifle-musket, description of the, 316; how dismounted, 332; trials of strength of, 340; cartridge, 349; bullet, penetrations of, 486.
Rifle-projectiles, different systems of, 545.
Rifle siege-gun, description of (*note*), 184.
Rifles, how classified, 168; invention and history of the, 275; original mode of loading, 308; as distinguished from the rifle-musket, 317; American sporting, 319.
Rimbases of cannon, 116; description of, 162.
Robins, elongated projectiles proposed by, 309; on the path of projectiles, 383; deviation by rotation discovered by, 431.
Rockets, theory and construction of, 94; motion of, 95; origin of movement in, 96; guiding principle of, 97; Hale's system for guiding, 98; Congreve's system for guiding, 99; how fired, 99; signal and war, 99; form of trajectory of, 100; effect of wind on the trajectory of, 101; history of, 102; advantages claimed for, over cannon, 102; kinds used in the United States service, 103; signal, principal parts of, 366.
Rodman, General, experiments of, with cake-powder (*note*), 49; experiments of, in relation to the density of gases (*note*), 58; plan of, for cooling cannon from the interior, 136; plan of, for determining the force of the charge in fire-arms, 152; experiments of, on the force of cake and grained powder (*note*), 153; his formula for calculating the exterior form of large cannon (*note*), 157; great improvements of, in casting large guns, 552.
Roller handspike, for new sea-coast carriages, 257.
Rolling fire, 458.
Rolling friction, 223.

Rolling musket-barrels, 321.
Roman candles, how made, 380.
Rope matting impenetrable to small-arm projectiles, 487.
Rosin in pyrotechny, 347.
Rotation, initial velocity of, in rifle-projectiles, 174; a principal cause of deviation of projectiles, 425; effect of, in producing deviation, 427; of the earth, influence of, in producing deviation, 434.
Round bullets, denomination of, 76.
Round powder, how made, 25.
Rules for firing to effect a breach, 484.
Rules for ricochet fire, 453.
Rumford, Count, experiments of, on the combustion of gunpowder, 54; eprouvette used by, for determining the force of gunpowder, 55.
Rupture of cannon, tendency to, greatest from tangential force, 156.
Rupture of shells, 84.

Sabot, description of the, 351.
Sabre, the ancient, 271; best construction of for handling, 283; description of, 284.
Saddle of the artillery horse, 218.
Saltpetre, description of, 9; whence obtained, 10; test of rough, 13; test of pure, 14; quantity of, in gunpowder, 31.
Sand, the best kind for moulding cannon, 196.
Sarissa, or Macedonian lance, 270.
Sawyer's rifle-projectile, 559.
Saxon, in movable fireworks, 379.
Scabbard, uses and material of the, 285; proof of, 290.
Schenkle's rifle-projectile, 558.
Schönbein, Prof., gun-cotton discovered by, 68.
Schwartz, Berthold, explosive properties of gunpowder discovered by, 36.
Science of gunnery, 382.
Scratches in cannon, 210.
Screw-jack, used in greasing carriage-wheels, 259.
Scott projectile, 557.
Sea-coast ammunition, 353.
Sea-coast cannon, 188; various kinds of, 190; rapidity of fire of, 449; employment of, 502.
Sea-coast carriages, classification of, 243; various kinds of, 247.
Sea-coast defence, fires advantageous in, 503.
Sea-coast fuze, description of, 363.
Sea-coast gun-carriages, 243.
Sea-coast guns, ranges of, 519.
Sea-coast howitzers, description of, 193; ranges of, 519.
Sea-coast howitzer shell, 79.
Sea-coast mortars, description of, 193; the new, 194; ranges with the 13-inch, 541.
Sea-coast shells, range and force of, 464.
Searcher, 201.
Seasoning and preserving timber, 265.
Seat of the charge of cannon, 120.
Second reinforce, thickness of, 158.
Self-priming percussion-lock, Maynard's, 297.
Serpents in fireworks, how composed, 368.
Sharp's system of "packing the joint" of breech-loading arms, 303.
Shear-steel, various kinds of, 141.
Shell-hooks, 254.
Shell-plug screw, 253.
Shells, material and denomination of, 77; spherical, description of, 78; charge of rupture of, 84; loss of gas by the fuze-holes of, 87; strapped to sabots, 354; service bursting-charges of, 355; loaded as carcasses, 372; paper, how made, 380; French mortar, table of times of flight of, 401; force and range of, 463.
Shod handspike, for mortars, 257.
Shot, inspection of, 90; for proving cannon, 206; wedged in cannon, how drawn out, 261.
Shrapnel, Colonel, spherical case-shot perfected by, 82.
Shrapnel firing, 465.
Siege and sea-coast ammunition, 353.
Siege-cannon, 184; rapidity of fire of, 449; employment of, 493.
Siege-carriages, various kinds of, 239.
Siege gun-carriage, construction of, 239.
Siege-guns, characteristics of, 184; ranges of, 518.
Siege-howitzer, description of, 186; fire of, in attack or defence, 497; ranges of, 519.
Siege-mortars, description of, 186; bed, how constructed, 241; ranges of, 520.
Siege-shells, uses of, 464.
Sights of portable fire-arms, 298.
Signal-rockets, 99; principal parts of, 366.
Signals, fireworks for, 366.
Sinope, destruction of the Turkish fleet at, by Russian shells, 189.
Skelp of a sword-blade, 287.
Sky-rockets, 378.
Sleeves, gunner's, 254.

Sling, the ancient, 271.
Sling-carts, wooden and iron, 249.
Slow-match, description of, 358.
Small-arm firing, 469.
Small-arm projectiles, 307.
Small-arms, classification of, 268; charge of powder for, 314; different kinds of, 316; manufacture of, 320; inspection of, 327; ammunition for, 348; trajectory of projectiles of, 414, 418; rapidity of fire of, 449; how pointed, 445.
Soda, nitrate of, 15.
Solid-shot, classification of, 75; penetrations of, 478; when broken, 478, 481.
Solid-shot firing, 462.
Solubility of nitre, 9.
Sparks in fireworks, how produced, 346.
Spherical case-shot, description of, 82; fabrication of, 88; for mortars, 188; late improvements in, 466; effect of, from rifle-cannon, 467.
Spherical chamber of fire-arms, 123.
Spherical projectiles, advantages of, 73.
Spherical shells, description of, 78.
Spiking cannon, process of, 260.
Spirits of turpentine in pyrotechny, 347.
Splinter-bar of the field-limber, 231.
Sponge, for cleaning out cannon, 251.
Sponge-bucket, for washing bore of cannon, 258.
Spontoon, or half-pike, 271.
Sporting rifle, American, 319.
Springfield, Mass., United States Armory at, 320.
Sprue, used in casting cannon, 196.
Stand of ammunition, how composed, 351.
Star-gauge, 200.
Stars of signal rockets, 367.
Steamer Princeton, why the large wrought-iron gun burst on, 144.
Steel as a material for cannon, 139; characteristics of, 140; properties of, 142; Damascus, how made, 287; how hardened and tempered, 326.
Steel plates and iron, comparative resistance of, to projectiles, 474.
Stick of signal rockets, 369.
Stock of an artillery carriage, 225.
Stock of portable fire-arms, 298; defects in, 338; inspection of, 329.
Stock trail system of artillery, 112.
Stone mortar, uses and dimensions of, 187; charge of, 465.
Stone projectiles, 71.
Storage of gunpowder, 34; arms, 331.
Storehouse for artillery harness, 219.
Straightening plates of musket-barrels, 321.
Straight sword, parts of, 278.
Strain upon cannon, various kinds of, 154.
Strapped ammunition, 353.
Strapping shells described, 354.
Straps of a stand of ammunition, 352.
Strength of cannon, effect of form of chamber on, 124.
Strength of material of cannon, 132, 134.
Strength of musket-barrels, 340.
Strength of wrought-iron, 143.
Structure of rockets, 94.
Sulphur, purification of, 20; properties of, 20; quantity of, in gunpowder, 32; the tenacity of cast-iron destroyed by, 149; in pyrotechny, 345.
Sulphuret of antimony in pyrotechny, 345.
System of artillery, 108.
Swaging, in the process of making musket-barrels, 323.
Sweden, peculiar form of cast-iron cannon cast in, 159.
Swell of the muzzle of cannon, 115; thickness of, 159.
Swiss service bullet, 314.
Sword, the ancient, 271.
Sword, the straight, parts of, 278.
Sword-bayonet, description and uses of the, 281.
Sword-blades, manufacture of, 287.

Table of initial velocities with service-charges, 387.
Tables of fire, 516; purpose of, 447.
Tables of multipliers, 505.
Tangential strain upon cannon, 154.
Tangent-scale, how used in pointing guns, 255.
Tar in pyrotechny, 347.
Tar-bucket, for carrying grease, 258.
Targets, construction of, 461.
Tarred-links, how made and used, 374.
Tartaglia on the path of projectiles, 382.
Telescopic sights of fire-arms, 299.
Temperature, importance of, in the manufacture of charcoal, 17.
Temperature of the gaseous products of gunpowder, 54.
Tempering steel, 326.
Tempering sword-blades, 287.
Tenacity of bronze, 139.
Tenacity of puddled steel, 142.
Theory and construction of rockets, 94.
Thickness of the metal of cannon, 151.

Thouvenin, form of projectile proposed by, 309.
Thrusting arms, 277, 284.
Thumbstall, for closing vent, 253.
Tige, or spindle, of Colonel Thouvenin, 310.
Tilted steel, 141.
Timber, for artillery carriages, 261–266; seasoning and preserving, 265.
Time-fuze, description of, 360.
Times, table of, for West Point ballistic machine, 515.
Tin, bronze cannon injured by the melting of the, 139.
Torches, preparation and use of, 375.
Tourbillion, description of the, 379.
Track of field carriages 233.
Trail handspike for field carriages, 256.
Trajectory of rockets, 100; of projectiles, ancient theory respecting, 382; in the air, 412–424; under high angles of projection, 420; of oblong projectiles, 423; true and calculated, 423.
Transportation of gunpowder, 35.
Transverse strain upon cannon, 155.
Travelling-forge, description of, 236.
Treadwell, Prof., plan of, for combining wrought-iron and cast-iron in cannon, 150.
Trees, selection of, for timber for artillery purposes, 263; defects of timber, 265.
Trench-cart, for use in trenches, 251.
Trigger of portable fire-arms, 300.
Trituration, influence of, on the combustibility of charcoal, 19; effect of, on the ingredients of gunpowder, 43.
Trunnion-gauge, 201.
Trunnion-rule, 201.
Trunnion-square, 201.
Trunnions of cannon, 115; description of, 160; influence of position of, on recoil, 160; size of, dependent on recoil, 160; importance of position of, on siege and sea-coast cannon, 162; verification of the axis of, 203.
Trunnions of the siege-howitzer, 186.
Truck for moving cannon in casemates, 250.
Truck handspike, for sea-coast carriages, 257.
Tube-pouch, worn by a cannonier, 252.
Turning cannon, 199; musket-barrels, 324.
Turning of field-carriages, 232.
Turpentine in pyrotechny, 346.
Twist, the inclination of a rifle-groove, 171.

Uniform groove in rifled fire-arms 171.
United States service bullet, 312.
Unspiking cannon, process of, 261.
Uses to which cannon are applied, 180.

Valière's sytem of artillery, 110.
Values, tables of, 510–514.
Variable groove in rifled fire-arms, 172.
Vauban's method of breaching walls, 482.
Velocities and times, tables of values of, 129, 512.
Velocity, initial, of projectiles, 383; maximum of, dependent on what, 129; loss of, by resistance of the air, 406; causes affecting, 424.
Velocity of combustion of gunpowder, 39.
Venice turpentine in pyrotechny, 347.
Vent, small loss of force by the escape of gas through, 119; table illustrating influence of position of, 119.
Vent of Armstrong gun, 528.
Vent of cannon, position of, 117; size of, 117; how bored, 200; inspection of, 204; wear of, how obviated, 208
Vent of fireworks, 358.
Vent of rockets, 95.
Vent-gauges, 201.
Vent-piece, copper, adopted for rifle guns (*note*), 117.
Vent-searcher, 201.
Vertical field of fire of field-guns, 181; of siege-guns, 185; of siege-mortars, 187; of the new columbiad, 192.

Wade, Major, experiments of, with eccentric shells, 430.
Wads (junk, hay, and ring), how used, 355.
Wagon, battery, 236.
War-club, ancient, 270.
War-powder, proportions of ingredients of, in the United States service, 53.
War-rockets, 99.
Water, penetration of projectiles into, 480.
Watering bucket, artillery, 258.
Weapons, ancient defensive, 272.
Weight, recoil diminished by, 137.
Weight of American small-arms, 316–319.
Weight of barrel of portable fire-arms, 291.
Weight of cannon, how determined, 165.
Weight of field-guns and howitzers, 180.

Weight of projectile and powder of American small-arms, 316–319.
Weight of siege-guns, 185.
Welding, description of the process of, 321.
Welding plates for musket-barrels, 321.
West Point ballistic machine, 390–395; table of times calculated for, 515.
Wheels of artillery carriages, names of parts of, 220; objects of, 221; influence of size of, 224; greased and ungreased, 227; should be of the same height, 233.
Wheels of gun-carriages, weight of, 224; should be few kinds of, 224.
White iron, characteristics of, 148.
White pine, effect of projectiles on, 475.
Whitworth rifle-gun, 544; projectiles, 556.
Willow, charcoal for gunpowder made from, 15.
Wind, deviation caused by, 433; effect of, on the trajectory of rockets, 101.
Windage, loss of force from, 124.
Wood, effect of projectiles on, 474; penetrations of projectiles into, 479.
Wootz, a natural steel from India, 140.
Worm, for withdrawing cartridges, 252.
Wounds made by thrusting-swords, 279.
Wrought-iron, projectiles of, 72; as a material for cannon, 143; for artillery carriages, 266; effect of projectiles on, 472.
Wrought-iron cannon made by the Phœnix Iron Co., 181.

Yataghan of the Arabs, 283.

Zorndorf, effect of a field-cannon ball at the battle of, 487.

Barre Duparcq's Military Art and History. Elements of Military Art and History: comprising the History and Tactics of the separate Arms; the Combination of the Arms; and the minor operations of War. By EDWARD DE LA BARRE DUPARCQ, Chef de Bataillon of Engineers in the Army of France; and Professor of the Military Art in the Imperial School of St. Cyr. Translated by Brig.-Gen. GEO. W. CULLUM, U. S. A., Chief of the Staff of Major-Gen. H. W. HALLECK, General-in-Chief U. S. Army. 1 vol. 8vo, cloth. $5.00.

'I read the original a few years since, and considered it the very best work I had seen upon the subject. Gen. Cullum's ability, and familiarity with the technical language of French military writers, are a sufficient guarantee of the correctness of his translation." H. W. HALLECK, Major-Gen. U. S. A.

Benet's Military Law. A Treatise on Military Law and the Practice of Courts-Martial. By Capt. S. V. BENÉT, Ordnance Department, U. S. A., late Assistant Professor of Ethics, Law, &c., Military Academy, West Point. Fifth edition, revised. 1 vol. 8vo, law sheep. $4.50.

"This book is manifestly well timed just at this particular period, and it is, without doubt, quite as happily adapted to the purpose for which it was written. It is arranged with admirable method, and written with such perspicuity, and in a style so easy and graceful, as to engage the attention of every reader who may be so fortunate as to open its pages. This treatise will make a valuable addition to the library of the lawyer or the civilian; while to the military man it seems to be indispensable."—*Philadelphia Evening Journal.*

Halleck's International Law. International Law; or, Rules Regulating the Intercourse of States in Peace and War. By Major-Gen. H. W. HALLECK, Commanding the Army. 1 vol. 8vo, law sheep. $6.00.

"The work will be found to be of great use to army and navy officers, to professional lawyers, and to all interested in the topics of which it treats—topics to which present events give a greatly enhanced importance—such as 'Declaration of War and its Effects;' 'Sieges and Blockades;' 'Visitation and Search;' 'Right of Search;' 'Prize Courts;' 'Military Occupation;' 'Treaties of Peace;' 'Sovereignty of States,' &c., and valuable information for consuls and ambassadors."—*N. Y. Evening Post.*

History of West Point. And its Military Importance during the American Revolution; and the Origin and Progress of the United States Military Academy. By Capt. EDWARD C. BOYNTON, A. M., Adjutant of the Military Academy. With numerous Maps and Engravings. 1 vol. 8vo, blue cloth, $6.00; half mor., $7.50; full mor., $10.00.

"It records the earliest attempt at instituting a Military School by the Continental Congress, in 1776. It conducts us through the life of the institution, arguing with terseness its constitutionality, defending its educational principles, and explaining the necessity for its preservation."—*United Service Magazine.*

History of the United States Naval Academy. With Biographical Sketches, and the Names of all the Superintendents, Professors, and Graduates; to which is added a Record of some of the earliest votes by Congress, of Thanks, Medals, and Swords to Naval Officers. By EDWARD CHAUNCEY MARSHALL, A. M. 1 vol. 12mo, cloth, plates. $1.00.

"Every naval man will find it not only a pleasant companion, but an invaluable book of reference. It is seldom that so much information is made accessible in so agreeable a manner in so small a space."—*New York Times.*

Scott's Military Dictionary. Comprising Technical Definitions; Information on Raising and Keeping Troops; Actual Service, including Makeshifts and improved *Matériel*, and Law, Government, Regulation, and Administration relating to Land Forces. By Colonel H. L. Scott, Inspector-General U. S. A. 1 vol. large octavo, fully illustrated, half morocco, $6; half russia, $8; full morocco, $10.

"This book is really an Encylopædia, both elementary and technical, and as such occupies a gap in military literature which has long been most inconveniently vacant. This book meets a present popular want, and will be secured not only by those embarking in the profession but by a great number of civilians, who are determined to follow the descriptions and to understand the philosophy of the various movements of the Campaign. Indeed, no tolerably good library would be complete without the work."—*New York Times.*

Benton's Ordnance and Gunnery. A Course of Instruction in Ordnance and Gunnery; compiled for the use of the Cadets of the United States Military Academy. By Capt. J. G. Benton, Ordnance Department, late Instructor of Ordnance and Gunnery, Military Academy, West Point; Principal Assistant to Chief of Ordnance, U. S. A. Third Edition, revised and enlarged. 1 vol. 8vo, cloth, cuts. $5.00.

"There is no one book within the range of our military reading and study, that contains more to recommend it upon the subject of which it treats. It is as full and complete as the narrow compass of a single volume would admit, and the reputation of the author as a scientific and practical artillerist is a sufficient guarantee for the correctness of his statements and deductions, and the thoroughness of his labors."—*N. Y. Observer.*

Gibbon's Artillerist's Manual. Compiled from various Sources, and adapted to the Service of the United States. Profusely illustrated with woodcuts and engravings on stone. Second edition, revised and corrected, with valuable additions. By Gen. John Gibbon, U. S. A. 1 vol. 8vo, half roan. $6.00.

This book is now considered the standard authority for that particular branch of the Service in the United States Army. The War Department, at Washington, has exhibited its thorough appreciation of the merits of this volume, the want of which has been hitherto much felt in the service, by subscribing for 700 copies.

Jomini's Life of the Emperor Napoleon I. Life of Napoleon. By Baron Jomini, General-in-Chief and Aide-de-Camp to the Emperor of Russia. Translated from the French. with Notes, by H. W. Halleck, LL. D., Major-General United States Army; author of "Elements of Military Art and Science," "International Law, and the Laws of War," etc., etc. In four volumes octavo, with an Atlas of Sixty Maps and Plans. Price, in red cloth, $25.00; half calf, or half morocco, $35.00; half russia, $37.50.

"It is needless to say any thing in praise of Jomini as a writer on the science of war.

"General Halleck has laid the professional soldier and the student of military history under equal obligations by the service he has done to the cause of military literature in the preparation of this work for the press. His rare qualifications for the task thus undertaken will be acknowledged by all.

"The Notes with which the text is illustrated by General Halleck are not among the least of the merits of the publication, which, in this respect, has a value not possessed by the original work. * * *" —*National Intelligencer.*

"The Atlas attached to this version of Jomini's *Napoleon* adds very materially to its value. It contains *sixty* Maps, illustrative of Napoleon's extraordinary military career, beginning with the immortal Italian Campaigns of 1796, and closing with the decisive Campaign of Flanders, in 1815, the last Map showing the Battle of Wavre. These Maps take the reader to Italy, Egypt, Palestine, Germany Moravia, Russia, Spain, Portugal, and Flanders; and their number and variety, and the vast and various theatres of action which they indicate, testify to the immense extent of Napoleon's operations, and to the gigantic character of his power. * * *" —*Boston Traveller.*

Jomini's Grand Military Operations. Treatise on Grand Military Operations; or, a Critical and Military History of the Wars of Frederick the Great, as contrasted with the Modern System, together with a few of the most important Principles of the Art of War. By Baron Jomini, General-in-Chief and Aide-de-Camp to the Emperor of Russia. Translated from the French by Col. S. B. Holabird, A. D. C. U. S. A. 2 vols. 8vo, cloth, with an Atlas of 40 Maps and Plans. $15.00.

Jomini's Campaign of Waterloo. The Political and Military History of the Campaign of Waterloo. Translated from the French of General Baron de Jomini, by Capt. S. V. Benét, Ordnance Department, U. S. Army. Third edition. 1 vol. 12mo, cloth, $1.25.

"Baron Jomini has the reputation of being one of the greatest military historians and critics of the century. His merits have been recognized by the highest military authorities in Europe, and were rewarded in a conspicuous manner by the greatest military power in Christendom. He learned the art of war in the school of experience, the best and only finishing school of the soldier. He served with distinction in nearly all the campaigns of Napoleon. and it was mainly from the gigantic military operations of this matchless master of the art that he was enabled to discover its true principles, and to ascertain the best means of their application to the infinity of combinations which actual war presents. Jomini criticises the details of Waterloo with great science, and yet in a manner that interests the general reader as well as the professional."—*New York World.*

Roemer's Cavalry; its History, Management, and Uses in War. By J. Roemer, LL. D., late an Officer of Cavalry in the Service of the Netherlands. Elegantly illustrated, with one hundred and twenty-seven fine wood engravings. In one large octavo volume, beautifully printed on tinted paper. Cloth, $6; half calf, $7.50.

"By far the best treatise upon Cavalry and its uses in the field, which has yet been published in this country, for the general use of officers of all ranks, is this elaborate and interesting work. Eschewing the elementary principles and tactics of cavalry, which may be learned from any hand-book, the author treats of the uses of cavalry in the field of strategy and tactics, and of its general discipline and management. The range of the work includes an admirable treatise upon rifled fire-arms, an historical sketch of cavalry, embodying many interesting facts, an account of the cavalry service in Europe and this country, and a treatise on horses, their equipment, management, &c. The work is copiously illustrated and elegantly printed. It is interesting not alone to military men but to the general reader, who will gain from its pages valuable historical facts and very clear ideas of some branches of the art of war, such as the employment of spies, gaining information in an enemy's country, advance movements, and other strategical manœuvres."—*Boston Journal.*

Nolan's System for Training Cavalry Horses. By Kenner Garrard, Captain Fifth Cavalry, U. S. A. 1 vol. 12mo, cloth. 24 lithographed plates. $2.00.

"This work is clearly written, is eminently practical, is fully illustrated, and contains numerous hints as applicable to the discipline and management of the draught-horse as that of his more showy and fiery brother of the cavalry."—*Boston Journal.*

Barnard and Barry's Peninsular Campaign. Report of the Engineer and Artillery Operations of the Army of the Potomac, from its organization to the close of the Peninsular Campaign. By Brig.-Gen. J. G. Barnard, and other Engineer Officers, and Brig.-Gen. W. F. Barry, Chief of Artillery. Illustrated by numerous Maps, Plans, etc. 1 vol. 8vo, cloth, $4.00.

"The title of this work sufficiently indicates its importance and value as a contribution to the history of the great rebellion. General Barnard's Report is a narrative of the Engineer operations of the Army of the Potomac from the time of its organization to the date it was withdrawn from the James River. Thus a record is given of an important part in the great work which the nation found before it when it was first confronted with the necessity of war; and perhaps on no other point in the annals of the rebellion will future generations look with a deeper or more admiring interest."—*Buffalo Courier.*

The "C. S. A.," and the Battle of Bull Run. (A Letter to an English friend), by J. G. Barnard, Lieut.-Colonel of Engineers, U. S. A., Brigadier-General and Chief Engineer, Army of the Potomac. With five Maps. 1 vol. 8vo, cloth. $2.00.

"The work is clearly written, and can but leave the impression upon every reader's mind that it is truth. We commend it to the perusal of every one who wants an intelligent, truthful, and graphic description of the 'C. S. A.,' and the Battle of Bull Run."—*New York Observer.*

Simpson's Ordnance and Naval Gunnery. A Treatise on Ordnance and Naval Gunnery, compiled and arranged as a Text-Book for the U. S. Naval Academy. By Lieut. Edward Simpson, U. S. N. Third edition, revised and enlarged. 1 vol. 8vo, plates and cuts, cloth. $5.00.

"It is scarcely necessary for us to say, that a work prepared by a writer so practically conversant with all the subjects of which he treats, and who has such a reputation for scientific ability, cannot fail to take at once a high place among the text-books of our naval service. It has been approved by the Secretary of the Navy, and will henceforth be one of the standard authorities on all matters connected with Naval Gunnery."—*New York Herald.*

Holley's Ordnance and Armor. Embracing Descriptions, Discussions, and Professional Opinions concerning the Material, Fabrication, Requirements, Capabilities, and Endurance of European and American Guns for Naval, Sea-Coast, and Iron-Clad Warfare, and their Rifling, Projectiles, and Breech-Loading; also, Results of Experiments against Armor, from Official Records. With an Appendix, referring to Gun-Cotton, Hooped Guns, etc., etc. By A. L. Holley, B. P. With 493 illustrations. 1 vol. 8vo, 948 pages. Half roan, $10.00.

Luce's Naval Light Artillery. Instructions for Naval Light Artillery, afloat and ashore, prepared and arranged for the U. S. Naval Academy, by Lieutenant W. H. Parker, U. S. N. Second edition, revised by Lieutenant S. B. Luce, U. S. N., Assistant Instructor of Gunnery and Tactics at the U. S. Naval Academy. 1 vol. 8vo, cloth, with twenty-two Plates. $3.00.

"The service for which this is the text-book of instruction is of special importance in the present war. The use of light boat-pieces is constant and important, and young officers are frequently obliged to leave their boats, take their pieces ashore, and manœuvre them as field artillery. Not unfrequently, also, they are incorporated, when ashore, with troops, and must handle their guns like the artillery soldiers of a battery. 'The Exercise of the Howitzer Afloat' was prepared and arranged by Captain Dahlgren, whose name gives additional sanction and value to the book. A Manual for the Sword and Pistol is also given. The Plates are numerous and exceedingly clear, and the whole typography is excellent."—*Philadelphia Inquirer.*

Ward's Naval Ordnance and Gunnery. Elementary Instruction in Naval Ordnance and Gunnery. By James H. Ward, Commander U. S. Navy; author of "Naval Tactics," and "Steam for the Million." New edition, revised and enlarged. 8vo, cloth. $2.00.

"It conveys an amount of information in the same space to be found nowhere else, and given with a clearness which renders it useful as well to the general as the professional inquirer."—*N. Y. Evening Post.*

"The whole detail of Ordnance, in its history, philosophy, and application, is given by Commander Ward in such a manner (with occasional diagrams) as to convey to the student accurate notions for practical use."—*New Yorker.*

Luce's Seamanship. Compiled from various authorities, and Illustrated with numerous original and selected Designs. For the use of the United States Naval Academy. By S. B. Luce, Lieut.-Commander U. S. N. In two parts. Second Edition. One royal octavo volume, cloth. $10.00.

Squadron Tactics Under Steam. By Foxhall A. Parker, Commander U. S. Navy. Published by authority of the Navy Department. 1 vol. 8vo, with numerous Plates. $5.00.

"In this useful work to Navy officers, the author demonstrates—by the aid of profuse diagrams and explanatory text—a new principle for manœuvring naval vessels in action. The author contends that the winds, waves, and currents of the ocean oppose no more serious obstacles to the movements of a steam fleet, than do the inequalities on the surface of the earth to the manœuvres of an army. It is in this light, therefore, that he views a vast fleet—simply as an army; the regiments, brigades, and divisions of which are represented by a certain ship or ships."—*Scientific American.*

Nautical Routine and Stowage. With Short Rules in Navigation. By John McLeod Murphy and Wm. N. Jeffers, Jr., U. S. Navy. 1 vol. 8vo, blue cloth. $2.50.

Osbon's Hand-Book of the United States Navy. Being a Compilation of all of the Principal Events in the History of every Vessel of the United States Navy, from April, 1861, to May, 1864. Compiled and arranged by B. S. Osbon. 1 vol. 12mo, blue cloth. $2.50.

"As a condensed and compact history, as well as a work containing a vast amount of information, this work cannot be surpassed."—*Boston Traveller.*

Brandt's Gunnery Catechism. Gunnery Catechism, as applied to the Service of Naval Ordnance. Adapted to the Latest Official Regulations, and approved by the Bureau of Ordnance, Navy Department. By J. D. Brandt, formerly of the U. S. Navy. 1 vol. 18mo, illustrated, blue cloth. $1.50.

"This manual is very full of information and instruction, and shows the 'chief end' of Gunnery, and the aim of those who follow that profession. It is indispensable to all those who are suddenly introduced to a gun-deck, and will be found a valuable aid also to experienced officers."—*Commercial Advertiser.*

Barrett's Gunnery Instructions. Gunnery Instructions, simplified for the Volunteer Officers of the United States Navy, with Hints to Executive and other Officers. By Lieut. Edward Barrett, U. S. N., Instructor of Gunnery, Navy Yard, Brooklyn. 1 vol. 12mo, cloth. $1.25.

"It is a thorough work, treating plainly on its subject, and contains also some valuable hints to executive officers. No officer in the volunteer navy should be without a copy"—*Boston Evening Traveller.*

Totten's Naval Text-Book. Naval Text-Book and Dictionary, compiled for the use of the Midshipmen of the U. S. Navy. By Com mander B. J. Totten, U. S. N. Second and revised edition. 1 vol. 12mo. $3.00.

Calculated Tables of Ranges for Navy and Army Guns. With a Method of Finding the Distance of an Object at Sea. By Lieut. W. P Buckner, U. S N. 1 vol. 8vo, cloth. $1.50.

Manual of Internal Rules and Regulations for Men-of-War. By Commodore U. P. Levy, U. S. N., late Flag Officer commanding U. S. Naval Force in the Mediterranean, &c. Flexible blue cloth. Third Edition, revised and enlarged. 50 cents.

"Among the professional publications for which we are indebted to the war, we willingly give a prominent place to this useful little Manual of Rules and Regulations to be observed on board of ships of war. Its authorship is a sufficient guarantee for its accuracy and practical value; and as a guide to young officers in providing for the discipline, police, and sanitary government of the vessels under their command, we know of nothing superior."—*N. Y. Herald.*

King's Lessons and Practical Notes on Steam, The Steam Engine, Propellers, &c., &c., for Young Marine Engineers, Students, and others. By the late W. H. King, U. S. Navy. Revised by Chief Engineer J. W. King, U. S. Navy. Ninth Edition, enlarged. 8vo, cloth. $2.00.

"This is the ninth edition of a valuable work of the late W. H. King, U. S. Navy. It contains lessons and practical notes on Steam and the Steam-Engine, Propellers, &c. It is calculated to be of great use to young marine engineers, students, and others. The text is illustrated and explained by numerous diagrams and representations of machinery. This new edition has been revised and enlarged by Chief Engineer J. W. King, U. S. Navy, brother to the deceased author of the work."—*Boston Daily Advertiser.*

Ward's Steam for the Million. A popular Treatise on Steam and its Application to the Useful Arts, especially to Navigation. By J. H. Ward, Commander U. S Navy. New and revised Edition. 1 vol. 8vo, cloth. $1.00.

The Naval Howitzer Ashore. By Foxhall A. Parker, Commander U. S. Navy. 1 vol. 8vo, with Plates. Cloth. $4.00.

Gillmore's Fort Sumter. Official Report of Operations against the Defences of Charleston Harbor, 1863. Comprising the Descent upon Morris Island, the Demolition of Fort Sumter, and the Siege and Reduction of Forts Wagner and Gregg. By Major-Gen. Q. A. GILLMORE, U. S. Volunteers, and Major U. S. Corps of Engineers. With Maps and Lithographic Plates, Views, &c. 1 vol. 8vo. cloth. $10.

Gillmore's Siege and Reduction of Fort Pulaski, Georgia. Papers on Practical Engineering. No. 8. Official Report to the U. S. Engineer Department of the Siege and Reduction of Fort Pulaski, Ga., February, March, and April, 1862. By Brig.-Gen. Q. A. GILLMORE, U. S. A. Illustrated by Maps and Views. 1 vol. 8vo, cloth. $2.50.

"This is an official history of the siege of Fort Pulaski, from the commencement, with all the details in full, made up from a daily record, forming a most valuable paper for future reference. The situation and construction of the Fort, the position of the guns, both of the rebels and the Federals, and their operation, are made plain by maps and engraved views of different sections. Additional reports from other officers are furnished in the appendix, and every thing has been done to render the work full and reliable."—*Boston Journal.*

Gillmore's Treatise on Limes, Hydraulic Cements, and Mortars. Papers on Practical Engineering, U. S. Engineer Department, No. 9, containing Reports of numerous Experiments conducted in New York City, during the years 1858 to 1861, inclusive. By Q. A. GILLMORE, Brig.-Gen. U. S. Volunteers, and Major U. S. Corps of Engineers. With numerous Illustrations. One vol. 8vo. $4.00.

"This work contains a record of certain experiments and researches made under the authority of the Engineer Bureau of the War Department from 1858 to 1861, upon the various hydraulic cements of the United States, and the materials for their manufacture. The experiments were carefully made, and are well reported and compiled."—*Journal Franklin Institute.*

The Volunteer Quartermaster. Containing a Collection and Codification of the Laws, Regulations, Rules, and Practices governing the Quartermaster's Department of the United States Army, and in force March 4, 1865. By Captain ROELIFF BRINKERHOFF, Assistant Quartermaster U. S. Volunteers, and Post Quartermaster at Washington. 1 vol. 12mo, cloth. $2.50.

This work embraces all the laws of Congress, and all the orders and circulars of the War Office and its bureaus, bearing upon the subject. It also embodies the decisions of the Second Comptroller of the Treasury, so far as they affect the Quartermaster's Department. These decisions have the force of law in the adjustment of accounts, and are therefore invaluable to all disbursing officers.

Cullum's Military Bridges. Systems of Military Bridges in Use by the United States Army; those adopted by the Great European Powers; and such as are employed in British India. With Directions for the Preservation, Destruction, and Re-establishment of Bridges. By Brig.-Gen. George W. Cullum, Lieut.-Col. Corps of Engineers, U. S. A. 1 vol. 8vo. With numerous Illustrations. cloth. $3.50.

"We have no man more competent to prepare such a work than Brig.-Gen. Cullum, who had the almost exclusive supervision, devising, building, and preparing for service of the various bridge-trains sent to our armies in Mexico during our war with that country. The treatise before us is very complete, and has evidently been prepared with scrupulous care. The descriptions of the various systems of military bridges adopted by nearly all civilized nations are very interesting even to the non-professional reader, and to those specially interested in such subjects must be very instructive, for they are evidently the work of a master of the art of military bridge-building."—*Washington Chronicle.*

Haupt's Military Bridges. For the Passage of Infantry, Artillery, and Baggage-Trains; with Suggestions of many new Expedients and Constructions for Crossing Streams and Chasms; designed to utilize the Resources ordinarily at command, and reduce the amount and cost of Army Transportation. Including also Designs for Trestle and Truss Bridges for Military Railroads, adapted especially to the wants of the Service of the United States. By Herman Haupt, Brig.-Gen. in charge of the Construction and Operation of the U. S. Military Railways, Author of "General Theory of Bridge Construction," &c. Illustrated by sixty-nine Lithographic Engravings. 8vo, cloth. $6.50.

"This elaborate and carefully prepared, though thoroughly practical and simple work, is peculiarly adapted to the military service of the United States. Mr. Haupt has added very much to the ordinary facilities for crossing streams and chasms, by the instructions afforded in this work."—*Boston Courier.*

Holley's Railway Practice. American and European Railway Practice, in the Economical Generation of Steam, including the Materials and Construction of Coal-burning Boilers, Combustion, the Variable Blast, Vaporization, Circulation, Superheating, Supplying and Heating Feed-water, &c., and the Adaptation of Wood and Coke-burning Engines to Coal-burning; and in permanent Way, including Road-bed, Sleepers, Rails, Joint Fastenings, Street Railways, &c., &c. By Alexander L. Holley, B. P. With 77 lithographed plates. 1 vol. folio, cloth. $12.00.

* * * "All these subjects are treated by the author in both an intelligent and intelligible manner. The facts and ideas are well arranged, and presented in a clear and simple style, accompanied by beautiful engravings, and we presume the work will be regarded as indispensable by all who are interested in a knowledge of the construction of railroads, and rolling stock, or the working of locomotives."—*Scientific American.*

Authorized U. S. Infantry Tactics. For the Instruction, Exercise, and Manœuvres of the Soldier, a Company, Line of Skirmishers, Battalion, Brigade, or Corps d'Armée. By Brig.-Gen. SILAS CASEY, U. S. A. 3 vols 24mo. Cloth, lithographed plates. $2.50.

Vol. I.—School of the Soldier; School of the Company; Instruction for Skirmishers.

Vol. II.—School of the Battalion.

Vol. III.—Evolutions of a Brigade; Evolutions of a Corps d'Armée.

"WAR DEPARTMENT, WASHINGTON, *August* 11, 1862.

"The System of Infantry Tactics prepared by Brig.-Gen. Silas Casey, U. S. A., having been approved by the President, is adopted for the instruction of the Infantry of the Armies of the United States, whether Regular, Volunteer, or Militia.

"EDWIN M. STANTON, *Secretary of War.*"

U. S. Tactics for Colored Troops. U. S. Infantry Tactics, for the Instruction, Exercise, and Manœuvres of the Soldier, a Company, Line of Skirmishers, and Battalion, for the use of the Colored Troops of the United States Infantry. Prepared under the direction of the War Department. 1 vol., plates. $1.50.

"WAR DEPARTMENT, WASHINGTON, *March* 9, 1863.

"This system of United States Infantry Tactics, prepared under the direction of the War Department, for the use of the Colored Troops of the United States Infantry, having been approved by the President, is adopted for the instruction of such troops.

"EDWIN M. STANTON, *Secretary of War.*"

Kelton's Bayonet Exercise. A New Manual of the Bayonet, for the Army and Militia of the United States. By Colonel J. C. KELTON, U. S. A. With forty beautifully engraved plates. Red cloth. $2.00.

"This Manual was prepared for the use of the Corps of Cadets, and has been introduced at the Military Academy with satisfactory results. It is simply the theory of the attack and defence of the sword applied to the bayonet, on the authority of men skilled in the use of arms."—*New York Times.*

Berriman's Sword-Play. The Militiaman's Manual and Sword-Play without a Master.—Rapier and Broad-Sword Exercises copiously Explained and Illustrated; Small-Arm Light Infantry Drill of the United States Army; Infantry Manual of Percussion Muskets; Company Drill of the United States Cavalry. By Major M. W. BERRIMAN, engaged for the last thirty years in the practical instruction of Military Students. Fourth edition. 1 vol. 12mo, red cloth. $1.00.

"This work will be found very valuable to all persons seeking military instruction; but it recommends itself most especially to officers, and those who have to use the sword or sabre. We believe it is the only work on the use of the sword published in this country."—*New York Tablet.*

Heavy Artillery Tactics.—1863. Instruction for Heavy Artillery; prepared by a Board of Officers, for the use of the Army of the United States. With service of a gun mounted on an iron carriage. In one volume, 12mo, with numerous illustrations. cloth. $2.50.

"WAR DEPARTMENT,
"WASHINGTON, D. C., *Oct.* 20, 1862.

"This system of Heavy Artillery Tactics, prepared under direction of the War Department, having been approved by the President, is adopted for the instruction of troops when acting as heavy artillery. EDWIN M. STANTON, *Secretary of War.*"

"The First Part consists of sixteen lessons relating to the service of the single piece, including the gun, howitzer, mortar, and columbiad; also, the formation of batteries, the art of aiming pieces and firing hot-shot. Part Second relates entirely to mechanical manœuvres, and appliances for handling, mounting, dismounting, and transporting heavy pieces. Part Third is of a miscellaneous character, containing directions for embarking and disembarking artillery and ordnance stores; also, tables of dimensions and weights of guns, carriages, shot, shell, machines, and implements, with charges for, and ranges of heavy artillery. These instructions are not only copious in detail, but aptly illustrated with thirty-nine elegant steel-plate engravings."—*Bulletin.*

Roberts's Hand-Book of Artillery. For the Service of the United States Army and Militia. New revised and greatly enlarged edition. By Major JOSEPH ROBERTS, U. S. A. 1 vol. 18mo, cloth. $1.25.

"A complete catechism of gun practice, covering the whole ground of this branch of military science, and adapted to militia and volunteer drill, as well as to the regular army. It has the merit of precise detail, even to the technical names of all parts of a gun, and how the smallest operations connected with its use can be best performed. It has evidently been prepared with great care, and with strict scientific accuracy." *New York Century.*

Duane's Manual for Engineer Troops. Consisting of Part I., Pontoon Drill; II., Practical Operations of a Siege; III, School of the Sap; IV., Military Mining; V., Construction of Batteries. By Major J. C. DUANE, Corps of Engineers, U. S. Army. 1 vol. 12mo, cloth. $2.50.

"I have carefully examined Capt. J. C. Duane's 'Manual for Engineer Troops,' and do not hesitate to pronounce it the very best work on the subject of which it treats.

"H. W. HALLECK, *Major-General, U. S. A.*"

Dufour's Principles of Strategy and Grand Tactics. Translated from the French of General G. H. Dufour. By WILLIAM P. CRAIGHILL, Capt. of Engineers, U. S. Army, and Assistant Professor of Engineering, U. S. Military Academy, West Point. From the last French Edition. Illustrated. In one volume 12mo. cloth. $3.00.

"In all military matters General Dufour is recognized as one of the first authorities in Europe, and consequently the translation of this very valuable work is a most acceptable addition to our military libraries."—*London Naval and Military Gazette.*

Instructions for Field Artillery. Prepared by a Board of Artillery Officers. 1 vol. 12mo, illustrated by 122 pages of Engravings. Cloth. $3.00.

"WAR DEPARTMENT, WASHINGTON, *March* 1, 1863.

"This system of Instruction for Field Artillery, prepared under direction of the War Department, having been approved by the President, is adopted for the instruction of troops when acting as field artillery.

"Accordingly, instruction in the same will be given after the method pointed out therein; and all additions to or departures from the exercise and manœuvres laid down in the system, are positively forbidden.

"EDWIN M. STANTON, *Secretary of War.*"

Anderson's Evolutions of Field Batteries of Artillery. Translated from the French, and arranged for the Army and Militia of the United States. By Gen. ROBERT ANDERSON, U. S. A. Published by order of the War Department. 1 vol., cloth, 32 plates. $1.

"WAR DEPARTMENT, *Nov. 2d*, 1859.

"The system of 'Evolutions of Field Batteries,' translated from the French, and arranged for the service of the United States, by Major Robert Anderson, of the 1st Regiment of Artillery, having been approved by the President, is published for the information and government of the army.

"All Evolutions of Field Batteries not embraced in this system are prohibited, and those herein prescribed will be strictly observed. J. B. FLOYD,

"*Secretary of War.*"

Mendell's Treatise on Military Surveying. Theoretical and Practical, including a Description of Surveying Instruments. By G. H. MENDELL, Captain of Engineers. 1 vol. 8vo, with numerous illustrations. cloth. $2.50.

"The author is a Captain of Engineers, and has for his chief authorities Salneuve, Lalobre, and Simms. He has presented the subject in a simple form, and has liberally illustrated it with diagrams, that it may be readily comprehended by every one who is liable to be called upon to furnish a military sketch of a portion of country."—*New York Evening Post.*

Viele's Hand-book. Hand-Book for Active Service, containing Practical Instructions in Campaign Duties. For the use of Volunteers. By Brigadier-General Egbert L. Viele, U. S. A. 12mo, cloth. $1.

"It is a thorough treatise, copiously illustrated, and embraces a complete drill by company, regiment, &c. It also embraces instructions in regard to the camp, fortifications, rations, and mode of cooking them, and has a manual for light and heavy artillery."—*New Haven Palladium.*

A Treatise on the Camp and March. With which is connected the Construction of Field-Works and Military Bridges; with an Appendix of Artillery Ranges, &c. For the use of Volunteers and Militia in the United States. By Captain Henry D. Grafton, U. S. A. 1 vol. 12mo, cloth. 75 cents.

The Automaton Regiment; or, Infantry Soldier's Practical Instructor.—For all Regimental Movements in the Field. By G. Douglas Brewerton, U. S. Army. Neatly put up in boxes, price $1; when sent by mail, $1.40.

The "Automaton Regiment" is a simple combination of blocks and counters, so arranged and designated, by a carefully considered contrast of colors, that it supplies the student with a perfect miniature regiment, in which the position in the battalion of each company, and of every officer and man in each division, company, platoon, and section, is clearly indicated. It supplies the studious soldier with the means whereby he can consult his "tactics," and at the same time join practice to theory by manœuvring a mimic regiment.

I hereby certify that I have examined the "Automaton Regiment," invented by G. Douglas Brewerton, late of the United States Regular Army, and now serving as a Volunteer Aide upon my military staff, and believe that his invention will prove a useful and valuable assistant to every student of military tactics. I take pleasure in recommending it accordingly.
B. Saxton,
Brigadier-General Volunteers.

The Automaton Company; or, Infantry Soldier's Practical Instructor.—For all Company Movements in the Field. By G. Douglas Brewerton, U. S. A. Price in boxes, $1.25; when sent by mail, $1.95.

The Automaton Battery; or, Artillerist's Practical Instructor.—For all Mounted Artillery Manœuvres in the Field. By G. Douglas Brewerton, U. S. A. Price in boxes, $1; when sent by mail, $1.40.

These productions are of a similar character, and cannot fail to be of great value to the military student. They are *object lessons*, that will teach more in an hour than mere verbal instruction could in a month. They cannot be too highly commended to both officers and men.—*Commercial Advertiser.*

Monroe's Company and Skirmish Drill.—The Company Drill of the Infantry of the Line, together with the Skirmish Drill of the Company and Battalion, after the method of General Le Louterel. Bayonet Fencing; with a Supplement on the Handling and Service of Light Infantry. By J. Monroe, Colonel 22d Regiment, N. G., N. Y. S. M., formerly Captain U. S. Infantry. 1 vol., 32mo, cloth. 75 cents.

This is a most valuable and timely little manual. It should be in the hands of every new recruit.—*Chicago Tribune.*

School of the Guides.—Designed for the use of the Militia of the United States. By Colonel Eugene Le Gal, 55th Regiment, N. Y. S. M. cloth, 60 cents.

"This excellent compilation condenses into a compass of less than sixty pages all the instruction necessary for the guides, and the information being disconnected with other matters, is more readily referred to and more easily acquired."—*Louisville Journal.*

Craighill's Army Officer's Pocket Companion. Principally designed for Staff Officers in the Field. Partly translated from the French of M. de Rouvre, Lieutenant-Colonel of the French Staff Corps, with Additions from standard American, French, and English Authorities. By Wm. P. Craighill, First Lieutenant U. S. Corps of Engineers, Assist. Prof. of Engineering at the U. S. Military Academy, West Point. 1 vol. 18mo, full roan. $2.00.

"I have carefully examined Capt. Craighill's Pocket Companion. I find it one of the very best works of the kind I have ever seen. Any Army or Volunteer officer who will make himself acquainted with the contents of this little book, will seldom be ignorant of his duties in camp or field. H. W. HALLECK, *Major-General U. S. A.*"

Hunter's Manual for Quartermasters and Commissaries. Containing Instructions in the Preparation of Vouchers, Abstracts, Returns, etc., embracing all the recent changes in the Army Regulations, together with instructions respecting Taxation of Salaries, etc. By Captain R. F. Hunter, late of the U. S. Army. 12mo, cloth, $1.25. Flexible morocco, $1.50.

"This is the only work of the kind extant. It is based on the latest regulations of the War Department, and will be regarded as authority by those officers for whose use it is designed."—*Saturday Evening Gazette.*

Ordronaux's Manual of Instructions for Military Surgeons, in the Examination of Recruits and Discharge of Soldiers. With an Appendix, containing the Official Regulations of the Provost-Marshal General's Bureau, and those for the formation of the Invalid Corps, etc., etc. Prepared at the request of the United States Sanitary Commission. By John Ordronaux, M. D., Professor of Medical Jurisprudence in Columbia College, New York. 12mo. Half morocco. $1.50.

"In a condensed form, it is an admirable treatise on the important subjects of which it treats. The author has aimed to be brief without being obscure, to omit nothing of real importance, and to draw his materials from the best sources. He treats of the physical disabilities which have relation to the military service, and of these alone. Medical Examiners are instructed in their duties, and the method of discovering feigned, artificially produced, and concealed diseases is pointed out. The book will prove valuable to all who are concerned in the manipulation of recruits or conscripts. An Appendix contains official regulations and instructions relative to the Provost-Marshal's office, the Invalid Corps, etc."—*Commercial Advertiser.*

Ordronaux's Hints on the Preservation of Health in Armies. For the use of Volunteer Officers and Soldiers. By John Ordronaux, M. D. New edition, 18mo, cloth. 75 cents.

Thomas's Rifled Ordnance. A Practical Treatise on the Application of the Principle of the Rifle to Guns and Mortars of every Calibre. To which is added, a new theory of the initial action and force of Fired Gunpowder. By LYNALL THOMAS, F. R. S. L. Fifth edition, revised. One volume, octavo, illustrated. cloth· $2.00.

"At a time when the manufacture of guns engrosses the attention of thousands on thousands, any practical treatise which may suggest desirable alterations or innovations is of importance, and deserves that attention we doubt not will be extended to the present volume."—*Boston Evening Gazette.*

Wilcox's Rifles and Rifle Practice. An Elementary Treatise on the Theory of Rifle Firing; explaining the causes of Inaccuracy of Fire and the manner of correcting it; with descriptions of the Infantry Rifles of Europe and the United States, their Balls and Cartridges. By Captain C. M. WILCOX, U. S. A. New edition, with engravings and cuts. Green cloth. $2.00.

"The book will be found intensely interesting to all who are watching the changes in the art of war arising from the introduction of the new rifled arms. We recommend to our readers to buy the book."—*Military Gazette.*

Lendy's Maxims and Instructions on the Art of War. Maxims, Advice, and Instructions on the Art of War; or, A Practical Military Guide for the use of Soldiers of all Arms and of all Countries. Translated from the French by Captain LENDY, Director of the Practical Military College, late of the French Staff, etc., etc. 1 vol. 18mo, cloth. 75 cents.

"This book treats generally of the art of war and the conduct of campaigns, and without going into all the details of a soldier's business, aims to explain the principles on which an army may be well established in camp, or successfully led and manœuvred on the field."—*Providence Journal.*

Andrews's Hints to Company Officers on their Military Duties. By Captain C. C. ANDREWS, Third Regiment Minnesota Volunteers. 1 vol. 18mo, cloth. 60 cents.

"This is a hand-book of good practical advice, which officers of all ranks may study with advantage."—*Philadelphia Press.*

Heth's System of Target Practice.—For the use of Troops when armed with the Musket, Rifle-Musket, Rifle, or Carbine. Prepared principally from the French, by Captain Henry Heth, 10th Infantry, U. S. A. 18mo, cloth. 75 cents.

Minifie's Mechanical Drawing. A Text-Book of Geometrical Drawing, for the Use of Mechanics and Schools, in which the Definitions and Rules of Geometry are familiarly explained; the Practical Problems are arranged from the most simple to the more complex, and in their description technicalities are avoided as much as possible. With Illustrations for Drawing Plans, Sections, and Elevations of Buildings and Machinery; an introduction to Isometrical Drawing, and an Essay on Linear Perspective and Shadows. Illustrated with over 200 Diagrams engraved on Steel. By WILLIAM MINIFIE, Architect. Seventh edition. With an Appendix on the Theory and Application of Colors. 1 vol. 8vo, cloth. $4.00.

"It is the best work on Drawing that we have ever seen, and is especially a text-book of Geometrical Drawing for the use of Mechanics and Schools. No young Mechanic, such as a Machinist, Engineer, Cabinet-Maker, Millwright, or Carpenter, should be without it." —*Scientific American.*

"One of the most comprehensive works of the kind ever published, and cannot but possess great value to builders. The style is at once elegant and substantial."—*Pennsylvania Inquirer.*

"We think this the *best* work on this subject, which is saying very much; as much attention has been given to the science of Drawing. There is nothing in the range of drawing that cannot be found in this book, and which is not well explained."—*Ohio Teacher.*

"Whatever is said is rendered perfectly intelligible by remarkably well executed diagrams on steel, leaving nothing for mere vague supposition; and the addition of an introduction to isometrical drawing, linear perspective, and the projection of shadows, winding up with a useful index to technical terms."—*Glasgow Mechanic's Journal.*

Minifie's Geometrical Drawing. Abridged from the Octavo edition, for the use of Schools. Illustrated with 48 Steel Plates. Fifth edition. 1 vol. 12mo, half roan. $1.50.

Peirce's System of Analytic Mechanics. Physical and Celestial Mechanics, by BENJAMIN PEIRCE, Perkins Professor of Astronomy and Mathematics in Harvard University, and Consulting Astronomer of the American Ephemeris and Nautical Almanac. Developed in four systems of Analytic Mechanics, Celestial Mechanics, Potential Physics, and Analytic Morphology. 1 vol. 4to, cloth. $10.00.

Dictionary of Weights and Measures. Universal Dictionary of Weights and Measures, Ancient and Modern, reduced to the standards of the United States of America. By J. H. ALEXANDER. New edition, enlarged. 1 vol. 8vo, cloth. $2.50

www.ingramcontent.com/pod-product-compliance
Lightning Source LLC
LaVergne TN
LVHW021106110826
845150LV00001B/187